Richard L. Carlin

Magnetochemistry

With 244 Figures and 21 Tables

Springer-Verlag
Berlin Heidelberg New York Tokyo

Professor Richard L. Carlin
Department of Chemistry
University of Illinois at Chicago
Chicago, Illinois 60680

ISBN 3-540-15816-2 Springer-Verlag Berlin Heidelberg New York Tokyo
ISBN 0-387-15816-2 Springer-Verlag New York Heidelberg Berlin Tokyo

Library of Congress Cataloging-in-Publication Data

Carlin, Richard L. (Richard Lewis), 1935–
  Magnetochemistry.
  Bibliography: p.
  Includes indexes.
  1. Magnetochemistry. I. Title.
QD591.C37 1986 541.3'78 85-26165
ISBN O-387-15816-2 (U.S.)

Typesetting, printing and bookbinding: Brühlsche Universitätsdruckerei, 6300 Giessen
2152/3020-543210

Preface

This is a book about things in magnetism that interest me. I think that these are important things which will interest a number of other chemists. The restriction is important, because it is difficult to write well about those things which are less familiar to an author. In general, the chemistry and physics of coordination compounds are what this book is about.

Magnetochemistry is the study of the ground states of metal ions. When the ions are not interacting, then the study of single-ion phenomena is called paramagnetism. When the metal ions interact, then we are concerned with collective phenomena such as occur in long-range ordering. Several years ago, Hans van Duyneveldt and I published a book that explored these subjects in detail. Since that time, the field has grown tremendously, and there has been a need to bring the book up to date. Furthermore, I have felt that it would be useful to include more subsidiary material to make the work more useful as a textbook. This book is the result of those feelings of mine.

The subject of magnetism is one of the oldest in science, and the magnetic properties of atoms and molecules have concerned physicists and chemists since at least the turn of the century. The Nobel Laureate J. H. Van Vleck has written a brief but comprehensive review of magnetism [1], and he has argued that quantum mechanics is the key to understanding magnetism. I would go further and argue that magnetism is the key to understanding quantum mechanics, at least in the pedagogical sense, for so many physical phenomena can be understood quantitatively in this discipline. I hope the book expresses this sense.

Units are a troublesome issue here, for people working in the field of magneto-chemistry have not yet adopted the SI system. I have converted literature values of applied magnetic fields from Oe to tesla (T) where feasible, and tried to use units of joules for energy throughout. In order to avoid total confusion, I have generally quoted parameters from the literature as given, and then my conversion to the SI units in parentheses.

I would once again like to dedicate this book to my wife, Dorothy, because she never answered my continual question, Why in heaven's name am I writing this book? I would like to thank Hans van Duyneveldt for reading the manuscript and offering perceptive criticism. He has also always been helpful in patiently explaining things to me. Jos de Jongh has also taught me a lot about magnetism. I would also like to thank my colleagues around the world who were kind enough to send me preprints and reprints, for allowing me to quote from their work, and for giving me permission to reproduce figures and tables from their publications. My research on the magnetic properties of transition metal complexes has been supported by a succession of grants from the Solid State Chemistry Program of the Division of Materials Research of the

National Science Foundation. Most of the typing was done by my good friends, Wally Berkowicz, Pat Campbell, and Regina Gierlowski.

Chicago, October 1985   Richard L. Carlin

Reference

1. Van Vleck J.H., Rev. Mod. Phys. **50**, 181 (1978)

Contents

List of symbols

M	= magnetization	H	= Hamiltonian
H	= magnetic field strength	S	= spin operator
κ	= magnetic susceptibility	10Dq	= cubic crystal field splitting
v	= molar volume	ν	= frequency
n	= principal quantum number	Q	= heat
l	= orbital quantum number	U	= internal energy
m	= magnetic quantum number	S	= entropy
m_s	= magnetic quantum number	E	= enthalpy
$\mathscr{S}$	= total spin	F	= free energy (Helmholtz)
$\mathscr{L}$	= total angular momentum	G	= free energy (Gibbs)
M_L	= total angular momentum quantum number	c	= specific heat
M_S	= total spin quantum number	$B_J(\eta)$	= Brillouin function
C	= Curie constant	τ	= relaxation time
T	= absolute temperature	ω	= frequency
μ	= magnetic moment	E	= energy
k	= Boltzmann constant	g_J	= splitting factor
g	= Landé constant	S	= total spin magnetic moment
N	= Avogadro's number	L	= total orbital magnetic moment
μ_B	= Bohr magneton	H	= magnetic field
h	= Planck's constant	λ	= spin-orbit coupling constant
c	= speed at light	J	= exchange constant
m	= mass of an electron	d	= distortion vector
e	= charge of an electron	T_c	= critical temperature
P	= pressure	λ	= Weiss-field constant
V	= volume	z	= magnetic coordination number
R	= molar gas constant	N	= demagnetization factor
J	= inner quantum number	p_c	= critical concentration
Z	= partition function	ε	= voltage
D, δ	= zero-field splitting parameter	n	= number of mols
		I	= current

1. Diamagnetism and Paramagnetism

1.1 Introduction

This is a book concerning the magnetic properties of transition metal complexes. The subject has been of interest for a long time, for it was realized by Pauling as long ago as the 1930's that there was a diagnostic criterion between magnetic properties and the bonding of metal ions in complexes. Indeed, over the years, magnetic properties have continued to be used in this fashion. With time and the influence of physicists working in this field, the emphasis has shifted so that chemists are becoming more interested in the magnetic phenomena themselves. The subject is no longer a subsidiary one. One result of this new emphasis is that chemists have continued to decrease the working temperature of their experiments, with measurements at liquid helium temperatures now being common.

These notes are designed as a supplement to such current texts of inorganic chemistry as that by Huheey [1]. We go beyond the idea of counting the number of unpaired electrons in a compound from a magnetic measurement to the important temperature-dependent behavior. We emphasize the structural correlation with magnetic properties, which follows from a development of magnetic ordering phenomena. A more quantitative treatment of certain aspects of this field may be found in the book by Carlin and van Duyneveldt [2], and more details of the descriptive chemistry of the elements may be found in the text by Cotton and Wilkinson [3].

A magnetic susceptibility is merely the quantitative measure of the response of a material to an applied (i.e., external) magnetic field. Some substances, called diamagnets, are slightly repelled by such a field. Others, called paramagnets, are attracted into an applied field; they therefore weigh more in the field, and this provides one of the classical methods for the measurement of magnetic susceptibilities, the Gouy method. Diamagnetic susceptibilities are temperature independent, but paramagnetic susceptibilities depend on the temperature of the sample, often in a rather complex fashion. In addition, many paramagnetic materials have interactions which cause them to become antiferromagnets or ferromagnets, and their temperature-dependent properties become even more complex. These notes will outline these different behaviors in detail, and correlate the magnetic behavior with the chemical nature of the materials.

Since the question of the units of susceptibilities is often confusing, let us emphasize the point here. For the susceptibility χ, the definition is $M = \chi H$, where M is the magnetization (magnetic moment per unit of volume) and H is the magnetic field strength. This χ is dimensionless, but is expressed as emu/cm^3. The dimension of emu is therefore cm^3. The molar susceptibility χ_N is obtained by multiplying χ with the molar volume, v (in cm^3/mol). So, the molar susceptibility leads to $M = H\chi_N/v$, or $Mv = \chi_N H$, where Mv is now the magnetic moment per mol. The dimension of molar susceptibility is thus emu/mol or cm^3/mol. We shall omit the subscript on χ in what follows.

Another physical quantity important for the understanding of magnetic systems is the specific heat. Indeed, no magnetic study, at least at low temperatures, is complete without the measurement of the specific heat. This point is emphasized in the text.

Finally, the magnetic properties of molecules, whether they interact with one another or not, depend on the local geometry and the chemical links between them. This means that a true understanding of a magnetic system requires a determination of the molecular geometry, which is usually carried out by means of an X-ray crystal structure determination. Chemists after all must be concerned about magneto-structural correlations, and this point is emphasized throughout.

1.2 Diamagnetism

Let us begin with diamagnetism, which of itself is not very interesting for transition metal chemistry. It is, nevertheless, something that cannot be ignored, for it is an underlying property of all matter.

Diamagnetism is especially important in the consideration of materials with completely filled electronic shells, that is, systems which do not contain any unpaired electrons. This cannot be taken as an operational definition of a diamagnet for, as we shall see, certain paramagnetic materials can become diamagnetic under certain conditions. So, we shall use the following definition [4, 5].

> If a sample is placed in a magnetic field H, the field within the material will generally differ from the free space value. The body has therefore become magnetized and, if the density of the magnetic lines of force within the sample is reduced, the substance is said to be *diamagnetic*. Since this is equivalent to the substance producing a flux opposed to the field causing it, it follows that the substance will tend to move to regions of lower field strength, or out of the field.

The molar susceptibility of a diamagnetic material is negative, and rather small, being of the order of -1 to -100×10^{-6} emu/mol. Diamagnetic susceptibilities do not depend on field strength and are independent of temperature. For our purposes, they serve only as a correction of a measured susceptibility in order to obtain the paramagnetic susceptibility.

Diamagnetism is a property of all matter and arises from the interaction of paired electrons with the magnetic field. Since transition metal substances with unpaired electrons also have a number of filled shells, they too have a diamagnetic contribution to their susceptibility. It is much smaller than the paramagnetic susceptibility, and can usually be separated out by the measurement of the temperature dependence of the susceptibility. Indeed, paramagnetic susceptibilities frequently become so large at low temperatures that it is scarcely necessary even to correct for them.

Diamagnetic susceptibilities of atoms in molecules are largely additive, and this provides a method for the estimation of the diamagnetic susceptibilities of ligand atoms and counter ions in a transition metal complex. The Pascal constants (Table 1.1)

Table 1.1. Pascal's constants[a]
(susceptibilities per gram atom $\times 10^6$ emu)

Cations		Anions	
Li^+	$-$ 1.0	F^-	$-$ 9.1
Na^+	$-$ 6.8	Cl^-	-23.4
K^+	-14.9	Br^-	-34.6
Rb^+	-22.5	I^-	-50.6
Cs^+	-35.0	NO_3^-	-18.9
Tl^+	-35.7	ClO_3^-	-30.2
NH_4^+	-13.3	ClO_4^-	-32.0
Hg^{2+}	-40.0	CN^-	-13.0
Mg^{2+}	$-$ 5.0	NCS^-	-31.0
Zn^{2+}	-15.0	OH^-	-12.0
Pb^{2+}	-32.0	SO_4^{2-}	-40.1
Ca^{2+}	-10.4	O^{2-}	-12.0

Neutral Atoms

H	$-$ 2.93	As (III)	$-$ 20.9
C	$-$ 6.00	Sb (III)	$-$ 74.0
N (ring)	$-$ 4.61	F	$-$ 6.3
N (open chain)	$-$ 5.57	Cl	$-$ 20.1
N (imide)	$-$ 2.11	Br	$-$ 30.6
O (ether or alcohol)	$-$ 4.61	I	$-$ 44.6
O (aldehyde or ketone)	$-$ 1.73	S	$-$ 15.0
P	$-$ 26.3	Se	$-$ 23.0
As (V)	-43.0		

Some Common Molecules

H_2O	-13	$C_2O_4^{2-}$	$-$ 25
NH_3	-18	acetylacetone	$-$ 52
C_2H_4	-15	pyridine	$-$ 49
CH_3COO^-	-30	bipyridyl	-105
$H_2NCH_2CH_2NH_2$	-46	o-phenanthroline	-128

Constitutive Corrections

C=C	5.5	N=N	1.8
C=C−C=C	10.6	C=N−R	8.2
C≡C	0.8	C−Cl	3.1
C in benzene ring	0.24	C−Br	4.1

a   From Ref. [5].

provide an empirical method for this procedure. One adds the atomic susceptibility of each atom, as well as the constitutive correction to take account of such factors as π-bonds in the ligands. For example, the diamagnetic contribution to the susceptibility of $K_3Fe(CN)_6$ is calculated as

$$K^+ \quad 3(-14.9 \times 10^{-6}) = -\ 44.7 \times 10^{-6}$$
$$CN^- \quad 6(-13.0 \times 10^{-6}) = -\ 78.0 \times 10^{-6}$$
$$\overline{\qquad\qquad\qquad -122.7 \times 10^{-6} \text{ emu/mol}}$$

This procedure is only of moderate accuracy, and the values given could change from compound to compound. Greater accuracy can sometimes be obtained by the direct measurement of the susceptibility of a diamagnetic analog of the paramagnetic compound which is of interest. On the other hand, since a paramagnetic susceptibility is in the range $(10^2 - 10^4) \times 10^{-6}$ emu/mol at room temperature, and increases with decreasing temperature, the exact evaluation of the diamagnetic contribution is often not important.

1.3 Atomic Term Symbols

Each electron has associated with it four quantum numbers, n, ℓ, m, and m_s. They are restricted to the following values: $0 \leq \ell \leq n$, $|m| \leq \ell$, and $m_s = \pm \frac{1}{2}$. The iron series ions have $n = 3$ and $\ell = 2$; this means that m may take on the values $2, 1, 0, -1$, and -2, and combining each of these states with $m_s = \pm \frac{1}{2}$, we see that there can be a maximum of 10 electrons in this shell. This result is consistent with the Pauli principle. The respective elements (Ti–Cu) are generally called the 3d or iron series. The naming of other shells, or states, is illustrated as follows:

$\ell =$	0	1	2	3
shell	s	p	d	f
orbital degeneracy	1	3	5	7
m	0	$0, \pm 1$	$0, \pm 1, \pm 2$	$0, \pm 1, \pm 2, \pm 3$.

In order to describe the states of an atom or ion, it is necessary to combine the quantum numbers of the electrons into what are called Russell-Saunders (R-S) term symbols; these are valid when spin-orbit coupling is relatively small. A general representation of a R-S term symbols is $^{2\mathscr{S}+1}L$, where $L = S, P, D, F,$ ----- as $\mathscr{L} = 0, 1, 2, 3,$ -----. An atom with filled shells has $\mathscr{S} = 0$ and $\mathscr{L} = 0$ and is therefore said to be in a ^1S state.

The derivation of term symbols is easily described with reference to the $(2p)^2$ configuration of the carbon atom. Each electron has $n = 2$ and $\ell = 1$, so one can construct a list (Table 1.2) of the values of m and m_s which are allowed by the Pauli principle. The quantum number m will take on the values of $0, +1$ and -1, and m_s will be $+\frac{1}{2}$ or $-\frac{1}{2}$; no two electrons can be assigned the same set of quantum numbers. As Table 1.2 illustrates, there are 15 such allowed combinations, or microstates. The value of $\mathscr{L}$ is $(M_L)_{max}$ and the first value found is 2. Five micro-states are assigned to this state $(M_{\mathscr{L}} = 2, 1, 0, -1, -2)$ and it is found that these micro-states all have $\mathscr{S} = (M_S)_{max} = 0$. The fivefold degenerate state is called ^1D.

After excluding the above micro-states from the table, one finds that there remains a set of nine micro-states belonging to the ^3P state. These correspond to $M_{\mathscr{L}} = +1$, $M_s = 0, \pm 1$; $M_{\mathscr{L}} = 0, M_s = 0, \pm 1$; and $M_{\mathscr{L}} = -1, M_s = 0, \pm 1$. Then $(M_L)_{max} = \mathscr{L} = 1$, and the term is a P state; $2\mathscr{S} + 1 = 3$, and so a ^3P state is ninefold degenerate. There is but one micro-state remaining, corresponding to a ^1S state. Hund's rule places the ^3P state as the ground state.

The degeneracies of some of the atomic states are partially resolved by weak crystalline fields, since the maximum orbital degeneracy allowed in this situation is

Table 1.2. Terms for $(2p)^2$: ^1D, ^3P, ^1S
Quantum nos.: $n\ell\, mm_s = 2\ell\, mm_s$; $m = \pm 1, 0$; $m_s = \pm\tfrac{1}{2}$.

$(mm_s)_1$	$(mm_s)_2$	$M_\mathscr{L}$	$M_\mathscr{S}$	
$1\ \ \tfrac{1}{2}$	$1\ \ -\tfrac{1}{2}$	2	0	
$1\ \ \tfrac{1}{2}$	$0\ \ -\tfrac{1}{2}$	1	0	
$1\ \ \tfrac{1}{2}$	$-1\ \ -\tfrac{1}{2}$	0	0	^1D
$0\ \ \tfrac{1}{2}$	$-1\ \ -\tfrac{1}{2}$	-1	0	
$-1\ \ \tfrac{1}{2}$	$-1\ \ -\tfrac{1}{2}$	-2	0	
$1\ \ \tfrac{1}{2}$	$0\ \ \tfrac{1}{2}$	1	1	
$1\ \ -\tfrac{1}{2}$	$0\ \ \tfrac{1}{2}$	1	0	
$1\ \ -\tfrac{1}{2}$	$0\ \ \tfrac{1}{2}$	1	-1	
$1\ \ \tfrac{1}{2}$	$-1\ \ \tfrac{1}{2}$	0	1	
$1\ \ -\tfrac{1}{2}$	$-1\ \ \tfrac{1}{2}$	0	0	^3P
$1\ \ -\tfrac{1}{2}$	$-1\ \ -\tfrac{1}{2}$	0	-1	
$0\ \ \tfrac{1}{2}$	$-1\ \ \tfrac{1}{2}$	-1	1	
$0\ \ -\tfrac{1}{2}$	$-1\ \ \tfrac{1}{2}$	-1	0	
$0\ \ -\tfrac{1}{2}$	$-1\ \ -\tfrac{1}{2}$	-1	-1	
$0\ \ \tfrac{1}{2}$	$0\ \ -\tfrac{1}{2}$	0	0	^1S

$\mathscr{L} = (M_L)_{max}$

$(M_L)_{max} = 2,\ M_\mathscr{S} = 0$

$(M_L)'_{max} = 1,\ M_\mathscr{S} = 1$

$(M_L)''_{max} = 0,\ M_\mathscr{S} = 0$

$\mathscr{S} = (M_S)_{max}$

$^1D\ (M_\mathscr{L} = 0, \pm 1, \pm 2;\ M_\mathscr{s} = 0)$

$^3P\ (M_\mathscr{L} = 0, \pm 1;\ M_\mathscr{S} = 0, \pm 1)$

$^1S\ (M_\mathscr{L} = 0;\ M_\mathscr{S} = 0)$

three. Neither S nor P states are affected, but they are frequently renamed as A_1 (or A_2) or T_1, respectively. A D state is resolved into $E + T_2$, and an F state to $A_2 + T_1 + T_2$.

1.4 Paramagnetism

These notes are primarily about paramagnets and the interactions they undergo. As has been indicated above, paramagnetism is a property exhibited by substances containing unpaired electrons. This includes the oxygen molecule, nitric oxide and a large number of organic free radicals, but we shall restrict our concerns to the transition metal and rare earth ions and their compounds.

A *paramagnet* concentrates the lines of force provided by an applied magnet and thereby moves into regions of higher field strength. This results in a measurable gain in weight. A paramagnetic susceptibility is generally independent of the field strength, but this is true only under the particular conditions which are discussed below.

Paramagnetic susceptibilities are temperature dependent, however. To a first (high-temperature) approximation, the susceptibility χ varies inversely with temperature, which is the Curie Law:

$$\chi = C/T. \tag{1.1}$$

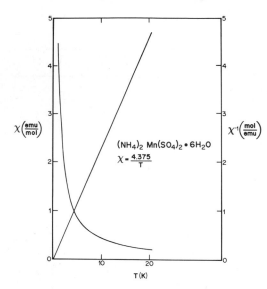

Fig. 1.1. Curie and inverse Curie plots for a salt containing manganese(II), for which $C = 4.375$ emu-K/mol.

Here, χ is the measured susceptibility (from now on, we assume that the proper diamagnetic correction is either unnecessary or else has been made), C is called the Curie constant, and T is the absolute temperature. We plot in Fig. 1.1 the Curie law and the inverse of χ for a representative ion. Since $\chi^{-1} = C^{-1} T$, a plot of χ^{-1} vs. T is a convenient procedure for the determination of the Curie constant; note that the line goes through the origin.

Since the magnitude of χ at room temperature is an inconvenient number, it is common among chemists to report the effective magnetic moment, μ_{eff}, which is defined as

$$\mu_{\text{eff}} = (3k/N)^{1/2} (\chi T)^{1/2}$$
$$= [g^2 s(s+1)]^{1/2} \mu_{\text{B}} . \tag{1.2}$$

Here, k is the Boltzmann constant, $1.38 \times 10^{-23} \, \text{JK}^{-1}$, N is Avogadro's number, $6.022 \times 10^{23} \, \text{mol}^{-1}$ and

$$\mu_{\text{B}} = |e|\hbar/2mc = 9.27 \times 10^{-24} \, \text{JT}^{-1}$$

is the Bohr magneton. Planck's constant h, divided by 2π, is denoted by $\hbar$. That is, the units of μ_{eff} is μ_{B}.

Compounds containing such ions as Cr^{3+} and Mn^{2+} have magnetic moments due to unpaired electron spins outside filled shells. The orbital motion is usually quenched by the ligand field, resulting in spin-only magnetism. Consider an isolated ion, acted on only by its diamagnetic ligands and an external magnetic field H (the Zeeman perturbation). The field will resolve the degeneracy of the various states according to the magnetic quantum number m_s, which varies from $-\mathcal{S}$ to $\mathcal{S}$ in steps of unity. Thus, the ground state of a free manganese(II) ion, which has $\mathcal{L} = 0$, has $\mathcal{S} = 5/2$, and yields six states with $m_s = \pm 1/2, \pm 3/2,$ and $\pm 5/2$. These states are degenerate (of equal energy) in the absence of a field, but the magnetic field H resolves this degeneracy. The

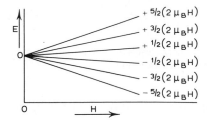

Fig. 1.2. Splitting of the lowest energy level of manganese(II) by a magnetic field into six separate energy levels

energy of each of the sublevels in a field becomes

$$E = m_s g\mu_B H, \tag{1.3}$$

where g is a (Landé) constant, characteristic of each system, which is equal to 2.0023 when $\mathscr{L} = 0$, but frequently differs from this value. The convention used here in applying Eq. (1.3) is that the ^6S ground state of the manganese ion at zero-field is taken as the zero of energy.

The situation is illustrated in Fig. 1.2. The separation between adjacent levels, ΔE, varies with field, and is easily calculated as $\Delta E = g\mu_B H = 2\mu_B H$. In a small field of 0.1 T

$$\Delta E = 2\mu_B H = 2 \times 9.27 \times 10^{-24} \times 0.1 = 1.85 \times 10^{-24} \, J$$

while at a temperature of 1 K, $kT = 1.38 \times 10^{-23} \times 1 = 1.38 \times 10^{-23} \, J$. Thus, at $H = 0.1 \, T$ and $T = 1 \, K$, $\Delta E < kT$, and the resulting levels have almost the same population, as may be found by calculating the distribution of magnetic ions among the various states from the Boltzmann relation,

$$N_i/N_j \propto \exp(-\Delta E_i/kT),$$

ΔE_i being the energy level separation between the levels i and the ground state j.

Since each state corresponds to a different orientation with regard to the external magnetic field, the net magnetic polarization or magnetization M of the substance would then be very small. Or, the field tends to align the spins with itself, but this is opposed by thermal agitation.

On the other hand, in a larger field of 2 T, $\Delta E = 4 \times 10^{-23} \, J$, and at 1 K, $\Delta E > kT$. Then, only the state of lowest energy, $-5/2g\mu_B H$, will be appreciably populated, having about 95% of the total. This corresponds to the moments lining up parallel to the external field, and the magnetization would almost have its largest or saturation value, M_{sat}. Real crystals that have been shown to exhibit behavior of this sort include potassium chrome alum, $KCr(SO_4)_2 \cdot 12 H_2O$, and manganese ammonium Tutton salt, $(NH_4)_2Mn(SO_4)_2 \cdot 6 H_2O$. Each of the magnetic ions in these salts is well-separated from the other magnetic ions, so it is said to be magnetically dilute. The behavior then is analogous to that of ideal gases in the weakness of the interactions. The statistical mechanics of weakly interacting, distinguishable particles is therefore applicable. This means that most properties can be calculated by means of the Boltzmann distribution, as already suggested.

At low temperatures, the vibrational energy and heat capacity of everything but the magnetic ions may be largely ignored. The spins and lattice do interact through the

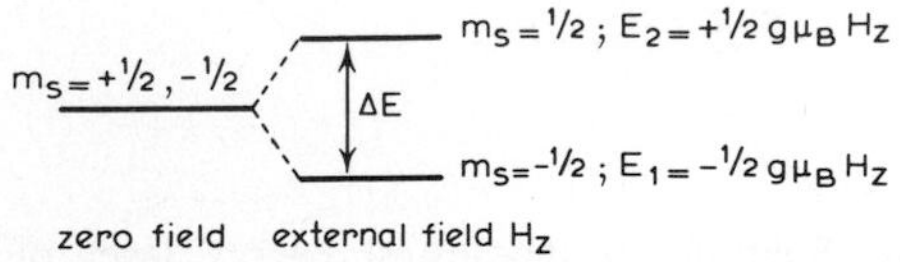

$m_S = +1/2, -1/2$ — zero field

$m_S = 1/2; E_2 = +1/2\, g\mu_B H_z$

ΔE

$m_S = -1/2; E_1 = -1/2\, g\mu_B H_z$

zero field   external field H_z

Fig. 1.3. Energy levels of an electron spin in an external magnetic field

time-dependent phenomenon of spin-lattice relaxation, a subject which will be described later. The magnetic ions form a subsystem with which there is associated a temperature which may or may not be the same as that of the rest of the crystal. The magnetization or total magnetic moment M is not correlated with the rest of the crystal, and even the external magnetic field has no effect on the rest of the crystal. Thus, the working hypothesis, which has been amply justified, is that the magnetic ions form a subsystem with its own identity, describable by the coordinates H, M, and *T*, independent of everything else in the system.

This model provides a basis for a simple derivation of the Curie law, which states that a magnetic susceptibility varies inversely with temperature. Although the same result will be obtained later by a more general procedure, it is useful to illustrate this calculation now. Consider a mol of $\mathscr{S} = 1/2$ particles. In zero field, the two levels $m_s = \pm 1/2$ are degenerate, but split as illustrated in Fig. 1.3 when a field $H = H_z$ is applied. The energy of each level is $m_s g\mu_B H_z$, which becomes $-g\mu_B H/2$ for the lower level, and $+g\mu_B H_z/2$ for the upper level; the separation between them is $\Delta E = g\mu_B H_z$, which for a g of 2, corresponds to about $1\,\text{cm}^{-1}$ at 1 T.

Now, the magnetic moment of an ion in the level n is given as $\mu_n = -\partial E_n/\partial H = -m_s g\mu_B$; the molar macroscopic magnetic moment M is therefore obtained as the sum over magnetic moments weighted according to the Boltzmann factor.

$$M = N \sum_n \mu_n P_n = N \frac{\sum\limits_{m_s = -1/2}^{1/2} (-m_s g\mu_B)\exp(-m_s g\mu_B H_z/kT)}{\sum\limits_{m_s = -1/2}^{1/2} \exp(-m_s g\mu_B H_z/kT)}, \tag{1.4}$$

where the summation in this case extends over only the two states, $m_s = -1/2$ and $+1/2$. Then,

$$M = \tfrac{1}{2}Ng\mu_B \left[\frac{\exp(g\mu_B H_z/2kT) - \exp(-g\mu_B H_z/2kT)}{\exp(g\mu_B H_z/2kT) + \exp(-g\mu_B H_z/2kT)}\right]$$

$$= \tfrac{1}{2}Ng\mu_B \tanh(g\mu_B H_z/2kT), \tag{1.5}$$

since the hyperbolic tangent (see the Appendix to this chapter) is defined as

$$\tanh y = (e^y - e^{-y})/(e^y + e^{-y}). \tag{1.6}$$

One of the properties of the hyperbolic tangent is that, for $y \ll 1$, $\tanh y = y$, as may be seen by expanding the exponentials:

$$\tanh y \simeq \frac{(1 + y + \text{---}) - (1 - y + \text{---})}{(1 + y + \text{---}) + (1 - y + \text{---})} \simeq y.$$

Thus, for moderate fields and temperatures, with $g\mu_B H_z/2kT \ll 1$,

$$\tanh(g\mu_B H_z/2kT) \simeq g\mu_B H_z/2kT$$

and

$$M = Ng^2\mu_B^2 H_z/4kT. \tag{1.7}$$

Since the static molar magnetic susceptibility is defined as $\chi = M/H$, in this case

$$\chi = M/H_z = \frac{Ng^2\mu_B^2}{4kT} = C/T \tag{1.8}$$

which is in the form of the Curie law where the Curie constant

$$C = Ng^2\mu_B^2/4k.$$

This is a special case of the more general and more familiar spin-only formula,

$$\chi = \frac{Ng^2\mu_B^2\mathscr{S}(\mathscr{S}+1)}{3kT} \tag{1.9}$$

$$= N\mu_{eff}^2/3kT, \tag{1.10}$$

where $\mu_{eff}^2 = g^2\mathscr{S}(\mathscr{S}+1)\mu_B^2$ is the square of the "magnetic moment" traditionally reported by inorganic chemists. This quantity is of less fundamental significance than the static susceptibility itself, particularly in those cases where μ_{eff} is not independent of temperature. Other definitions of the susceptibility will be introduced later, since in practice one often measures the differential susceptibility, dM/dH, which is not always identical to the static one.

It is also of interest to examine the behavior of Eq. (1.5) in the other limit, of large fields and very low temperatures. In Eq. (1.6), if $y \gg 1$, one may neglect e^{-y} compared to e^y, and

$$\tanh y = 1. \tag{1.11}$$

Then,

$$M_{sat} = Ng\mu_B/2, \tag{1.12}$$

where the magnetization becomes independent of field and temperature, and, as discussed earlier, becomes the maximum or saturation magnetization M_{sat} which the spin system can exhibit. This situation corresponds to the complete alignment of magnetic dipoles by the field. The more general, spin-only version of (1.12) reads as

$$M_{sat} = Ng\mu_B\mathscr{S}. \tag{1.13}$$

1.5 Some Curie Law Magnets

The Curie constant

$$C = Ng^2 \mu_B^2 \mathscr{S}(\mathscr{S}+1)/3k \qquad (1.14)$$

takes the following form

$$C = \frac{[6.02 \times 10^{23}\,\text{mol}^{-1}]\,[9.27 \times 10^{-24}\,\text{JT}^{-1}]^2\,g^2\mathscr{S}(\mathscr{S}+1)}{3[1.38 \times 10^{-23}\,\text{JK}^{-1}]}$$

which, since it can be shown that the unit T^2 is equivalent to J/cm³, becomes

$$C = 0.125\,g^2\mathscr{S}(\mathscr{S}+1)\,\text{cm}^3\text{-K/mol}$$

or dividing this expression by T in K, one has the volume susceptibility χ in units of cm³/mol.

It should be apparent that a good Curie law magnet will be found only when there are no thermally accessible states whose populations change with changing temperature. Four salts which offer good Curie-law behavior are listed in Table 1.3. Each is assumed to have $g = 2.0$, which is consistent with a lack of mixing of the ground state with nearby states with non-zero orbital angular momentum. A plot of inverse susceptibility vs. temperature for chrome alum in Fig. 1.4 illustrates how well the Curie law holds over a wide region of temperature for this substance.

Table 1.3. Several Curie law magnets.

	$\mathscr{S}$	$\mathscr{S}(\mathscr{S}+1)$	C (expt)	C (calc)
			emu-K/mol	
$KCr(SO_4)_2 \cdot 12\,H_2O$	$\frac{3}{2}$	3.75	1.84	1.88
$(NH_4)_2Mn(SO_4)_2 \cdot 6\,H_2O$	$\frac{5}{2}$	8.75	4.38	4.38
$(NH_4)Fe(SO_4)_2 \cdot 12\,H_2O$	$\frac{5}{2}$	8.75	4.39	4.38
$Gd_2(SO_4)_3 \cdot 8\,H_2O$	$\frac{7}{2}$	15.75	7.80	7.87

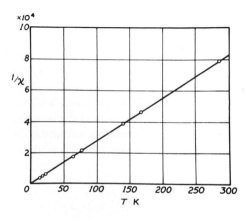

Fig. 1.4. The inverse susceptibility per g of chromium potassium alum as a function of temperature showing the very good agreement with Curie's law (after de Haas and Gorter, 1930)

1.6 Curie-Weiss Law

The Curie law is really just the magnetic analog of the ideal gas law, which is expressed in terms of the variables pVT (pressure, volume, temperature). For magnetic systems, one uses HMT (magnetic field, magnetization, temperature) and the thermodynamic relations derived for a perfect gas can be translated to a magnetic system by replacing p by H and V by $-$M.

When the pressure of a gas becomes too high or molecular interactions occur, deviations from ideal gas behavior occur. Then, one turns to the van der Waals equation to describe the situation, or perhaps an even more elaborate correction is required. In the same fashion, there are many situations in which the Curie law is not strictly obeyed; much of this book is devoted to those cases. One source of the deviations can be the presence of an energy level whose population changes appreciably over the measured temperature interval; another source is the magnetic interactions which can occur between paramagnetic ions. These interactions will be discussed in later chapters, but we can assume their existence for now. To the simplest approximation, this behavior is expressed by a small modification of the Curie law, to the Curie-Weiss law,

$$\chi = C/(T-\theta),\qquad\qquad(1.15)$$

where the correction term, θ, has the units of temperature. Negative values of θ are common, but this should not be confused with unphysical negative temperatures. The θ is obtained empirically from a plot of χ^{-1} vs. T, as for the Curie law, but now the intercept with the abscissa is not at the origin. When θ is negative it is called *antiferromagnetic* in sign; when θ is positive, it is called *ferromagnetic*. These terms will be explained in later chapters. The constant, θ, characteristic of any particular sample, is best evaluated when $T \geq 10\theta$, as curvature of χ^{-1} usually becomes apparent at smaller values of T.

The Curie and Curie-Weiss laws are compared for a hypothetical situation in Fig. 1.5.

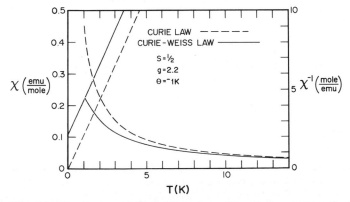

Fig. 1.5. The Curie and Curie-Weiss laws. The Curie constant C=0.454 emu-K/mol ($\mathscr{S}=\frac{1}{2}$, g=2.2) has been used for both curves, while $\theta=-1$ K has been used

1.7 Temperature Independent Paramagnetism

It sometimes happens that systems with a spin singlet ground state, which from the development presented above would be expected to be diamagnetic, in fact exhibit a weak paramagnetic behavior. This paramagnetism is found to be temperature independent and, since it is only of the order of 10^{-4} emu/mol, is generally more important when considering measurements made at 80 K and above. This temperature independent paramagnetism (TIP) arises from a mixing into the ground state wave function the wave function of the excited states that are *not* thermally populated. They may be connected with an orbital operator to the ground state, however; a spin operator will not suffice, for the problem assumes that $\mathscr{S}=0$ for the ground state, and all spin matrix elements will therefore be zero.

An ion such as octahedral cobalt(III), with a $^1A_1$ ground state, is typical of those which exhibits a TIP. The first excited singlet state ($^1T_1$) usually lies some $16\,000$–$21\,000$ cm^{-1} above this, varying with the particular compound, and it can be shown [2] that

$$\chi_{TIP} \simeq 4/\varDelta(T_1),$$

where $\varDelta(T_1)$ is the energy of the $^1T_1$ state, in wave numbers, above the ground state. For $[Co(NH_3)_6]^{3+}$, the $^1T_{1g}$ state is at $21\,000$ cm^{-1}, so one calculates

$$\chi_{TIP} = 1.95 \times 10^{-4} \text{ emu/mol},$$

which is the order of magnitude of the experimental result.

Temperature independent paramagnetism has been observed in such other systems as chromate and permanganate ions and with such paramagnetic systems as octahedral cobalt(II) complexes which have low-lying orbital states.

1.8 References

1. Huheey J.E., Inorganic Chemistry, (1983), 3rd Ed.: Harper & Row, New York
2. Carlin R.L. and van Duyneveldt A.J., (1977) Magnetic Properties of Transition Metal Compounds, Springer, Berlin Heidelberg New York
3. Cotton F.A. and Wilkinson G., (1980), 4th Ed.: Advanced Inorganic Chemistry, Wiley, New York
4. Earnshaw A., Introduction to Magnetochemistry, (1968) Academic Press, New York
5. Mabbs F.E. and Machin D.J., Magnetism and Transition Metal Complexes, (1973) Chapman and Hall, London

1.9 General References

Casimir H.B.G., (1961) Magnetism and Very Low Temperatures, Dover Publications, New York
Earnshaw A., (1968) Introduction to Magnetochemistry, Academic Press, New York
Garrett C.G.B., (1954) Magnetic Cooling, Harvard University Press, Cambridge
Gopal E.S.R., (1966) Specific Heats at Low Temperatures, Plenum Press, New York
van den Handel J., (1956) Handbuch der Physik, Bd XV, Springer, Berlin Göttingen Heidelberg

Hudson R.P., (1972) Principles and Application of Magnetic Cooling, North-Holland, Amsterdam

Kittel C., (1974), 4th Ed.: Introduction to Solid State Physics, J. Wiley and Sons, New York

Mabbs F.E. and Machin D.J., (1973) Magnetism and Transition Metal Complexes, Chapman and Hall, London

van Vleck J.H., (1932) The Theory of Electric and Magnetic Susceptibilities, Oxford University Press, Oxford

Zemansky M.W., (1968), 5th Ed.: Heat and Thermodynamics, McGraw-Hill, New York

1.10 Appendix

1.10.1 Physical Constants and Units

Molar gas constant	$R = 8.3144\,\mathrm{J\,mol^{-1}\,K^{-1}}$
Avogadro constant	$N = 6.0220 \times 10^{23}\,\mathrm{mol^{-1}}$
Boltzmann constant	$k = 1.3807 \times 10^{-23}\,\mathrm{J\,K^{-1}}$
Bohr magneton	$\mu_B = 9.274 \times 10^{-24}\,\mathrm{J\,T^{-1}}$

Easy to remember

$$N\mu_B^2/k = 0.375 \text{ emu-K/mol}$$

and for the translation of energy 'units'

$$1\,\mathrm{cm^{-1}} \simeq 30\,\mathrm{GHz} \simeq 1.44\,\mathrm{K} \simeq 1.24 \times 10^{-4}\,\mathrm{eV} \simeq 1.99 \times 10^{-23}\,\mathrm{J}.$$

At 1 K, $(kT/hc) = 0.695\,\mathrm{cm^{-1}}$. We do not distinguish magnetic field from magnetic induction, and take 1 Tesla (T) as 1 kOe.

1.10.2 Hyperbolic Functions

The hyperbolic functions occur repeatedly in the theory of magnetism. Though they are described in most elementary calculus texts, some properties are summarized here.

The two basic hyperbolic functions are defined in terms of exponentials as follows

$$\sinh x = \tfrac{1}{2}[\exp(x) - \exp(-x)],$$

$$\cosh x = \tfrac{1}{2}[\exp(x) + \exp(-x)]$$

and, by analogy to the common trigonometric functions, there are four more hyperbolic functions defined in terms of $\sinh x$ and $\cosh x$, as follows:

$$\tanh x = \sinh x/\cosh x$$

$$\mathrm{sech}\, x = 1/\cosh x$$

$$\coth x = \cosh x/\sinh x$$

$$\mathrm{csch}\, x = 1/\sinh x.$$

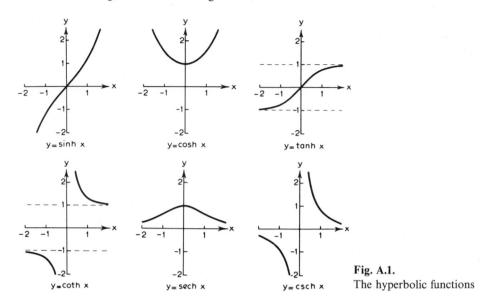

Fig. A.1.
The hyperbolic functions

These functions are sketched in Fig. A.1. It is evident that, unlike the trigonometric functions, none of the hyperbolic functions is periodic. $\sinh x$, $\tanh x$, $\coth x$, and $\operatorname{csch} x$ are odd functions, while $\cosh x$ and $\operatorname{sech} x$ are even.

Lastly, the derivatives of the hyperbolic functions may be shown to be:

$$d(\sinh x) = (\cosh x)dx$$
$$d(\cosh x) = (\sinh x)dx$$
$$d(\tanh x) = (\operatorname{sech}^2 x)dx$$
$$d(\coth x) = -(\operatorname{csch}^2 x)dx$$
$$d(\operatorname{sech} x) = -(\operatorname{sech} x)(\tanh x)dx$$
$$d(\operatorname{csch} x) = -(\operatorname{csch} x)(\coth x)dx .$$

1.10.3 Magnetic Moment of a Magnetic Ion Subsystem

It is illuminating to calculate the magnetic moment of a magnetic system with arbitrary spin-quantum number. In order to allow an orbital contribution, we use the total quantum number $\mathscr{J}$, where $\mathscr{J} = \mathscr{L} + \mathscr{S}$. With $\mathscr{J}\hbar$ the total angular momentum, the energy is $E = -\boldsymbol{\mu} \cdot \boldsymbol{H}$, where $\boldsymbol{\mu} = g\mu_B\mathscr{J}$ and $E_m = m g\mu_B H_z$. The Boltzmann factor is

$$P_m = \exp(-m g\mu_B H_z/kT)/ \sum_m \exp(-m g\mu_B H_z/kT)$$

so that $\langle\mu_z\rangle$, the average magnetic moment of one atom is

$$\langle\mu_z\rangle = \frac{\sum_{m=-\mathscr{J}}^{\mathscr{J}} (-m g\mu_B)\exp(-m g\mu_B H_z/kT)}{\sum_{m=-\mathscr{J}}^{\mathscr{J}} \exp(-m g\mu_B H_z/kT)} .$$

But,

$$\sum_{m=-\mathscr{I}}^{\mathscr{I}} (-mg\mu_B)\exp(-mg\mu_B H_z/kT) = kT\frac{\partial Z_a}{\partial H_z},$$

where

$$Z_a = \sum_{m=-\mathscr{I}}^{\mathscr{I}} \exp(-mg\mu_B H_z/kT)$$

is the magnetic partition function of one atom. Hence,

$$\langle\mu_z\rangle = \frac{kT}{Z_a}\frac{\partial Z_a}{\partial H_z} = kT\frac{\partial\ln Z_a}{\partial H_z}.$$

Now, define a dimensionless parameter η,

$$\eta = g\mu_B H_z/kT,$$

which measures the ratio of the magnetic energy $g\mu_B H_z$, which tends to align the magnetic moments, to the thermal energy kT, that tends to keep the system oriented randomly. Then,

$$Z_a = \sum_{m=-\mathscr{I}}^{\mathscr{I}} e^{-\eta m} = e^{\eta\mathscr{I}} + e^{\eta(\mathscr{I}-1)} + \ldots + e^{-\eta\mathscr{I}}$$

which is a finite geometric series that can be summed to yield

$$Z_a = (e^{-\eta\mathscr{I}} - e^{\eta(\mathscr{I}+1)})/(1-e^\eta).$$

On multiplying both top and bottom by $e^{-\eta/2}$, Z_a becomes

$$Z_a = \frac{e^{-\eta(\mathscr{I}+1/2)} - e^{\eta(\mathscr{I}+1/2)}}{e^{-\eta/2} - e^{\eta/2}} = \frac{\sinh(\mathscr{I}+1/2)\eta}{\sinh\eta/2},$$

since $(\sinh y) = \dfrac{e^y - e^{-y}}{2}$. Thus, $\ln Z_a = \ln\sinh(\mathscr{I}+1/2)\eta - \ln\sinh\eta/2$, and

$$\langle\mu_z\rangle = \frac{kT\partial\ln Z_a}{\partial H_z} = \frac{kT\partial\ln Z_a}{\partial\eta}\frac{\partial\eta}{\partial H_z} = g\mu_B\frac{\partial\ln Z_a}{\partial\eta}.$$

Hence

$$\langle\mu_z\rangle = g\mu_B\left[\frac{(\mathscr{I}+\tfrac{1}{2})\cosh(\mathscr{I}+\tfrac{1}{2})\eta}{\sinh(\mathscr{I}+\tfrac{1}{2})\eta} - \frac{\tfrac{1}{2}\cosh\eta/2}{\sinh\eta/2}\right]$$

or

$$\langle \mu_z \rangle = g\mu_B \mathscr{J} B_J(\eta),$$

where the Brillouin function $B_J(\eta)$ is defined as

$$B_J(\eta) = \frac{1}{\mathscr{J}} \left[\left(\mathscr{J} + \frac{1}{2} \right) \coth \left(\mathscr{J} + \frac{1}{2} \right) \eta - \frac{1}{2} \coth \eta/2 \right].$$

Now,

$$(\coth y) = \frac{1 + \dfrac{1}{2} y^2 + \cdots}{y + \dfrac{1}{6} y^3 + \cdots} \simeq \frac{\left(1 + \dfrac{1}{2} y^2 \right) \left(\dfrac{1}{y} \right)}{\left(1 + \dfrac{1}{6} y^2 \right)}$$

$$\simeq \left(\frac{1}{y} \right) \left(1 + \frac{1}{2} y^2 \right) \left(1 - \frac{1}{6} y^2 \right)$$

$$\simeq \left(\frac{1}{y} \right) \left(1 + \frac{1}{3} y^2 \right),$$

or for small y,

$$(\coth y) = 1/y + y/3.$$

Thus, in the two cases, the limiting behaviors of the Brillouin function are:
a) For $\eta \gg 1$,

$$B_J(\eta) = (1/\mathscr{J})[(\mathscr{J} + \tfrac{1}{2}) - \tfrac{1}{2}] = 1.$$

(Using the more exact expression for the coth, $B_J(\eta) = 1 - e^{-\eta}/\mathscr{J}$.)

$\eta = g\mu_B H/kT \gg 1$ means

$$H/T \gg \frac{k}{g\mu_B} = \frac{1.38 \times 10^{-23} \, \text{J/K}}{2 \times 9.27 \times 10^{-24} \, \text{J/T}}$$

$$\simeq 0.7 \, \text{T/K}.$$

b) For $\eta \ll 1$,

$$B_J(\eta) = (1/\mathscr{J})\{(\mathscr{J} + \tfrac{1}{2})[((\mathscr{J} + \tfrac{1}{2})\eta)^{-1}$$
$$+ (\tfrac{1}{3})(\mathscr{J} + \tfrac{1}{2})\eta] - \tfrac{1}{2}[2/\eta + \eta/6]\}$$
$$= 1/\mathscr{J}\{\tfrac{1}{3}(\mathscr{J} + \tfrac{1}{2})^2 \eta - 1/12\eta\}$$
$$= (\eta/3\mathscr{J})(\mathscr{J}^2 + \mathscr{J} + \tfrac{1}{4} - \tfrac{1}{4}) = (\mathscr{J} + 1)\eta/3$$

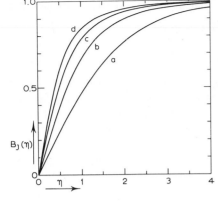

Fig. A.2. The Brillouin function, $B_J(\eta)$, plotted vs. η for several values of $\mathscr{J}$; curve a: $\mathscr{J} = \frac{1}{2}$, b: $\mathscr{J} = \frac{3}{2}$, c: $\mathscr{J} = \frac{5}{2}$, and d: $\mathscr{J} = \frac{7}{2}$

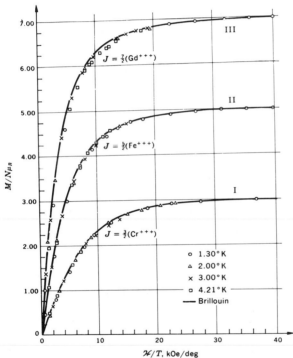

Fig. A.3. Plot of magnetization per magnetic ion, expressed in Bohr magnetons, against H/T for (I) chromium potassium alum ($\mathscr{J} = \frac{3}{2}$); (II) iron ammonium alum ($\mathscr{J} = \frac{5}{2}$); and (III) gadolinium sulfate ($\mathscr{J} = \frac{7}{2}$). The points are experimental results of W.E. Henry (1952), and the solid curves are graphs of the Brillouin equation. (Recall that $10\,\mathrm{kOe}$ is $1\,\mathrm{T}$)

and the initial slope of a plot of $B_J(\eta)$ vs. η will be $(\mathscr{J} + 1)/3$. Such a plot for several values of $\mathscr{J}$ is illustrated in Fig. A.2.

For N non-interacting atoms, the mean magnetic moment, or magnetization, is

$$M = N\langle \mu_z \rangle = Ng\mu_B \mathscr{J} B_J(\eta) \tag{A1.1}$$

and so for η small, M is proportional to η, that is to say, $M \propto H/T$. In fact, for $g\mu_B H_z/kT \ll 1$, $M = \chi H_z$, where

$$\chi = Ng^2 \mu_B^2 \mathscr{J}(\mathscr{J}+1)/3kT$$

which once again is simply the Curie law. As one should now expect, in the limit of large η, $g\mu_B H_z/kT \gg 1$, M becomes $Ng\mu_B \mathscr{J}$, which is again the saturation moment. The physical ideas introduced here are not new, but of course the theory allows a comparison of theory with experiment over the full range of magnetic fields and temperatures. The theory was tested by Henry [1] amongst others for three salts of different $\mathscr{J}$ value with no orbital contribution (i.e., $\mathscr{L}=0$, $\mathscr{S}=\mathscr{J}$, and $g=2.0$) at four temperatures (1.30, 2.00, 3.00, 4.21 K). As illustrated in Fig. A.3, at 1.3 K a field of 5 T produces greater than 99.5% magnetic saturation.

1.10.4 Reference

1. Henry W.E., Phys. Rev. **88**, 559 (1952)

2. Paramagnetism: Zero-Field Splittings

2.1 Introduction

The subject of zero-field splittings requires a detailed description, for it is central to much that is of interest in magnetism. Zero-field splittings are often responsible for deviations from Curie law behavior, they give rise to a characteristic specific heat behavior, and they limit the usefulness of certain substances for adiabatic demagnetization. Furthermore, zero-field splittings cause a single-ion anisotropy which is important in characterizing anisotropic exchange, and are also one of the sources of canting or weak ferromagnetism (Chapt. 7).

An energy level diagram which illustrates the phenomenon of zero field splitting is presented in Fig. 2.1, where for convenience we set $\mathscr{L}=0$ for an $\mathscr{S}=\frac{3}{2}$ system. In an axial crystalline field, the fourfold degeneracy, $m_s = \pm\frac{1}{2}, \pm\frac{3}{2}$, is partially resolved with the $\pm\frac{3}{2}$ states separated an amount 2 D (in units of energy) from the $\pm\frac{1}{2}$ states. We shall use below the notation $|m_s\rangle$ to denote the wave function corresponding to each of these states. The population of the levels then depends, by the Boltzmann principle, on the relative values of the zero-field splitting parameter D and the thermal energy, kT. The fact that D not only has magnitude but also sign is of large consequence for many magnetic systems. As drawn, $D>0$; if $D<0$, then the $\pm\frac{3}{2}$ levels have the lower energy. The Zeeman energy of a state labeled by the quantum number m_s is

$$E = m_s g \mu_B H_z, \tag{2.1}$$

where the orientation of the field is denoted by the subscript z (or equivalently, $\parallel$). This is done because an axial crystalline field which is capable of causing a zero-field splitting establishes a principal or symmetry axis in the metal-containing molecule, and it is important to specify whether that axis of the molecule is oriented parallel to the field (H_z), perpendicular ($H_\perp$) or at some intermediate direction. The splittings caused by H_z are illustrated in Fig. 2.1.

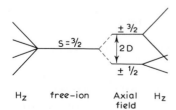

Fig. 2.1. The behavior of an $\mathscr{S}=\frac{3}{2}$ state (center). The usual Zeeman splitting is illustrated on the left, while on the right the effect of zero-field splitting is illustrated. The degeneracies are partially resolved before the magnetic field affects the levels

The name "zero-field splitting" arises from the fact that the splitting occurs in the absence of a magnetic field. As we learn from crystal field theory, it can be ascribed to the electrostatic field of the ligands. For an axial distortion of an $\mathscr{S}=\frac{3}{2}$ system, the conventional zero-field splitting parameter is called $2D$. For chromium (III), this splitting of the $^4A_2$ state is typically much less than a Kelvin (i.e., a few tenths of a cm^{-1}); tetrahedral cobalt (II), however, also has a $^4A_2$ ground state and typically exhibits zero-field splittings of 10–15 K.

2.2 Van Vleck's Equation

Before we can calculate the effect of zero-field splittings on magnetic susceptibilities, we require a more general method for calculating these susceptibilities. If the energy levels of a system are known, the magnetic susceptibility may always be calculated by application of Van Vleck's equation. The standard derivation [1] follows, along with several typical applications.

Let the energy, E_n, of a level be developed in a series in the applied field:

$$E_n = E_n^0 + HE_n^{(1)} + H^2 E_n^{(2)} + \ldots ,$$

where, in the standard nomenclature, the term linear in H is called the first-order Zeeman term, and the term in H^2 is the second order Zeeman term. Since $\mu_n = -\partial E_n/\partial H$, the total magnetic moment, M, for the system follows as before:

$$M = \frac{N \sum_n \mu_n \exp(-E_n/kT)}{\sum_n \exp(-E_n/kT)} .$$

Now,

$$\exp(-E_n/kT) = \exp\{-(E_n^0 + HE_n^{(1)} + \ldots)/kT\}$$
$$\simeq (1 - HE_n^{(1)}/kT) \exp(-E_n/kT)$$

by expansion of the exponential, and

$$\mu_n = -\frac{\partial E_n}{\partial H} = -E_n^{(1)} - 2HE_n^{(2)} + \cdots .$$

To this approximation we obtain

$$M = N \frac{\sum_n (-E_n^{(1)} - 2HE_n^{(2)})(1 - HE_n^{(1)}/kT) \exp(-E_n^0/kT)}{\sum_n \exp(-E_n^0/kT)(1 - HE_n^{(1)}/kT)} .$$

We limit the derivation to paramagnetic substances, as distinct from ferromagnetic ones so that the absence of permanent polarization in zero magnetic field (i.e., M = 0 at

H=0) requires that

$$\sum_n -E_n^{(1)} \exp(-E_n^0/kT) = 0.$$

Retaining only terms linear in H,

$$M = N \frac{H \sum_n [(E_n^{(1)})^2/kT - 2E_n^{(2)}] \exp(-E_n^0/kT)}{\sum_n \exp(-E_n^0/kT)}.$$

Since the static molar susceptibility is $\chi = M/H$, the final result is:

$$\chi = \frac{N \sum_n [(E_n^{(1)})^2/kT - 2E_n^{(2)}] \exp(-E_n^0/kT)}{\sum_n \exp(-E_n^0/kT)}. \tag{2.2}$$

The degeneracy of any of the levels has been neglected here, but of course the r-degeneracy of any level must be summed r times.

The general form of the spin-only susceptibility is obtained as follows. Consider an orbital singlet with $2\mathscr{S}+1$ spin degeneracy. The energy levels are at $m_\mathscr{S} g\mu_\mathrm{B}H$, where $m_\mathscr{S}$ spans the values from $+\mathscr{S}$ to $-\mathscr{S}$. Note that the energy levels correspond to

$$E_n^0 = E_n^{(2)} = 0, \qquad E_n^{(1)} = m_\mathscr{S} g\mu_\mathrm{B}$$

since the zero of energy can be taken as that of the level of lowest energy in the magnetic field. Then, applying these results to Eq. (2.2), i.e.,

$$\chi = \frac{Ng^2\mu_\mathrm{B}^2}{kT} \frac{(-\mathscr{S})^2 + (-\mathscr{S}+1)^2 + \cdots + (+\mathscr{S})^2}{2\mathscr{S}+1}$$

or

$$\chi = Ng^2\mu_\mathrm{B}^2\mathscr{S}(\mathscr{S}+1)/3kT \tag{2.3}$$

since

$$\sum_{-\mathscr{S}}^{\mathscr{S}} m_\mathscr{S}^2 = \frac{1}{3} \mathscr{S}(\mathscr{S}+1)(2\mathscr{S}+1).$$

This result was presented earlier as Eq. (1.9).

2.3 Paramagnetic Anisotropy

Zero-field splittings are one of the most important sources of paramagnetic anisotropies, that is, of susceptibilities that differ as the external (measuring) magnetic field

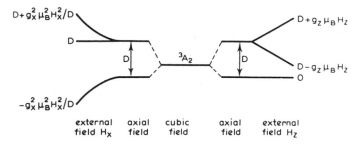

Fig. 2.2. Nickel(II) with an internal axial field and an external magnetic field

is rotated with respect to the principal ($\parallel$, $\perp$) axis of the molecule. With the aid of the energy level diagram of nickel(II) in a weak tetragonal field, Fig. 2.2, and Van Vleck's equation, we are now able to calculate the susceptibilities with the measuring field parallel or perpendicular to the principal axis. The parameter D measures the zero-field splitting of the ground state. Our procedure yields for the parallel susceptibility,

$$\chi_\parallel/N = \frac{(g_z^2\mu_B^2/kT)\exp(-D/kT) + 0\cdot\exp(0/kT) + (g_z^2\mu_B^2/kT)\exp(-D/kT)}{[1+2\exp(-D/kT)]}.$$

Assuming D may be large (i.e., the exponentials are not expanded)

$$\chi_\parallel = \frac{2Ng_z^2\mu_B^2}{kT}\frac{\exp(-D/kT)}{[1+2\exp(-D/kT)]}. \tag{2.4}$$

Frequently, however, $D \ll kT$ (with $T \simeq 300$ K at room temperature, and D/k is often of the order of a few Kelvins, and rarely is as large as 25 K), and then this reduces to

$$\chi_\parallel \simeq \frac{2Ng_z^2\mu_B^2(1-D/kT)}{kT(1+2-2D/kT)} \simeq \frac{2Ng_z^2\mu_B^2}{3kT}(1-D/3kT). \tag{2.5}$$

Many susceptibility measurements are made on powdered paramagnetic samples, so that only the average susceptibility, $\langle\chi\rangle$, is obtained. The quantity $\langle\chi\rangle$ is defined as

$$\langle\chi\rangle = (\chi_\parallel + 2\chi_\perp)/3, \tag{2.6}$$

where $\chi_\perp$ is the susceptibility measured with the field perpendicular to the principal axis. We require $\chi_\perp$ in order to calculate $\langle\chi\rangle$ and the calculation is done in the following fashion.

The set of energy levels in a field used above for the calculation of $\chi_\parallel$ is of course valid only when the applied field is parallel to the axis of quantization. If an anisotropic single crystal is oriented such that the external field is normal to the principal molecular or crystal field axis, we need to consider the effect of this field H_x (H_y is equivalent in trigonal and tetragonal fields) on the energy levels. The problem is to calculate the energies of the levels labeled $m_s = \pm 1$ in zero-field which are D in energy above the level with $m_s = 0$. The Hamiltonian is

$$H_x = g_x \mu_B H_x S_x + D\left[S_z^2 - \frac{1}{3}\mathscr{S}(\mathscr{S}+1)\right]$$

$$= \frac{g_x \mu_B H_x}{2}(S_+ + S_-) + D\left[S_z^2 - \frac{1}{3}\mathscr{S}(\mathscr{S}+1)\right], \qquad (2.7)$$

where $S_\pm = S_x \pm iS_y$, and S_x, S_y are the operators for the x, y components of electron spin, respectively, and $\mathscr{S}$ is 1. We require the eigenvalues of the 3×3 matrix made up from the set of matrix elements $\langle m_s | H_x | m_s' \rangle$. We make use of the standard formulas

$$\langle m_s | S_z | m_s \rangle = m_s$$

$$\langle m_s \pm 1 | S_\pm | m_s \rangle = [\mathscr{S}(\mathscr{S}+1) - m_s(m_s \pm 1)]^{1/2}.$$

The only nonzero off-diagonal elements are

$$\langle \pm 1 | H_x | 0 \rangle = \langle 0 | H_x | \pm 1 \rangle = g_x \mu_B H_x / \sqrt{2}$$

and the matrix takes the form

$$
\begin{array}{ccc}
+1 & -1 & 0
\end{array}
$$
$$
\begin{bmatrix}
D & 0 & g_x \mu_B H_x / \sqrt{2} \\
0 & D & g_x \mu_B H_x / \sqrt{2} \\
g_x \mu_B H_x / \sqrt{2} & g_x \mu_B H_x / \sqrt{2} & 0
\end{bmatrix}
$$

Subtracting the eigenvalues W from each of the diagonal terms and setting the determinant of the matrix equal to zero results in the following cubic equation

$$W^3 + W^2(-2D) + W(D^2 - g_x^2 \mu_B^2 H_x^2) + g_x^2 \mu_B^2 H_x^2 D = 0$$

which has roots

$$W_1 = \tfrac{1}{2}[D - (D^2 + 4g_x^2 \mu_B^2 H_x^2)^{1/2}]$$
$$W_2 = D$$
$$W_3 = \tfrac{1}{2}[D + (D^2 + 4g_x^2 \mu_B^2 H_x^2)^{1/2}].$$

Making use of the expression $(1+r)^{1/2} \approx 1 + (\tfrac{1}{2})r$ for small r, which corresponds in this case to small magnetic fields, the energy levels become

$$W_1 = -g_x^2 \mu_B^2 H_x^2 / D$$
$$W_2 = D$$
$$W_3 = D + g_x^2 \mu_B^2 H_x^2 / D. \qquad (2.8)$$

These energy levels are plotted, schematically, as a function of external field in Fig. 2.2, on the left side. They are readily inserted into Van Vleck's equation to yield

$$\chi_x/N = \frac{2g_x^2\mu_B^2/D - (2g_x^2\mu_B^2/D)\exp(-D/kT)}{[1+2\exp(-D/kT)]}$$

or, since the notation x and $\perp$ are equivalent here,

$$\chi_x = \chi_\perp = \frac{2Ng_\perp^2\mu_B^2}{3kT}\frac{6kT}{D}\left[\frac{1-\exp(-D/kT)}{1+2\exp(-D/kT)}\right]. \tag{2.9}$$

Neglecting the difference between $g_\parallel$ and $g_\perp$, we also obtain, after averaging,

$$\langle\chi\rangle = \frac{2Ng^2\mu_B^2}{3kT}\left[\frac{2/x - 2\exp(-x)/x + \exp(-x)}{1+2\exp(-x)}\right],$$

where $x = D/kT$.

Although it was assumed that $D > 0$, the equations are equally valid if $D < 0$. The complete solutions for $\chi_\parallel$ and $\chi_\perp$ are plotted in Fig. 2.3 for a typical set of parameters. It will be noticed that Curie-Weiss-like behavior is found at temperatures high with respect to D/k, but marked deviations occur at $D/kT \leq 1$. The perpendicular susceptibility approaches a constant value at low temperatures, while the parallel susceptibility goes to zero. A measurement of $\langle\chi\rangle$ alone in this temperature region would offer few clues as to the nature of the situation.

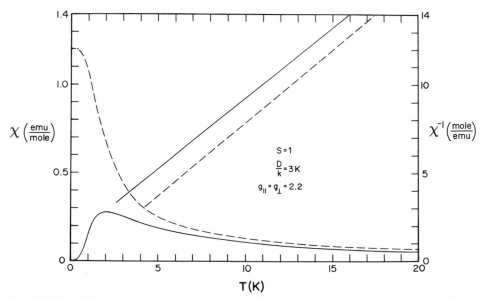

Fig. 2.3. Parallel and perpendicular susceptibilities and their inverses, for a (hypothetical) nickel(II) ion with $g_\parallel = g_\perp = 2.2$ and $D/k = 3$ K; drawn lines: $\chi_\parallel$ and $\chi_\parallel^{-1}$; dotted-drawn lines: $\chi_\perp$ and $\chi_\perp^{-1}$

Plots of χ^{-1} vs. T are also seen to be linear at high temperatures and to deviate when $D/kT \to 1$. If the straight-line portion is extrapolated to $\chi^{-1}=0$, non-zero intercepts are obtained. This behavior is often treated empirically by means of the Curie-Weiss law, when χ is written

$$\frac{T-\theta}{C} = \chi^{-1}, \qquad (2.10)$$

where $1/C$ is the slope of the curve (and C is the supposed Curie constant) and θ is the intercept of the curve with the T axis (positive or negative). This result illustrates merely one of the sources of Curie-Weiss behavior, exchange (interactions between the magnetic ions) being another, and shows that caution is required in ascribing the non-zero value of θ to any one particular source.

The ground-state energy-levels of Fig. 2.2 are applicable particularly to vanadium(III) and nickel(II). The parameter D/k is relatively larger for vanadium, often being of the order of 5 to 10 K [2]. Interestingly, $D>0$ for every vanadium compound investigated to date. Nickel usually has a smaller D/k, some 0.1 to 6 K, and both signs of D have been observed. There seems to be no rational explanation for these observations at present. As will be discussed later, there are several compounds of nickel studied recently that seem to have very large zero-field splittings, so the observations described here may no longer be valid when a larger number and variety of compounds have been studied. A typical set of data is illustrated in Fig. 2.4; only two parameters, the g values and D, are required in order to fit the measurements on $[C(NH_2)_3]V(SO_4)_2 \cdot 6H_2O$ [3].

Another simple illustration of the use of Van Vleck's equation is provided by chromium(III) in an octahedral crystal field, e.g., in $[Cr(H_2O)_6]^{3+}$. The ion has a $^4A_2$ ground state, which was illustrated in Fig. 2.1 and $^4T_2$ and $^4T_1$ excited states at, respectively, around $18\,000\,\text{cm}^{-1}$ and $26\,000\,\text{cm}^{-1}$. The excited states therefore make no contribution to the paramagnetic susceptibility, and may be ignored. The ground state is orbitally nondegenerate and exhibits only a first-order Zeeman effect, corresponding to $\mathscr{S}=\frac{3}{2}$, $m_s=\pm\frac{1}{2}$, $\pm\frac{3}{2}$. Upon application of Van Vleck's equation, the spin-only result for a system with three spins is obtained.

If, now, there is a slight axial field, the zero-field splitting may partially resolve the fourfold degeneracy of $^4A_2$ into the $m_s=\pm\frac{1}{2}$, and the $m_s=\pm\frac{3}{2}$ levels, with the latter 2D (in energy) above the former. (Alternately, the sign of D may be changed.) In this

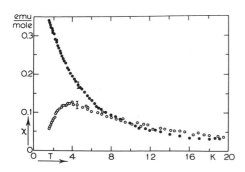

Fig. 2.4. Parallel (o) and perpendicular (●) susceptibilities of $[C(NH_2)_3]V(SO_4)_2 \cdot 6H_2O$. From Ref. [3]

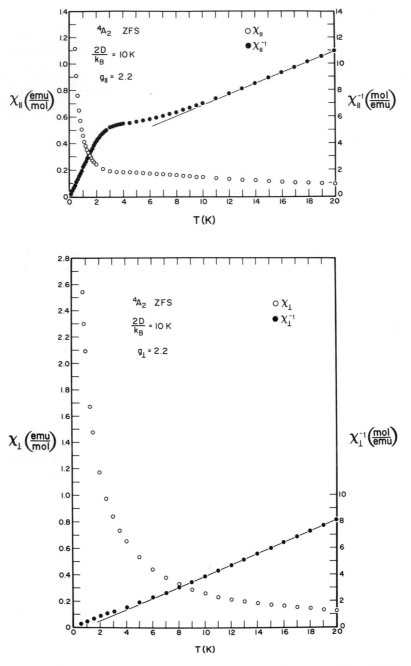

Fig. 2.5. The susceptibilities of Eqs. (2.11) and (2.12) with $g_\parallel = g_\perp = 2.2$ and $2D/k = 10$ K

situation one calculates, for H_z parallel to the principal molecular axis,

$$\chi_{\parallel} = \frac{Ng_z^2\mu_B^2}{4kT}\left[\frac{1+9\exp(-2D/kT)}{1+\exp(-2D/kT)}\right].$$
(2.11)

For $2D/kT \ll 1$, corresponding to $2D \rightarrow 0$ or T becoming large, $\chi = (\frac{5}{4})(Ng^2\mu_B^2/kT)$, the isotropic spin-only formula. For $2D/kT \gg 1$, corresponding to a large zero-field splitting, the exponential terms go to zero, yielding $\chi = Ng^2\mu_B^2/4kT$, which is the spin-only formula for isotropic $\mathscr{S} = \frac{1}{2}$. That is, only the $m_s = \pm\frac{1}{2}$ state is populated in this case.

The calculation of $\chi_\perp$ for Cr(III) is more complicated [4]; the result is

$$\chi_\perp = \frac{Ng_\perp^2\mu_B^2}{kT}[1+\exp(-2D/kT)]^{-1} + \frac{3Ng_\perp^2\mu_B^2\tanh(D/kT)}{4D}.$$
(2.12)

Equations (2.11) and (2.12) are plotted for a set of representative parameters in Figs. 2.5 and 2.6.

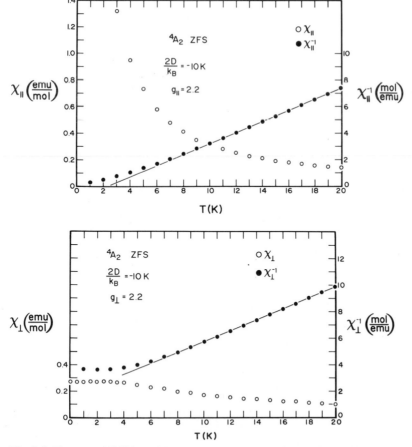

Fig. 2.6. The susceptibilities of Eqs. (2.11) and (2.12) with $g_\parallel = g_\perp = 2.2$ and $2D/k = -10$ K

2.4 Spin-Orbit Coupling

It is difficult to discuss spin-orbit coupling without mathematics and matrix elements. But, it is an important concept and effects many magnetic quantities, so we shall attempt to give a qualitative discussion of it.

In Russell-Saunders coupling, the angular momenta, m, of each electron of an atom are summed to give the total orbital angular momentum, L. Similarly, the m_s values of the spin components are summed to give the quantum number $\mathscr{S}$. The interaction between L and S is assumed to be so small that the values of $\mathscr{L}$ and $\mathscr{S}$ are not affected by such an interaction. This interaction, called spin-orbit coupling, is often represented by an energy term, $\lambda L \cdot S$, where λ then has energy units. The most commonly used units for λ is cm^{-1}; the parameter ζ is also often used, and defined as $\lambda = \pm(\zeta/2S)$, where the $+$ sign is used for an electronic shell less than half full, and the $-$ sign if the shell is more than half full. Spin-orbit coupling is then not an important phenomenon for systems with filled shells (for then, $S=0$) and is of only second-order importance for systems with half-filled shells [e.g., spin-free manganese(II)]. The parameter λ has a characteristic (empirical!) value for each oxidation state of each free ion, and is often found (or at least represented) to have a slightly smaller value when an ion is put into a compound. The value of λ increases strikingly with atomic number, so that spin-orbit coupling becomes more and more important for the heavier elements. Indeed, for the rare earth ions, for example $\mathscr{L}$ and $\mathscr{S}$ no longer are good quantum numbers (that is, they are strongly coupled) and we must turn to a different formalism, called $j-j$ coupling, in order to understand the electronic structure.

Spin-orbit coupling is an important contributor to the zero-field splitting effects discussed above. For example, the 2 D splitting of the $^4A_2$ state of chromium(III) arises from the spin-orbit coupling acting in concert with the axial crystal field distortion. Or, the g value of chromium(III) is found to be $g = 2.0023 - 8\lambda/10\,Dq$, which accounts quite nicely for the fact that g is typically about 1.98 for this ion. The parameter $10\,Dq$ is the cubic crystal field splitting.

More significantly, the ground state of titanium(III) is altered strikingly by spin-orbit coupling. The $^2T_{2g}$ ground state of cubic Ti(III) is three-fold degenerate orbitally and two-fold degenerate in spin. The total degeneracy is therefore six, but spin-orbit coupling, acting again in concert with an axial crystalline field, is capable of resolving this degeneracy into 3 doublets. The separations among the doublets are a function of the ratio λ/δ and its sign, where δ is a zero-field splitting parameter [13b]. The result of

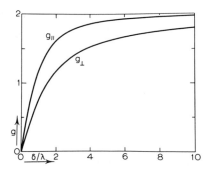

Fig. 2.7. Variation of $g_\parallel$ and $g_\perp$ as a function of tetragonal field parameter δ/λ for the E_- doublet of Ti^{3+}

Table 2.1. Spin-orbit coupling parameter ζ as estimated from analysis of experimental data.[a]

	3d	ζ (cm^{-1})	4d	ζ (cm^{-1})
d^1	Ti^{3+}	154	Nb^{4+}	750
d^2	V^{3+}	209	Mo^{4+}	950
d^3	Cr^{3+}	276	Mo^{3+}	800
	V^{2+}	168		
d^4	Mn^{3+}	360	Mo^{2+}	695
	Cr^{2+}	236	Ru^{4+}	(1350)
d^5	Fe^{3+}	280[b]	Ru^{3+}	(1180)
	Mn^{2+}	335[b]		
d^6	Fe^{2+}	404	Ru^{2+}	1000
d^7	Co^{2+}	528	Rh^{2+}	1220
d^8	Ni^{2+}	644	Pd^{2+}	1600
d^9	Cu^{2+}	829	Ag^{2+}	1840

[a]   Source – Abragam and Bleaney [13].
[b]   Values for strong-field t$_{2g}^5$ configuration.

this is that the g values are found to vary dramatically with δ/λ as is illustrated in Fig. 2.7; when the doublet called E$_-$ is the ground state, the g-values *approach* the value of two only when δ/λ becomes very large.

Values of ζ are listed in Table 2.1.

2.5 Effective Spin

This is a convenient place to introduce the concept of the effective spin of a metal ion, which is not necessarily the same as the true spin.

The two assignments of spin are the same for a copper(II) ion. This ion has nine d electrons outside the argon core and so has a $\mathscr{S} = \frac{1}{2}$ configuration no matter what geometry the ion is placed in. The net spin does not depend on or vary with the strength of the crystal or ligand field. Thus it always has a spin ("Kramers") doublet as the ground state, which is well-isolated from the optical states. One consequence of this is that EPR spectra of Cu(II) may usually be obtained in any lattice (i.e., liquid or solid) at any temperature.

Consider divalent cobalt, however, which is 3d^7. With a ^{4}T$_1$ ground state in an octahedral field, there are three unpaired electrons, and numerous Co(II) compounds exhibit spin-$\frac{3}{2}$ magnetism at 77 K and above [5]. However, the spin-orbit coupling constant of Co(II) is large ($\lambda \sim -180$ cm^{-1} for the free ion), and so the true situation here is better described by the energy level diagram in Fig. 2.8. It will be seen there that, under spin-orbit coupling, the ^{4}T$_1$ state splits into a set of three levels, with degeneracies as noted on the figure in parentheses. At elevated temperatures, the excited states are occupied and the $\mathscr{S} = \frac{3}{2}$ configuration, with an important orbital contribution, obtains. Fast electronic relaxation also occurs, and EPR is not observed. At low temperatures – say, below 20 K – only the ground state is occupied, but this is a doubly-degenerate level, and even though this is a spin-orbit doublet we may

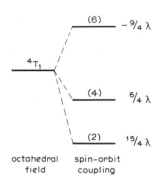

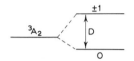

Fig. 2.8. Fine-structure splitting of the lowest levels of cobalt(II) under the action of the combined octahedral crystalline field, spin-orbit interaction, and a magnetic field. Degeneracies are given in parentheses. The spin-orbit coupling constant, λ is negative for cobalt(II)

Fig. 2.9. Zero-field splitting (positive) of a nickel(II) [or vanadium(III)] ion

characterize the situation with an effective spin, $\mathscr{S}' = \frac{1}{2}$. This is consistent with all available EPR data, as well as with magnetic susceptibility results. One further consequence of this situation is that the effective, or measured, g values of this level deviate appreciably from the free spin value of 2; in fact [6] for octahedral cobalt(II) the three g-values are so unusual that they sum to 13!

Let us return to the example of nickel(II), Fig. 2.9, but change the sign of D to a negative value and thereby invert the figure. Note that if $|D| \gg kT$, only the doubly-degenerate ground state will be populated, a spin doublet obtains, and $\mathscr{S}' = \frac{1}{2}$. The g-values will be quite anisotropic for, using primes to indicate effective g-values, $g'_{\parallel} = 2g_{\parallel}$ and $g'_{\perp} = 0$. The calculation differs from the cobalt situation in that the spin-orbit coupling is not as important.

Similar situations occur with tetrahedral Co^{2+} in both Cs_3CoCl_5 [7] and Cs_2CoCl_4 [8] and with Ni^{2+} in tetrahedral $NiCl_4^{2-}$ [9]. The case for spin $\mathscr{S} = \frac{3}{2}$ with large positive ZFS was described in connection with Eq. (2.11).

2.6 Direct Measurement of D

The magnetization as a function of field, frequency, and temperature has also been measured for several compounds with large zero-field splittings [10–12]. The isothermal magnetization is given as

$$M = Ng_{\parallel}\mu_B \langle S_z \rangle \tag{2.13}$$

with the field H||z, where

$$\langle S_z \rangle = (e^h - e^{-h})(e^d + e^h + e^{-h})^{-1}.$$

The quantity h is $g_{\parallel}\mu_B H/kT$ and d is D/kT. Equation (2.13) is plotted for a representative choice of parameters in Fig. 2.10.

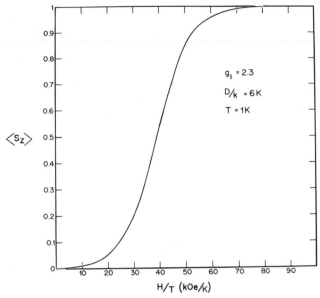

Fig. 2.10. The isothermal magnetization of a system with zero-field splitting, as calculated from Eq. (2.13). The parameters used are $D/k=6\,\mathrm{K}$ and $g_\parallel=2.3$, at $T=1\,\mathrm{K}$

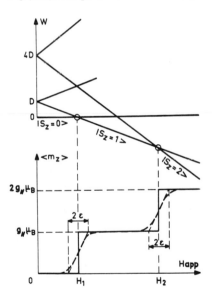

Fig. 2.11. Electronic level scheme and calculated magnetization at $0\,\mathrm{K}$ (full line) and finite temperature (dashed line) of Fe(II) in a magnetic field along the z-axis, $D \gg \lambda$

There has recently [10] been a direct measurement of D in $FeSiF_6 \cdot 6\,H_2O$, a trigonal crystal which is easy to orient. The technique employed illustrates some of the results that are becoming available through the use of high-field magnets.

The ferrous ion, Fe^{2+}, has a $3d^6$ electronic configuration, which gives rise to a ^5D ground state. The energy level scheme, illustrated in Fig. 2.11, has two doubly degenerate levels respectively D and 4D in energy above the non-degenerate ground state. Application of a magnetic field, as illustrated, causes the usual Zeeman interaction, and it will be noticed that a doubly-degenerate ground state occurs at each of the fields labeled H_1 and H_2. There can be no magnetization at $T \ll D/k$ when the external field, H_{ext}, is less than H_1, but a measurement of M when $H_{ext} = H_1$ or H_2 should be non-zero. The large fields required are necessarily pulsed fields, which in turn requires a rapid measurement of M; one must also be careful about spin-lattice relaxation effects (energy transfer) in such experiments.

A quick glance at Fig. 2.11 suggests that $D = g_{\parallel}\mu_B H_1 = (\frac{1}{3})g_{\parallel}\mu_B H_2$. Assuming $g_{\parallel} = 2$, an increase in M and therefore a cross-over was found at $H_1 = 132$ kOe (13.2 T). The derived value of D, 12.2 ± 0.2 cm^{-1}, may be compared to that obtained from susceptibility measurements (10.5–10.9 cm^{-1}). The field H_2 is then calculated at about 400 kOe (40 T), and indeed was measured as 410 kOe (41.0 T). Similar measurements have recently been reported on $[Ni(C_5H_5NO)_6](ClO_4)_2$ and $[C(NH_2)_3]V(SO_4)_2 \cdot 6H_2O$ [11] as well as $Cs_3VCl_6 \cdot 4H_2O$ [12].

2.7 Electron Paramagnetic Resonance (EPR)

Electron paramagnetic resonance, also called electron spin resonance, will not be treated in detail in these notes, for there are so many other excellent sources available [13]. Nevertheless, EPR is so implicitly tied up with magnetism that is is important just for the sake of completeness to include this short section on the determination of crystal field splittings by EPR. While the magnetic measurements that are the major topic of this book measure a property that is thermally averaged over a set of energy levels, EPR measures properties of the levels individually.

Thus, return in Fig. 1.3, where we see that two levels are separated by an energy $\Delta E = g_{\parallel}\mu_B H_z$. Following the Planck relationship, if we set

$$\Delta E = h\nu = g_{\parallel}\mu_B H_z \tag{2.14}$$

we then have the basic equation of paramagnetic resonance spectroscopy. The quantities h and μ_B are fundamental constants, $g_{\parallel}$ is a constant for a particular orientation of a given substance, and so we see that the basic experiment of EPR is to measure the magnetic field H_z at which radiation of frequency ν is absorbed. For ease of experiment, the usual procedure is to apply radiation of a constant frequency and then the magnetic field is scanned. For a substance with $g = 2$, the common experiment is done at X-band, with a frequency of some 9 GHz, and absorption occurs at approximately 3400 Oe (0.34 T). The energy separation involved is approximately 0.3 cm^{-1}. Q-band experiments, at some 35 GHz are also frequently carried out, with the energy separation that can be measured then of the order of 1 cm^{-1}.

As we have seen, zero-field splittings or other effects can sometimes cause energy levels to be separated by energies larger than these relatively small values. In those cases, EPR absorption cannot take place between those energy levels. If that particular electronic transition should be the only allowed transition, then there will no EPR spectrum at all, even in an apparently normal paramagnetic substance. This can happen with non-Kramers ions, which are ions with an even number of electrons. Zero-field splittings can resolve the energy levels of such an ion so as to leave a spin-singlet ground state (cf. Fig. 2.9).

The usual selection rule for electron resonance is $\Delta m_s = \pm 1$, although a variety of forbidden transitions are frequently observed in certain situations. Paramagnetic ions are usually investigated as dopants in diamagnetic host lattices. This procedure dilutes the magnetic ion concentration, minimizes dipole-dipole interaction effects, and so sharper lines are thereby observed. Exchange effects upon the spectra are also minimized by this procedure. Since paramagnetic relaxation is temperature dependent, a decrease in temperature lengthens the relaxation times and thereby also sharpens the lines (cf., below).

Now, return to Fig. 2.1, the energy level diagram for an $\mathscr{S} = \frac{3}{2}$ ion with and without zero-field splitting. Because of the $\Delta m_s = \pm 1$ selection rule, only one triply-degenerate line is observed at $hv = g_{\parallel}\mu_B H_z$ when the zero-field splitting is identically zero.

But, when the zero-field splitting is $2|D|$, the $+\frac{1}{2} \leftrightarrow -\frac{1}{2}$ transition remains at $hv = g_{\parallel}\mu_B H_z$, and the $+\frac{3}{2} \leftrightarrow -\frac{3}{2}$ transition is formally forbidden. The transitions $+\frac{1}{2} \leftrightarrow +\frac{3}{2}$ and $-\frac{1}{2} \leftrightarrow -\frac{3}{2}$ are allowed, the first occurring at $hv = g_{\parallel}\mu_B H_z + 2|D|$, the latter at $hv = g_{\parallel}\mu_B H_z - 2|D|$. Three lines are observed, then, when the magnetic field is parallel to the axis of quantization: a central line, flanked by lines at $\pm 2|D|$.

Thus, zero-field splittings which are of the size of the microwave quantum may be determined with ease and high precision, often even at room temperature, when the conditions described above are realized. As has been implied, however, this simple experiment does *not* determine the sign of D, but only its magnitude. The sign of D has occasionally been obtained by measuring the spectrum over a wide temperature interval.

It frequently happens that g is anisotropic, and the values of $g_{\parallel}$ and $g_{\perp}$ must be evaluated independently. In rhombic situations, there may be three g-values, g_x, g_y, and g_z.

Now, if there are two resonance fields given by

$$hv = g_{\parallel}\mu_B H_{\parallel} \quad \text{and} \quad hv = g_{\perp}\mu_B H_{\perp},$$

and if $g_{\parallel}$ is measurably different from $g_{\perp}$, then $H_{\parallel}$ will differ from $H_{\perp}$ since hv is assumed to be held constant by the experimental equipment. For the sake of argument, let $g_{\parallel} = 2$ and $g_{\perp} = 1$, corresponding, in a particular spectrometer, to typical values of resonance fields of $H_{\parallel} = 3400$ Oe (0.34 T) and $H_{\perp} = 6800$ Oe (0.68 T). Define θ as the angle between the principal (parallel) axis of the sample and the external field, and so the above results correspond, respectively, to $\theta = 0°$ and $90°$. At intermediate angles, it can be shown that

$$g_{\text{eff}}^2 = g_{\parallel}^2 \cos^2\theta + g_{\perp}^2 \sin^2\theta, \tag{2.15}$$

where g_{eff} is merely the effective or measured g-value. A typical set of data for this system would than appear as in Fig. 2.12, where both g_{eff}^2 and the resonant field are plotted for the above example as functions of θ. A fit of the experimental data over a variety of angles leads then to the values of both $g_{\parallel}$ and $g_{\perp}$.

The experiment as described requires all relevant axes to be parallel, which in turn requires a high-symmetry crystal. Commonly, systems of interest are of lower symmetry, and lines will overlap or several spectra will be observed simultaneously. Methods of solving problems such as these are described elsewhere [13].

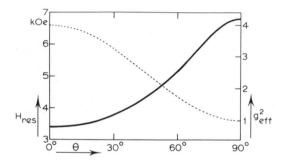

Fig. 2.12. g_{eff}^2 [Eq. (2.15), dotted curve] and resonant field H_{res}, plotted vs. the angle θ, with $g_{\parallel}=2$ and $g_{\perp}=1$

Let us return to the question of linewidths in the EPR spectra; we follow the discussion of Orton (13b). As a consequence of the Uncertainty Principle, the absorption of energy extends over a range of fields centered at the resonance field H_{res} and extending a finite amount on either side. The width of any energy level is related to the lifetime of the corresponding state by the relation

$$\Delta E \Delta t \simeq h/2\pi,$$

where h is Planck's constant, and gives rise to an EPR line width in frequency units of

$$\Delta v = 1/(2\pi \Delta t).$$

That is, the shorter the lifetime of the state, the broader is the resonance. Any interaction between the magnetic ion and its surroundings may broaden the line if it results in energy being transferred from the ion following excitation. We are concerned here primarily with spin-lattice relaxation, with the characteristic time called T_1, in which energy is transferred from the excited ion to the crystal lattice. The word "lattice" in this case refers to the entire sample as the system (e.g., ligands and other nuclei) and is not meant to imply that the "lattice" is necessarily a crystal lattice. (The same term is used to describe relaxation in both liquids and gases.) In many cases very short values of T_1 occur at room temperature and the resulting line widths are so large as to make resonance unobservable. This happens frequently when the metal ion has a low-lying (not necessarily populated) energy level, as with octahedral cobalt(II) and many of the rare earth ions. However, T_1 is usually sharply temperature-dependent so that cooling the specimen to 20 K or 4 K increases T_1 sufficiently to render this contribution negligible [13b].

Spin-lattice relaxation can also be important when magnetic susceptibilities are measured by the *ac* mutual inductance method in the presence of an applied field. This will be described in later chapters.

2.8 References

1. Van Vleck J.H., (1932) The Theory of Electric and Magnetic Susceptibilities, Oxford University Press, Oxford
2. McElearney J.N., Schwartz R.W., Siegel A.E. and Carlin R.L., J. Am. Chem. Soc. **93**, 4337 (1971)

3. McElearney J.N., Schwartz R.W., Merchant S. and Carlin R.L., J. Chem. Phys. **55**, 466 (1971)
4. Chirico R.D. and Carlin R.L., Inorg. Chem. **19**, 3031 (1980)
5. Carlin R.L., (1965) Transition Metal Chemistry, Ed. R.L. Carlin, Marcel Dekker, New York, Vol. 1, p. 1
6. Abragam A. and Pryce M.H.L., Proc. Roy. Soc. (London) **A206**, 173 (1951). See also Bates C.A. and Wood P.H., Contemp. Phys. **16**, 547 (1975) for an extended discussion of this point
7. Mess K.W., Lagendijk E., Curtis D.A., and Huiskamp W.J., Physica **34**, 126 (1967); Wielinga R.F., Blöte H.W.J., Roest J.A. and Huiskamp W.J., Physica **34**, 223 (1967)
8. Algra H.A., de Jongh L.J., Blöte H.W.J., Huiskamp W.J., and Carlin R.L., Physica **82B**, 239 (1976); McElearney J.N., Merchant S., Shankle G.E., and Carlin R.L., J. Chem. Phys. **66**, 450 (1977). See also Carlin R.L., J. Appl. Phys. **52**, 1993 (1981) and Science **227**, 1291 (1985)
9. Inman, Jr. G.W., Hatfield W.E., and Jones, Jr. E.R., Inorg. Nucl. Chem. Lett. **7**, 721 (1971)
10. Varret F., Allain Y., and Miedan-Gros A., Solid State Comm. **14**, 17 (1974); Varret F., J. Phys. Chem. Solids **37**, 257 (1976)
11. Smit J.J., de Jongh L.J., de Klerk D., Carlin R.L., and O'Connor C.J., Physica **B86–88**, 1147 (1977)
12. Smit J.J., van Wijk H.J., de Jongh L.J., and Carlin R.L., Chem. Phys. Lett. **62**, 158 (1979)
13. The most complete discussion of EPR is provided by: Abragam A. and Bleaney B., (1970) Electron Paramagnetic Resonance of Transition Ions, Oxford University Press. Popular texts on the subject are:
  a) McGarvey B.R., Transition Metal Chem. **3**, 90 (1966)
  b) Orton J.W., (1968) Electron Paramagnetic Resonance, Iliffe Books, London
  c) Pake G.E. and Estle T.L., (1973), 2nd Ed.: Physical Principles of Electron Paramagnetic Resonance, W.A. Benjamin, Reading, Mass.
  d) Carrington A. and McLachlan A.D., (1967) Introduction to Magnetic Resonance, Harper and Row

3. Thermodynamics

3.1 Introduction

It is clear from the preceding chapters that the relative population of the lowest energy levels of paramagnetic ions depends on both the temperature and the strength of the external magnetic field. The equilibrium states of a system can often be described by three variables, of which only two are independent. For the common example of the ideal gas, the variables are pVT (pressure, volume, temperature). For magnetic systems, one obtains HMT (magnetic field, magnetization, temperature) as the relevant variables and the thermodynamic relations derived for a gas can be translated to a magnetic system by replacing p by H and V by $-M$. In the next section we review a number of the thermodynamic relations. Then, two sections are used to demonstrate the usefulness of these relations in analyzing experiments. In fact the simple thermodynamic relations are often applicable, even to magnetic systems that require a complicated model to describe the details of their behavior [1].

3.2 Thermodynamic Relations

The first law of thermodynamics states that the heat dQ added to a system is equal to the sum of the increase in internal energy dU and the work done by the system. For a magnetic system work has to be done on the system in order to change the magnetization, so the first law of thermodynamics may be written as

$$dQ = dU - HdM,$$

where H is the applied field and M is the magnetization. Remembering that the entropy S is related to Q by $TdS = dQ$, the first law can be written as

$$dU = TdS + HdM. \qquad (3.1)$$

The energy U is an exact differential, so

$$\left(\frac{\partial T}{\partial M}\right)_S = \left(\frac{\partial H}{\partial S}\right)_M. \qquad (3.2)$$

The enthalpy E is defined as $E = U - HM$, thus

$$dE = dU - HdM - MdH = TdS - MdH,$$

and, as E is also an exact differential

$$\left(\frac{\partial T}{\partial H}\right)_S = -\left(\frac{\partial M}{\partial S}\right)_H.$$ (3.3)

The Helmholtz free energy F is defined as $F = U - TS$, so

$$dF = dU - TdS - SdT = -SdT + HdM,$$

and the exact differentiability leads to

$$\left(\frac{\partial S}{\partial M}\right)_T = -\left(\frac{\partial H}{\partial T}\right)_M.$$ (3.4)

The Gibbs free energy is $G = E - TS$, thus

$$dG = dE - TdS - SdT = -SdT - MdH$$

and

$$\left(\frac{\partial M}{\partial T}\right)_H = \left(\frac{\partial S}{\partial H}\right)_T.$$ (3.5)

Equations (3.2)–(3.5) are the Maxwell relations in a form useful for magnetic systems.

The specific heat of a system is usually defined as dQ/dT, but it also depends on the particular variable that is kept constant when the temperature changes. For magnetic systems, one has to consider both c_M and c_H, the specific heats at constant magnetization and field, respectively. From $dQ = TdS$ and the definitions of U and E one obtains

$$c_M = \left(\frac{\partial Q}{\partial T}\right)_M = T\left(\frac{\partial S}{\partial T}\right)_M = \left(\frac{\partial U}{\partial T}\right)_M$$ (3.6)

and

$$c_H = \left(\frac{\partial Q}{\partial T}\right)_H = T\left(\frac{\partial S}{\partial T}\right)_H = \left(\frac{\partial E}{\partial T}\right)_H.$$ (3.7)

Now let the entropy, which is a state function, be a function of temperature and magnetization, $S = S(T, M)$. Then, an exact differential may be written

$$dS = \left(\frac{\partial S}{\partial T}\right)_M dT + \left(\frac{\partial S}{\partial M}\right)_T dM.$$

Multiplying through by T, and using the Maxwell relation given by Eq. (3.4),

$$TdS = T\left(\frac{\partial S}{\partial T}\right)_M dT - T\left(\frac{\partial H}{\partial T}\right)_M dM$$

in which the coefficient of dT is just the specific heat at constant magnetization. Thus

$$TdS = c_M dT - T \left(\frac{\partial H}{\partial T}\right)_M dM. \tag{3.8}$$

In a similar way the entropy may be considered as a function of temperature and field, $S = S(T, H)$. Then

$$dS = \left(\frac{\partial S}{\partial T}\right)_H dT + \left(\frac{\partial S}{\partial H}\right)_T dH,$$

and multiplying through by T and using the Maxwell relation (3.5) yields

$$TdS = c_H dT + T \left(\frac{\partial M}{\partial T}\right)_H dH. \tag{3.9}$$

Specific heats are of interest in experimental work, and it is important to have expressions for the difference as well as for the ratio between the two specific heats, c_H and c_M, respectively. By subtracting Eq. (3.8) from Eq. (3.9), one obtains

$$(c_H - c_M)dT = -T \left(\frac{\partial M}{\partial T}\right)_H dH - T \left(\frac{\partial H}{\partial T}\right)_M dM,$$

and because

$$dT = \left(\frac{\partial T}{\partial H}\right)_M dH + \left(\frac{\partial T}{\partial M}\right)_H dM$$

a comparison of the coefficients of dM shows

$$\left(\frac{\partial T}{\partial M}\right)_H = -T(\partial H/\partial T)_M/(c_H - c_M).$$

From this equation one can resolve the quantity $c_H - c_M$. The fact that only two variables out of HMT are independently variable is expressed also by

$$\left(\frac{\partial H}{\partial T}\right)_M \left(\frac{\partial T}{\partial M}\right)_H \left(\frac{\partial M}{\partial H}\right)_T = -1, \tag{3.10}$$

a relation that may be used to eliminate $(\partial H/\partial T)_M$ from the expression for the difference between the two specific heats. So,

$$c_H - c_M = T \left(\frac{\partial M}{\partial T}\right)_H^2 \left(\frac{\partial H}{\partial M}\right)_T. \tag{3.11}$$

In a similar way an expression for the ratio the two specific heats is obtained. Solving for dT from Eq. (3.9), then

$$dT = \frac{T dS}{c_H} - \frac{T}{c_H}\left(\frac{\partial M}{\partial T}\right)_H dH.$$

But also

$$dT = \left(\frac{\partial T}{\partial S}\right)_H dS + \left(\frac{\partial T}{\partial H}\right)_S dH$$

and the coefficients of dH must be equal. This leads to

$$c_H = -T\left(\frac{\partial M}{\partial T}\right)_H \left(\frac{\partial H}{\partial T}\right)_S.$$

In an identical way, the use of Eq. (3.8) leads to

$$c_M = T\left(\frac{\partial H}{\partial T}\right)_M \left(\frac{\partial M}{\partial T}\right)_S.$$

For the quotient c_H/c_m, we derive

$$c_H/c_M = -\frac{(\partial M/\partial T)_H (\partial H/\partial T)_S}{(\partial M/\partial T)_S (\partial H/\partial T)_M} = -\frac{(\partial T/\partial M)_S (\partial H/\partial T)_S}{(\partial T/\partial M)_H (\partial H/\partial T)_M}$$

an expression that can be simplified by the use of relation (3.10) to

$$c_H/c_M = (\partial M/\partial H)_T/(\partial M/\partial H)_S. \tag{3.12}$$

The variation of M upon H is called the differential susceptibility. The constancy of either S or T determines whether it is the adiabatic or the isothermal susceptibility, respectively. It is usually found that relaxation effects occur in the adiabatic regime. The measurements that inorganic chemists are most generally interested in are the isothermal susceptibilities.

3.3 Thermal Effects

The specific heat of a magnetic system is, as we shall see repeatedly, one of its most characteristic and important properties. Magnetic ordering in particular, is evidenced by such thermal effects as anomalies in the specific heat. Single ion anisotropies also offer characteristic heat capacity curves, as explained below. We show here that there can even be a specific heat contribution by a Curie law paramagnet under certain conditions. But first, it is necessary to discuss lattice heat capacities.

Every substance, whether it contains ions with unpaired spins or not, exhibits a lattice heat capacity. This is because of the spectrum of lattice vibrations, which forms the basis for both the Einstein and Debye theories of lattice heat capacities. For our purposes, it is sufficient to be aware of the phenomenon and, in particular, that the lattice heat capacity decreases with decreasing temperature. It is this fact that causes so much of the interest in magnetic systems to concern itself with measurements at low temperatures, for then the magnetic contribution constitutes a much larger fraction of the whole.

In the Debye model, the lattice vibrations (phonons) are assumed to occupy the $3N$ lowest energies of an harmonic oscillator. The Debye lattice specific heat, derived in any standard text of solid state physics, is

$$c_L = 9R(T/\theta_D)^3 \int_0^{\theta_D/T} \frac{e^x x^4}{(e^x - 1)^2}\, dx,$$

where $x = \hbar\omega/kT$ and $\theta_D = \hbar\omega_{max}/k$ is called the Debye characteristic temperature. At low temperatures, where x is large, the integral in the above expression becomes a constant; thus c_L may be approximated as

$$c_L \propto (T/\theta_D)^3$$

which is useful up to temperatures of the order of $\theta_D/10$. Each substance has its own value of θ_D, but as a practical matter, many of the insulating salts which are the subject of these notes obey a T^3 law up to approximately 20 K. The specific heat of aluminum alum, a diamagnetic salt which is otherwise much like many of the salts of interest here, has been measured [2] and does obey the law

$$c_L = 0.801 \times 10^{-3} T^3\,\text{cal/mol-K} = 1.91 \times 10^{-4} T^3\,\text{J/mol-K}$$

at temperatures below 20 K, as illustrated in Fig. 3.1. On the other hand, substances which have clear structural features that are one or two dimensional in nature do not necessarily obey a T^3 law over wide temperature intervals, and caution must be used in assuming such a relationship.

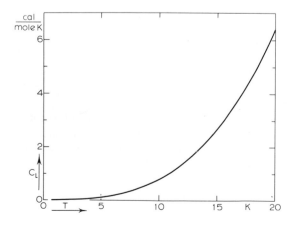

Fig. 3.1. Specific heat of $KAl(SO_4)_2 \cdot 12\,H_2O$. Data from Ref. [2].

Naturally, the lattice heat capacity must be evaluated in order to subtract it from the total to find the desired magnetic contribution. Several procedures are in common use:

1. Many of the magnetic contributions have, as we shall see below, a T^{-2} dependence for the specific heat in the high temperature limit. If the lattice specific heat follows the T^3 law in the measured temperature region, the total specific heat in a situation such as this should obey the relationship

$$c - a T^3 + b T^{-2} \qquad (3.13)$$

and a plot of cT^2 vs. T^5 will, if it is linear, show the applicability of the procedure in the particular situation at hand over a certain temperature interval, as well as allow the evaluation of the constants a and b. Extrapolation to lower temperatures then allows an empirical evaluation of the lattice contribution.

2. Many substances are not amenable to such a procedure, especially those which exhibit short-range order effects (Chap. 7). This is because the magnetic contribution extends over a wide temperature interval, too wide and too high in temperature for Eq. (3.13) to be applicable. A procedure introduced by Stout and Catalano [3] is then often of value. The method depends on the law of corresponding states, which in the present situation states that the specific heats of similar substances will be similar, if weighted by the differences in molecular weights. In practice, one measures the total heat capacity of a magnetic system over a wide temperature region, and compares it to the heat capacity of an isomorphic but nonmagnetic substance. By use of several relationships [3] one then *calculates* what the specific heat of the lattice of the magnetic compound is, and then subtracts this from the total. Though this procedure is used frequently it often fails just in those cases where it is needed the most. A careful analysis [4] suggests that the accuracy of this procedure is limited, especially in the application to layered systems, and a consistent evaluation of the lattice term can be made only in conjunction with an evaluation of the magnetic contribution.

3. Occasionally a diamagnetic isomorph will not be available. This is not a serious problem if the magnetic phenomena being investigated occur at sufficiently low temperatures but, again, if short range order effects are present, other procedures must be resorted to. Such a situation occurs for example with $CsMnCl_3 \cdot 2 H_2O$, in which the broad peak in the magnetic heat capacity has a maximum at about 18 K but extends even beyond 50 K, where the lattice term is then the major contributor [5]. In this case, the magnetic contribution could be calculated with some certainty, and so an empirical procedure could be used to fit the experimental data to the theoretical magnetic contribution and fitted lattice contribution. A similar problem is posed by $[(CH_3)_4N]MnCl_3$, and similar procedures led to the estimation of the several contributing terms [6].

Let us return now to the paramagnetic system described in the Appendix to Chap. 1, whose magnetization M is

$$M = N\langle \mu_z \rangle = Ng\mu_B \mathscr{J} B_J(\eta)$$

and recall that interactions between the ions are considered to be negligible. Thus the internal energy $U = 0$ and the enthalpy becomes simply the energy of the system in the

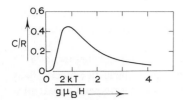

Fig. 3.2. Magnetic specific heat of a paramagnet, Eq. (3.14)

field, which is the product of M with H_z. The heat capacity at constant field is obtained by differentiating the enthalpy with respect to temperature, Eq. (3.9), so that, for example, in the case of $\mathscr{J} = \frac{1}{2}$, $B_{1/2}(\eta) = \tanh(\eta/2)$, $\eta = g\mu_B H_z/kT$,

$$E = \frac{Ng\mu_B H_z}{2} \tanh(g\mu_B H_z/2kT)$$

and

$$c_H = \left(\frac{\partial E}{\partial T}\right)_H = \frac{Ng^2\mu_B^2 H_z^2}{4kT^2} \operatorname{sech}^2\left(\frac{g\mu_B H_z}{2kT}\right). \qquad (3.14)$$

Equation (3.14) is plotted in Fig. 3.2 where it can be seen that a broad maximum occurs. The temperature range of the maximum can be shifted by varying the magnetic field strength. This curve, the shape of which is common for other magnetic phenomena as well, illustrates again the noncooperative ordering of a paramagnetic system caused by the combined action of both magnetic field and low temperatures. In a number of cases, $g\mu_B H_z/2kT \ll 1$ so the hyperbolic secant in Eq. (3.14) is equal to one, and the specific heat becomes

$$c_H = \frac{Ng^2\mu_B^2 H_z^2}{4kT^2} = \frac{CH_z^2}{T^2}. \qquad (3.15)$$

This simplification, using the Curie constant C, is in fact correct for more complicated systems also.

3.4 Adiabatic Demagnetization

Until recently, adiabatic demagnetization served as the best procedure for obtaining temperatures below 1 K. The subject is introduced here because the exploration of appropriate salts for adiabatic demagnetization experiments was the initial impetus for much of the physicists' interest in paramagnets. Recall, in this regard, that kT changes by the same ratio whether the temperature interval be 0.1–1 K, 1–10 K, or even 10–100 K, and that the ratio of kT with some other quantity such as a ZFS parameter is often more significant than the particular value of T.

A schematic plot of the entropy of a magnetic system as a function of temperature for two values of the field, $H = 0$ and a nonzero H is illustrated in Fig. 3.3. If for no other reason than the existence of the lattice heat capacity, every substance will have an

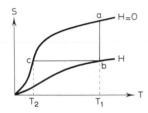

Fig. 3.3. Entropies (schematic) of a paramagnetic system with and without an external magnetic field

entropy increasing with temperature in some fashion. A magnetic system will have a lower entropy, as a function of temperature, when the field is applied than when the field is zero, simply from the paramagnetic alignment caused by a field.

Consider that the system is already at temperatures of the order of 1 K, say at the point a on the S vs. T plot. Let the system be magnetized isothermally by increasing the field to the point b (Fig. 3.3). Application of the field aligns the spins, decreasing the entropy, and so heat is given off. The sample must be allowed to remain in contact with a heat sink at say, 1 K, so that the process is isothermal. The next step in the process requires an adiabatic demagnetization. The system is isolated from its surroundings and the magnetic field is removed adiabatically. The system moves horizontally from point b to point c, and the temperature is lowered significantly.

It is useful to examine these procedures in more detail with the help of the thermodynamic relations. Equation (3.9)

$$T dS = c_H dT + T \left(\frac{\partial M}{\partial T} \right)_H dH$$

simplifies for an isothermal process to

$$T dS = T \left(\frac{\partial M}{\partial T} \right)_H dH .$$

For a paramagnet, $M = \chi H = CH/T$, so $(\partial M/\partial T)_H$ will necessarily be negative, and thus dS will be negative for the first step. During the adiabatic demagnetization step $(dS = 0)$, the above equation becomes

$$0 = c_H dT + T \left(\frac{\partial M}{\partial T} \right)_H dH$$

and since $(\partial M/\partial T)_H$ remains negative, for $dH < 0$, we find that $dT < 0$, and the system cools. (One should keep in mind that the heat capacity, c_H, is a positive quantity.) In general, a finite adiabatic change in field thus produces a temperature change given by

$$\Delta T = - \{ (T/c_H)(\partial M/\partial T)_H \} \Delta H$$

if the adiabatic field change is "ideal", that is, there is no heat exchange between the paramagnet and its surroundings. This effect is often called the magnetocaloric effect; it is the basis of adiabatic cooling, but also may cause unwanted temperature changes during experiments with pulsed (large) magnetic fields. The method of adiabatic demagnetization stands on such a firm thermodynamic foundation that the first experiments by Giauque [7], one of the originators of the method, were not to test the method but in fact to use it for other experiments once the cooling had occurred!

Let us now consider the effect of zero-field splittings upon adiabatic demagnetization. Letting Z be the partition function for a system, the relation for the entropy of a magnetic system [8] is

$$S = RT \left(\frac{\partial \ln Z}{\partial T} \right)_H + R \ln Z. \qquad (3.16)$$

This relation can be written down completely when Z is known, which as a practical matter is true only when there are no interactions between ions. If only a zero-field splitting δ is considered the entropy behaves simply in the limits

1) $kT \gg \delta$, $S = S_1(H/T)$,

2) $kT \leq \delta$, $S = S_2(H, T)$.

Several curves are displayed in Fig. 3.4 for various values of the field H.

The requirement of finding a good refrigerant crystal for adiabatic demagnetization work was the initial impetus for the measurement of zero-field splittings. Consider a two level system. The smaller the splitting, the more the populations of the two levels will be equalized. This corresponds to more disorder or larger entropy of the spin-system, and so, as illustrated in Fig. 3.5, the entropy curve at zero field for the system with smaller zero-field splitting lies above that for the system with the larger zero-field splitting. Application of the adiabatic demagnetization cycle, as was illustrated earlier, shows that the final temperature after the adiabatic demagnetization is in fact lower for that system with the smaller zero-field splitting.

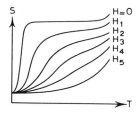

Fig. 3.4. Entropy of a magnetic ion subsystem as a function of temperature for several values of the magnetic field. In the region where $S(H=0)$ is constant, the energy U is a constant, and thus $c_M = 0$

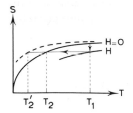

Fig. 3.5. Effect of zero-field splitting on the final temperature

3.5 Schottky Anomalies

One of the most important signatures of a zero-field splitting is the broad maximum that it causes in the specific heat, which for historical reasons is called a Schottky anomaly. Consider the $\mathscr{S}=1$ nickel ion with $^3A_2$ ground state with positive parameter D as illustrated in Fig. 2.9. There is a doubly-degenerate level with energy D above a non-degenerate level. The single-ion partition function is

$$Z=1+2e^{-D/kT} \tag{3.17}$$

and straightforward application of the thermodynamic relation

$$c=\frac{\partial}{\partial T}\left(RT^2\frac{\partial \ln Z}{\partial T}\right) \tag{3.18}$$

leads immediately to the magnetic specific heat,

$$c=\frac{2R(D/kT)^2\exp(-D/kT)}{[1+2\exp(-D/kT)]^2} \tag{3.19}$$

which has the shape illustrated in Fig. 3.6. This curve has a maximum at approximately $T_{max}=0.4\,D/k$, where D/k is the zero-field splitting expressed in Kelvins. It is essential to recall that this specific heat is the magnetic contribution alone, which is of course superimposed on the lattice heat capacity. As an example, Fig. 3.7 illustrates the specific heat of α-NiSO$_4\cdot6$H$_2$O [9]. A zero-field splitting of about 7 K provides the magnetic contribution, which falls on top of the T^3 lattice term. Clearly, the smaller the zero-field splitting, the lower the temperature at which it will give rise to a maximum, which can in turn be measured more accurately.

As an example of this situation, the magnetic specific heat of [V(urea)$_6$] Br$_3\cdot3$H$_2$O is shown in Fig. 3.8, together with the fit [10] to the Schottky curve, Eq. (3.19). The lattice contribution was estimated by the corresponding states procedure, making use of the specific heat of the isomorphic [Fe(urea)$_6$]Cl$_3\cdot3$H$_2$O. Although this is not a

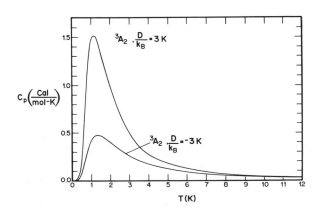

Fig. 3.6. Schottky specific heat for a $^3A_2$ ground state for $D/k=3$ K and $D/k=-3$ K

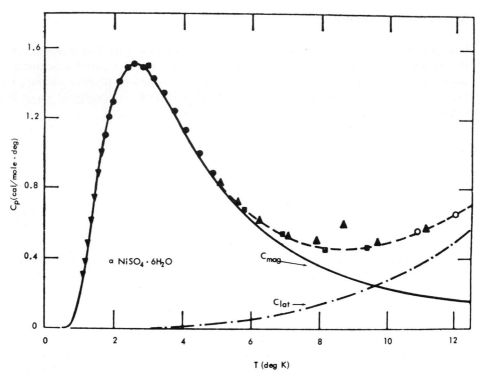

Fig. 3.7. Heat capacity of α-NiSiO$_4$ · 6 H$_2$O. From Ref. [9]

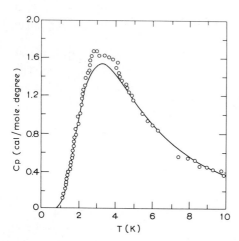

Fig. 3.8. Magnetic heat capacity of [V(urea)$_6$]Br$_3$ · 3 H$_3$O, along with theoretical fit to the data. From Ref. [10]

diamagnetic compound, the measured specific heat above 1 K gave no evidence for any magnetic effects, as one would expect because the zero-field splitting of the ^{6}S ground state of this compound is much smaller than kT in this region.

More generally, consider a level of degeneracy ξ_1 which is δ in energy above a level of degeneracy ξ_0. (It is of course not necessary to restrict the analysis to a two level

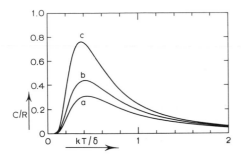

Fig. 3.9. Schottky heat capacity of a two-level system for several values of the relative degeneracy, ξ_1/ξ_0. Curve a: $\xi_1/\xi_0=\frac{1}{2}$, b: $\xi_1/\xi_0=1$, c: $\xi_1/\xi_0=2$

system.) Then, following the procedure described above,

$$c = \frac{R(\delta/kT)^2(\xi_0/\xi_1)\exp(\delta/kT)}{[1+(\xi_0/\xi_1)\exp(\delta/kT)]^2}. \tag{3.20}$$

Equation (3.20) is plotted in Fig. 3.9 with reduced parameters for several values of ξ_0 and ξ_1. The height of the maximum depends only on the relative values of ξ_0 and ξ_1, while the parameter δ determines the position of the maximum on the T axis.

It is common to find substances, particularly of Cr(III), Mn(II), and Fe(III), with δ of the order of only 0.1 K, a situation that is more accurately studied by means of EPR rather than by specific heat measurements. At those temperatures for which $\delta \ll kT$, the exponentials in Eq. (3.20) may be expanded to yield

$$c/R = \frac{(\xi_0/\xi_1)}{(1+\xi_0/\xi_1)^2}(\delta/kT)^2 = b'/T^2. \tag{3.21}$$

That is, the high-temperature tail of the Schottky specific heat varies as T^{-2}. This result hold at higher temperatures not only for a two-level system, but for a system with any number of closely-spaced levels. Magnetic ions can interact (Chap. 6) by both magnetic-dipole-dipole and magnetic exchange interactions, and these interactions also cause a characteristic specific heat maximum which varies as T^{-2} at high temperatures. Furthermore, nuclear spin-electron spin hyperfine interactions become important at very low temperatures, and can cause a Schottky-like contribution. All these possible contributions can be included in the one parameter b if we write the high-temperature magnetic specific heat as

$$c/R = b/T^2 \tag{3.22}$$

and b is a (high-temperature) measure of the importance of the interactions combined with the resolution of the energy levels. Equation (3.22) is applied to $K_2Cu(SO_4)_2 \cdot 6H_2O$ in Fig. 3.10. Some typical values of b for salts which have been discussed so far are:

chrome alum	$b=0.018\,K^2$
iron alum	0.013
$Gd_2(SO_4)_3 \cdot 8H_2O$	0.35
$Ce_2Mg_3(NO_3)_{12} \cdot 24H_2O(CMN)$	7.5×10^{-6}.

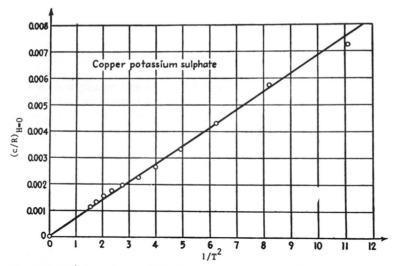

Fig. 3.10. T^{-2} dependence of c/R for copper potassium sulfate

The very small value of b for CMN (also sometimes called CeMN) indicates that all the factors that split the lowest energy level are also very small. Qualitatively then, we see that the quantity b is a guide to the usefulness of a particular salt for adiabatic demagnetization, and in fact the lowest temperatures are reached in this way using salts in which b is small. Clearly, CMN is one of the best salts in this regard.

Notice finally, that an evaluation of a zero-field splitting by susceptibility measurements requires at least two parameters – the g-values and the zero-field splitting parameter, as in Eq. (2.11). On the other hand, only one fitting parameter is required in the analysis of Schottky anomalies in the specific heat in terms of the zero-field splitting.

The compound $(CH_3NH_3)Fe(SO_4)_2 \cdot 12\,H_2O$ provides an example of the utility of combining EPR data with specific heat measurements of magnetic systems. In particular, a Schottky specific heat was observed [11] and allowed an unambiguous assignment of the energy levels of the ground state.

The alums, of which this compound is an example, have a cubic crystal structure and contain hexaquometal(III) ions. The long-standing interest in these compounds arises from the fact that they have low ordering temperatures, and thus have been explored extensively below 1 K as possible magnetic cooling salts and magnetic thermometers. In the current example, it has been shown by EPR that the ^6S ground state of the iron was split into the three doublets, $|\pm\frac{1}{2}\rangle, |\pm\frac{3}{2}\rangle$, and $|\pm\frac{5}{2}\rangle$, with successive splittings of 0.40 and 0.73 cm^{-1}. The power of EPR lies with the ease and accuracy that zero-field splittings of this magnitude can be obtained; the fault of EPR lies with the fact that the *sign* of the zero-field splitting cannot be directly determined, and so it was not known which state, $|\pm\frac{1}{2}\rangle$ or $|\pm\frac{3}{2}\rangle$, was the ground state of the system.

The calculation of the anticipated Schottky behavior is simple for a three level system, for the partition function is merely

$$Z = 1 + \exp(-\delta_1/kT) + \exp(-\delta_2/kT), \tag{3.23}$$

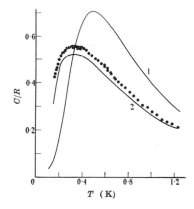

Fig. **3.11.** Magnetic specific heat of $(CH_3NH_3)Fe(SO_4)_2 \cdot 12\,H_2O$ at temperatures between 0.17 and 1.2 K. •, experimental points; curve 1, calculated specific heat, assuming $|\pm\frac{3}{2}\rangle$ level lowest; curve 2, calculated specific heat, assuming $|\pm\frac{1}{2}\rangle$ level lowest. From Ref. [11]

where δ_1 and δ_2 may be either both positive or both negative. The measured specific heat is illustrated in Fig. 3.11, along with the specific heat calculated from Eq. (3.23) for both signs of δ_1 and δ_2. Curve 1 corresponds to the $|\pm\frac{5}{2}\rangle$ state low and curve 2, which agrees quite well with experiment, corresponds to the $|\pm\frac{1}{2}\rangle$ state as the ground state.

3.6 Spin-Lattice Relaxation

In Chap. 1 the influence of a magnetic field on a system of magnetic ions was considered in equilibrium situations only. In the paragraph about adiabatic demagnetization it was demonstrated that the paramagnetic system behaves considerably differently if it is isolated from its surroundings. Under experimental conditions isothermal or adiabatic changes often do not occur, so that the system will reach an equilibrium situation only after some time.

The recovery of a perturbed magnetic system to a "new" equilibrium can be described phenomenologically with the help of the relation:

$$\frac{dM}{dt} = (M_0 - M)/\tau,$$

where M_0 is the equilibrium magnetization, M is the magnetization at time t, and τ is the relaxation time. In fact, the response times are very short, but, nevertheless, as soon as we study a magnetic material by means of techniques that require an oscillating field, the magnetization may no longer be able to follow the field changes instantaneously. One of the phenomena that has to be considered under such circumstances is the spin lattice relaxation. This relaxation process described the transfer of energy between the magnetic spin subsystem and the lattice vibrations. One needs to consider several possible processes [12]:

1. *Direct process.* This is the relaxation process in which one magnetic ion flips to another energy level under the absorption or emission of the energy of one phonon. The frequency ω of the required phonon is determined by $\hbar\omega = \Delta$ if Δ is the energy change of the magnetic ion.

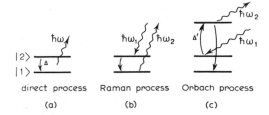

direct process   Raman process   Orbach process

   (a)   (b)   (c)

Fig. 3.12. The three spin-lattice relaxation processes indicated schematically

2. *Raman process.* In this non-resonant scattering process a phonon with frequency ω_1 is absorbed, causing the magnetic ion to reach a so-called virtual or non-stationary state from which it instantaneously decays by emission of a new phonon with frequency ω_2. The phonon frequencies are related to each other by $h(\omega_1 - \omega_2) = 2\pi\Delta$.

3. *Orbach process.* There is a possibility that a direct, resonant two-phonon process occurs via a real intermediate state, if the paramagnetic ion has, in addition to the two ground state levels, another level at such a position that the phonons can excite the ion to this state. In this case the phonon frequencies are also determined by $h(\omega_1 - \omega_2) = 2\pi\Delta$.

The three different types of relaxation processes are schematically indicated in Fig. 3.12, where the curly arrows represent the phonons. In the following equations we give the temperature and field dependences of the different kinds of processes [12]. One must realize that these relations are far from complete, and that the quoted dependences are simplified. The coefficients in each of the equations are different, and also vary from one magnetic substance to another.

$$1/\tau = A_1 H^2 T + B_1 T^7 + C_1 \exp(-\Delta/kT), \tag{3.24}$$

$$1/\tau = A_2 H^4 T + B_2 T^9 + C_2 \exp(-\Delta/kT), \tag{3.25}$$

$$1/\tau = A_3 H^2 T + B_3 T^5 \tag{3.26}$$

$$\begin{pmatrix} \text{Direct} \\ \text{process} \end{pmatrix} \begin{pmatrix} \text{Raman} \\ \text{process} \end{pmatrix} \begin{pmatrix} \text{Orbach} \\ \text{process} \end{pmatrix}.$$

Equation (3.24) refers to non-Kramers ions, Eq. (3.25) to Kramers ions with an "isolated" doublet and Eq. (3.26) to Kramers ions with various doublets (energy difference between the doublets small compared to kT).

These equations show the characteristic features of spin-lattice relaxation processes. The terms representing the direct process depend on the magnetic field H, while the others do not. The reason for this different behavior is found in the assumed phonon distribution. In the direct process phonons of one particular frequency are involved, and the number of such phonons depends not only on temperature but also on the energy or, in other words, on the applied magnetic field. In the Raman and Orbach processes, phonons of all available energies participate as only the difference between two of them is important and the total number of phonons does not vary with the external magnetic field. A similar argument in which, apart from the features of the Debye model for the phonons, the characteristics of the Bose-Einstein statistics are also considered, gives a direct explanation for the more pronounced temperature dependences of the two-phonon relaxation processes.

3.7 References

1. General references for this chapter include Morrish A.H., (1965) Physical Principles of Magnetism, J. Wiley and Sons, New York; de Klerk D., Handbuch der Physik, Vol. XV, p. 38
2. Kapadnis D.G. and Hartmans R., Physica **22**, 173 (1956)
3. Stout J.W. and Catalano E., J. Chem. Phys. **23**, 2013 (1955); see also, Boo W.O.J. and Stout J.W., J. Chem. Phys. **65**, 3929 (1976)
4. Bloembergen P. and Miedema A.R., Physica **75**, 205 (1974)
5. Kopinga K., de Neef T., and de Jonge W.J.M., Phys. Rev. B**11**, 2364 (1975)
6. de Jonge W.J.M., Swüste C.H.W., Kopinga K., and Takeda K., Phys. Rev. B**12**, 5858 (1975); see also, Kopinga K., van der Leeden P., and de Jonge W.J.M., Phys. Rev. B**14**, 1519 (1976)
7. Giauque W.F. and MacDougall D.P., J. Am. Chem. Soc. **57**, 1175 (1935)
8. Zemansky M.W., (1968), 5th Ed.: Heat and Thermodynamics, McGraw-Hill, New York
9. Stout J.W. and Hadley W.B., J. Chem. Phys. **40**, 55 (1964); see also Fisher R.A., Hornung E.W., Brodale G.E., and Giauque W.F., J. Chem. Phys. **46**, 4945 (1967)
10. McElearney J.N., Schwartz R.W., Siegel A.E., and Carlin R.L., J. Am. Chem. Soc. **93**, 4337 (1971)
11. Cooke A.H., Meyer H., and Wolf W.P., Proc. Roy. Soc. (London) A**237**, 404 (1956)
12. More details are provided in the books cited earlier on EPR spectroscopy as well as in Chapt. 2 of Carlin R.L. and van Duyneveldt A.J., (1977) Magnetic Properties of Transition Metal Compounds, Springer, Berlin, Heidelberg, New York

4. Paramagnetism and Crystalline Fields: The Iron Series Ions

4.1 Introduction

The discussion thus far has centered on the lowest-lying energy levels – the ground state – of a transition metal ion. These are the states which determine the magnetic susceptibility at low temperatures – say, the temperature of boiling liquid ^{4}He (4.2 K). Yet, there are many magnetic measurements that do not extend to temperatures below that of boiling liquid nitrogen. Many measurements of interest to chemists have been reported over the temperature interval 80 K to 300 K; for example, that is generally an adequate temperature range in which one is able to determine whether the divalent cobalt in a given sample is present in either octahedral or tetrahedral stereochemistry. Even more straightforwardly, such a method easily allows one to distinguish whether a given sample is paramagnetic or diamagnetic! This can be important with nickel(II) for example, which is paramagnetic at room temperature in both octahedral and tetrahedral stereochemistry, but is diamagnetic in four-coordinate square planar geometry. Though the color of nickel compounds often changes dramatically as its stereochemistry changes, this has long been known to be an unreliable guide. Magnetic measurements can be conclusive in this regard.

So, this chapter will concentrate on what might be called high-temperature magnetism, and the empirical correlations which have been developed between magnetism and structure.

4.2 Magnetic Properties of Free Ions

The first-order Zeeman energy of a state was presented in Eq. (1.3) as

$$E = m_s g \mu_B H$$

with m_s the magnetic quantum number and g equal to 2.0023 when the orbital quantum number $\mathscr{L}$ was zero. Much of the following discussion will be seen to come under the headings of situations when $\mathscr{L}$ is not zero. In that case, the quantity m_s in the above equation is replaced by m_J, the component of $\boldsymbol{J} = \boldsymbol{L} + \boldsymbol{S}$, the total angular momentum. The related splitting factor g_J then takes the more complete form

$$g_J = 1 + \frac{[\mathscr{S}(\mathscr{S}+1) - \mathscr{L}(\mathscr{L}+1) + \mathscr{J}(\mathscr{J}+1)]}{2\mathscr{J}(\mathscr{J}+1)}. \tag{4.1}$$

We see that this reduces to $g=2$ when $\mathscr{L}=0$ and therefore $\mathscr{J}=\mathscr{S}$. Then, the effective moment μ_{eff} becomes

$$\mu_{\mathrm{eff}}=2[\mathscr{S}(\mathscr{S}+1)]^{1/2}\mu_{\mathrm{B}}, \tag{4.2}$$

though the more general result is

$$\mu_{\mathrm{eff}}=g_{\mathrm{J}}[\mathscr{J}(\mathscr{J}+1)]^{1/2}\mu_{\mathrm{B}}. \tag{4.3}$$

The terminology of "spin-only magnetism" should now be clear; it refers to Eq. (4.2). When $\mathscr{L}=0$ or there is no orbital angular momentum, the angular momentum is said to be "quenched" [1].

If n is the number of unpaired electrons on the ion, $\mathscr{S}=n/2$ and then Eq. (4.2) can be rewritten as $\mu_{\mathrm{eff}}=[n(n+2)]^{1/2}\mu_{\mathrm{B}}$. One can thereby see how a determination of μ_{eff} directly provides a measurement of the number of unpaired electrons (when $\mathscr{L}$ is zero).

4.3 Quenching of Orbital Angular Momentum

The moment for a free ion with ground state quantum numbers $\mathscr{L}$ and $\mathscr{S}$ is

$$\mu=[\mathscr{L}(\mathscr{L}+1)+4\mathscr{S}(\mathscr{S}+1)]^{1/2}\mu_{\mathrm{B}}. \tag{4.4}$$

This result is readily derived using elementary quantum mechanics [2]. This differs from Eq. (4.3) only in that spin-orbit coupling was included in obtaining that earlier equation.

We list in Table 4.1 the values of the magnetic moment as calculated according to both Eqs. (4.2) and (4.4). The last column lists typical values that have been observed. In general, the experimental values lie closer to those calculated from the spin-only formula than from the one that sets $\mathscr{L}$ non-zero. In general then, the orbital angular momentum is quenched, at least for the iron series ions. What does this mean?

Table 4.1. Magnetic moments of first row transition metal spin-free configurations.

No. of d electrons	$\mathscr{L}$	$\mathscr{S}$	Free ion ground term	$\mu=$ $[4\mathscr{S}(\mathscr{S}+1)$ $+\mathscr{L}(\mathscr{L}+1)]^{1/2}$, μ_{B}	$\mu_{\mathrm{eff}}=$ $[4\mathscr{S}(\mathscr{S}+1)]^{1/2}$, μ_{B}	μ_{eff} observed at 300 K, μ_{B}
1	2	$\frac{1}{2}$	^{2}D	3.00	1.73	1.7–1.8
2	3	1	^{3}F	4.47	2.83	2.8–2.9
3	3	$\frac{3}{2}$	^{4}F	5.20	3.87	3.7–3.9
4	2	2	^{5}D	5.48	4.90	4.8–5.0
5	0	$\frac{5}{2}$	^{6}S	5.92	5.92	5.8–6.0
6	2	2	^{5}D	5.48	4.90	5.1–5.7
7	3	$\frac{3}{2}$	^{4}F	5.20	3.87	4.3–5.2
8	3	1	^{3}F	4.47	2.83	2.9–3.9
9	2	$\frac{1}{2}$	^{2}D	3.00	1.73	1.7–2.2
10	0	0	^{1}S	0.00	0.00	0

Since we are dealing with electrons in d orbitals, this corresponds to assigning the orbital quantum number $\ell = 2$ to such an electron. Then, according to the methods of atomic spectroscopy, as illustrated in Chap. 1, one can combine such assignments to derive the free ion atomic states of a system. For a d^1 ion, then, with $\mathscr{L} = 2$ and $\mathscr{S} = \frac{1}{2}$, a ^2D (Russell-Saunders) state is found to be the ground state; this state is 10-fold degenerate, which means that there are 10 micro-states which are of equal energy. The electron may be equivalently in any one of the five d orbitals, and may have either spin component value, $m_s = +\frac{1}{2}$ or $-\frac{1}{2}$. A d^2 free ion has several Russell-Saunders states, the twelve-fold degenerate ^3F being the term symbol of the ground state, and so on.

Putting such an ion into a complex ion tends to remove some of the degeneracy, because of the crystalline field. In the simplest case of the d^1 ion, the five d orbitals split into two sets, called the t_{2g} and e_g orbitals. The d_{xy}, d_{xz}, and d_{yz} orbitals are the (degenerate) t_{2g} orbitals, while the $d_{x^2-y^2}$ and d_{z^2} orbitals are now called e_g orbitals. An electron in either set is said to give rise to either a $^2T_{2g}$ state or an ^2E_g state, with the ^2E_g state being 10 Dq in energy (typically about 20 000 cm^{-1}) above the $^2T_{2g}$ state. Since the ^2E_g state is too high in energy to be populated, the unpaired electron resides in the $^2T_{2g}$ state, which is a 6-fold degenerate state. An electron can now be in any one of the t_{2g} set, but the rotational symmetry of the free ion which had previously allowed such an electron also to be in one of the e_g orbitals is now lost. This is the physical meaning of the term orbital quenching.

Since the d^1 ion now has a $^2T_{2g}$ ground state, some but not all of the orbital angular momentum is quenched. The d^2 ion, with a cubic field $^3T_{1g}$ state is similar, but a d^3 ion such as chromium(III) has a t_{2g}^3 configuration, which gives rise to a $^4A_{2g}$ ground state. As the label A implies, such a state is orbitally nondegenerate: There is one electron assigned to each of the t_{2g} orbitals, and since electrons are indistinguishable, there is only one such electron assignment. Therefore the orbital angular momentum should be quenched for such a system, and this is indeed found to be true.

4.4 Coordination Compounds

We are now in a position to survey the paramagnetic behavior of the iron series ions. There are a number of characteristics that we shall describe, but it is impossible to review all the data in the literature. We limit ourselves to several of the features of the important oxidation states and suggest that the reader consult the incredible volumes [3] put together by the Königs for more thorough literature surveys. More selective reviews are also available [4–12] and we give literature references only to data not taken from one of these sources.

An important point is that the transition metal ions are not isolated but reside in a coordination compound. This is the source of the crystalline field effects described above. Let us digress for a moment on that point.

A complex ion, $[Cr(NH_3)_6]^{3+}$ or $[NiCl_4]^{2-}$, for example, consists of a central metal ion surrounded by, in these cases, ammonia molecules or chloride ions, respectively. These latter groups are called ligands. The net charge on the complex ion is the algebraic sum of the charges on the metal ions and its ligands. Depending on the system, the complex ion will therefore be a cation or anion, but neutral molecules are also common when the charges balance properly. Examples of the latter situation are provided by, for example, $[Ni(thiourea)_4Cl_2]$, $[Cr(NH_3)_3Cl_3]$ and $Fe(acac)_3$, where acac is the acetylacetonate ion, $[CH_3C(O)CHC(O)CH_3]^-$. These materials are all also called coordination compounds (as, e.g., $[Cr(NH_3)_6]Cl_3$), a term which we shall use interchangeably with complex ions.

Complex ions exhibit a number of stereochemistries. The most important geometry for a six-coordinate compound is octahedral, as illustrated:

All the positions are equivalent, so that there is only one compound of stoichiometry $[Cr(NH_3)_5Cl]^{2+}$, for example. Isomers exist when there is further substitution, as:

trans   cis

The (idealized) geometry of the *trans* isomer is such that the Cl–Cr–Cl angle is linear and the Cl–Cr–NH$_3$ angles are all 90°.

There are a number of polynuclear compounds in which some of the ligands serve as bridges between the metal ions, as is illustrated below:

$$[(H_3N)_4Co(OH)_2Co(NH_3)_4]^{4+} \qquad Cr_2Cl_9^{3-}$$

In the compound on the left, the octahedra share an edge, while a face is shared by the octahedra in the Cr compound. Infinite polymeric structures are common, as in $CoCl_2 \cdot 2H_2O$:

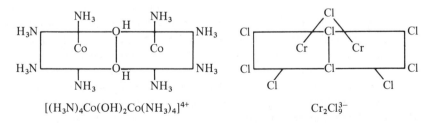

Two four-coordinate geometries are important: the tetrahedral (e.g., $[CoCl_4]^{2-}$) and the planar [e.g., $Cu(acac)_2$]. Five-coordinate geometries are less common, and the limiting structures are based either on the tetragonal pyramid

or the trigonal bipyramid

Further information on transition metal complexes may be found in [5].

4.5 Jahn-Teller Behavior [13]

A few words about the Jahn-Teller effect seem necessary, since many transition metal systems are influenced by the phenomenon. We have mentioned that certain complexes, particularly those with E_g ground states, are subject to Jahn-Teller influences. A theorem states explicitly that the energy in nonlinear molecules is minimized when distortions occur to remove orbital degeneracy. We illustrate the Jahn-Teller effect for octahedral orbital degeneracy.

Consider a d^9 ion in an octahedral complex. Using a crystal field model, the half-vacant d orbital may be either $d_{x^2-y^2}$ or d_{z^2} (both e_g) or their linear combination, giving a ^2E_g ground state. If we imagine that the $d_{x^2-y^2}$ orbital contains the "hole" we assume that the in-plane ligand atoms are less effectively shielded from the metal than the two axial ligands and hence bond more strongly to it. An elongated tetragonal complex of D_{4h} symmetry results. The opposite is true when the hole is in d_{z^2}. In reality, however, d_{z^2} and $d_{x^2-y^2}$ are equivalent in O_h and the complex cannot gain any more energy by tetragonal distortion along x than along y or z. However, it will gain energy by a symmetry reduction to D_{4h} with the E_g ground state split into two nondegenerate states (Fig. 4.1). The T_{2g} excited state also loses its degeneracy on distortion, but the splitting is much smaller.

Since tetragonal distortion along either x, y or z is equally feasible, the molecule may undergo a minima exchange between the three D_{4h} stereochemistries pictured in Fig. 4.2. The average geometry of the complex is octahedral, but its instantaneous geometry is likely to be D_{4h}. Since electronic transitions occur within a time much more rapid than nuclear motions, the electronic spectra may display features consistent with a tetragonal geometry about the metal ion. In solution or in the solid state, molecules cannot undergo nuclear motions without interacting in some way with other molecules. These interactions may produce barriers to the pseudo-rotation and

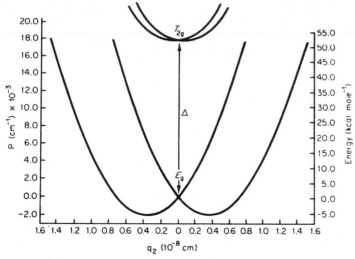

Fig. 4.1. Splitting of E_g and T_{2g} states as q_2, the coordinate describing the Jahn-Teller distortion of an octahedral molecule, is varied. The splitting of T_{2g} is not drawn to scale. From Ref. [13]

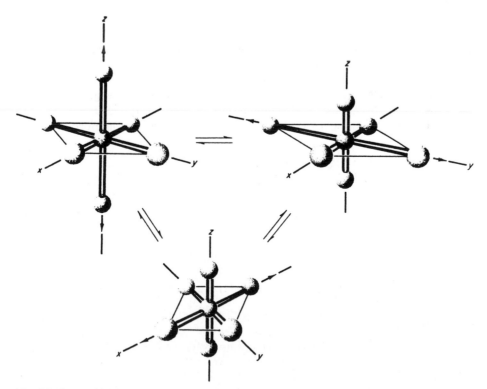

Fig. 4.2. Stereochemical rearrangement of a tetragonally distorted octahedral complex between the three equally energetic minima. From Ref. [13]

"freeze" distortions which are then observed by some physical technique such as X-ray crystallography. Thus, the $[M(H_2O)_6]^{2+}$ ion in $(NH_4)_2M(SO_4)_2 \cdot 6H_2O$ with $M = Cr(II)$ or $Cu(II)$ exists as a grossly distorted MO_6 octahedron, while with $M = Ni(II)$ or $Co(II)$ (non-E_g ground states) distortions are considerably smaller.

4.6 The Iron Series Ions

4.6.1 Titanium(III), d^1, $^2T_2$

With but one unpaired electron, titanium(III) is a Kramers ion and the ground state in a cubic field is $^2T_{2g}$. This has therefore been the prototype ion for all crystal field calculations, for the complication caused by electron-electron repulsions need not be considered. Despite this large body of theoretical work, the air-sensitive nature of many compounds of this ion has prevented a wide exploration of its magnetic behavior; more importantly, in those compounds where it has been studied, it rarely acts in a simple fashion or in any sensible agreement with theory. The problem lies with the six-fold degeneracy of the $^2T_{2g}$ state, and the nature of its resolution by spin-orbit coupling and crystal field distortions.

The results of calculations of the g values and susceptibility of trivalent titanium will be outlined here. A trigonal or tetragonal field is always present, which splits the $^2T_{2g}$ state (in the absence of spin-orbit coupling) into an orbital singlet and doublet, separated by an amount δ. After adding spin-orbit coupling, one finds that for negative δ, $g_\parallel = g_\perp = 0$, the orbital and spin contributions cancelling. There appear to be no examples known of this situation, though $g_\parallel$ can be non-zero if the trigonal field is not very much weaker than the cubic one. Thus, $g_\parallel = 1.067$ and $g_\perp < 0.1$ for Ti^{3+} in Al_2O_3. In the situation of positive δ, the g values depend on the relative strengths of the axial field distortion (δ) and spin-orbit coupling (λ), as was illustrated in Fig. 2.7. The Hamiltonian for the d^1 ion perturbed by both a magnetic field and spin-orbit coupling is:

$$H = \lambda L \cdot S + \mu_B(L + gS) \cdot H.$$

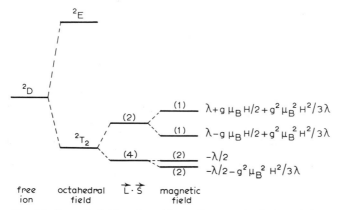

Fig. 4.3. Crystal field, spin-orbit, and magnetic field splitting of the ^2D energy level of titanium(III)

The resulting energy level diagram is given in Fig. 4.3. The numbers in parentheses indicate the degeneracy of a particular level. We ignore the contribution to paramagnetism from the ^2E states, for they are some $20\,000\,\mathrm{cm}^{-1}$ higher in energy. (The contribution of any energy level n is proportional to $\exp(-E_n/kT)$, and $kT \simeq 205\,\mathrm{cm}^{-1}$ at room temperature.)

Therefore, applying Eq. (2.2) to the energy levels sketched in Fig. 4.3.

$$\chi/N = [20^2/kT - 2 \times 0) \exp(\lambda/2kT) + 2(0^2/kT + 2g^2\mu_B^2/2\lambda)$$
$$\cdot \exp(\lambda/2kT) + 2(g^2\mu_B^2/4kT - 2g^2\mu_B^2/3\lambda) \exp(-\lambda/kT)]/$$
$$\cdot 2[\exp(\lambda/2kT) + \exp(\lambda/2kT) + \exp(-\lambda/kT)]$$

which reduces to

$$\chi = Ng^2\mu_{\mathrm{eff}}^2/3kT$$

with

$$\mu_{\mathrm{eff}}^2 = \frac{8 + (3\lambda/kT - 8)\exp(-3\lambda/2kT)}{4(\lambda/kT)\,[2 + \exp(-3\lambda/2kT)]} \, . \tag{4.5}$$

Note that the Curie Law does not hold in this case. In fact, in the limit as $T \to \infty$, the exponential may be expanded, the first term retained, and one finds $\mu^2 \to \frac{5}{4}$. In the other limit, $\mu^2 \to 0$ as $T \to 0$. How can a system with an unpaired electron have zero susceptibility at $0\,\mathrm{K}$? The spin and orbital angular momenta cancel each other out. Note also that $\mu^2 \to \frac{5}{4}$ as $\lambda \to 0$. This susceptibility is sketched in Fig. 4.4.

As the following discussion will show, this calculation has never been applied to any experimental data. The calculated susceptibility does not allow for any anisotropy,

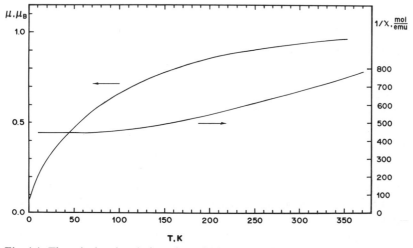

Fig. 4.4. The calculated variation of μ and $1/\chi$ versus T for octahedral titanium(III). The free-ion value of λ is used

even in the g-values, nor for any crystal field splittings. These are important for titanium.

Cubic cesium titanium alum, $CsTi(SO_4)_2 \cdot 12 H_2O$, is probably the most thoroughly studied titanium(III) compound, largely because it is the least air-sensitive, but it is only recently that an understanding of its magnetic properties seems to be emerging. The initial impetus arose because it was assumed that its electronic structure would prove easy to determine, as well as attempts were made to discover its feasibility as a cooling salt for adiabatic demagnetization experiments. The salt is decidedly not ideal at low temperatures, however, and interest in the substance, at least as a cooling salt, never developed very far.

Thus, it was assumed that the $^2T_2(D)$ state of $[Ti(H_2O)_6]^{3+}$ was resolved by a static crystal field into three doublets separated by an energy of $100\,cm^{-1}$ or so. At low temperatures, to first order, the system should then have acted as an effective $\mathscr{S} = \frac{1}{2}$ ion with a g of about 2 but interactions with the nearby excited statees would cause some modification of the behavior. Measurements of the powder susceptibility led to a $\langle g \rangle$ of only 1.12, and EPR on the diluted material provided the values $g_\| = 1.25, g_\perp = 1.14$, and it proved impossible to rationalize these values with simple crystal field theory. It was suggested that the excited states in fact lay only some tens of cm^{-1} above the ground state, and the most recent measurements of the magnetic susceptibility of $CsTi(SO_4)_2 \cdot 12 H_2O$ suggested that in fact the lowest lying doublet was some $30\,cm^{-1}$ above the ground state.

Over the years, Ti(III) has been studied by EPR in many alum lattices although apparently the spectra have been assigned incorrectly to the presence of too many sites. The reason that Ti(III) does not behave in a straightforward fashion has been ascribed to a number of factors. Trigonal distortions have usually been assumed, due in part to size. After all, this is a relatively large ion with a radius estimated as 0.067 nm, and many of the EPR experiments with titanium have been carried out on samples doped into isomorphous aluminum diluents, with the radius of the host Al(III) estimated as 0.053 nm. Size mismatch, followed with distortion of the lattice and a consequent effect on the electronic structure of the ion were thus frequently invoked. Indeed, a study of $Ti^{3+} : [C(NH_2)_3]Al(SO_4)_2 \cdot 6 H_2O$ suggested that the size mismatch was so important that titanium entered the guest lattice in effectively a random fashion, creating impurity centers as distinct from isomorphous replacement. Both signs of the trigonal field splitting parameter were observed within the one system. Static Jahn-Teller effect distortions have also been mentioned as a source of the problem.

Nevertheless, recent work suggests that the above arguments may not be correct, or at least not applicable to the pure, undiluted $CsTi(SO_4)_2 \cdot 12 H_2O$. For one thing, the crystal structure analysis of the compound shows the structure to be highly regular and not distorted. Furthermore, important dynamic Jahn-Teller effects have been implicated as the likely source of the complexity of the problem.

In a series of papers on both $CsTi(SO_4)_2 \cdot 12 H_2O$ and $(CH_3NH_3)Ti(SO_4)_2 \cdot 12 H_2O$, Walsh and his co-workers show [14] that the energy level diagram of Ti^{3+} illustrated in Fig. 4.5 accounts for all the observations described above. In addition, spin-lattice relaxation studies on both compounds showed the presence at low temperatures of the Orbach relaxation process, which requires that there be energy levels accessible at rather low energies. In particular, for the Cs^+ compound, the Γ_{5g} electronic state is strongly coupled to the Γ_{3g} vibrational mode by

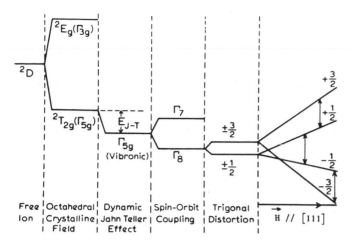

Fig. 4.5. Schematic energy-level diagram of Ti^{3+}

means of a dynamic Jahn-Teller interaction. This fortuitously has the effect of canceling the major trigonal distortion, resulting in a quasi-Γ_8 quartet ground state, separated from Γ_7 by a mere $7 \pm 1\,cm^{-1}$. A residual trigonal distortion gives rise to a zero-field splitting and three EPR lines, which disposes of the earlier model of twelve magnetic complexes for Ti^{3+} in CsAl alum. The best values of the parameters on this model for Ti^{3+} : $CsAl(SO_4)_2 \cdot 12\,H_2O$ are, $g_{\parallel} = 1.1937 \pm 0.001$, $g_{\perp} = 0.6673 \pm 0.005$, and a zero-field splitting of some 0.002 K.

The resolution of this long-standing problem illustrates the fact that, at least for trivalent titanium, static crystal field calculations such as those mentioned earlier are useful pedagogical devices, but bear little relationship to the true nature of the problem.

Optical spectra, EPR spectra, as well as spin-lattice relaxation studies suggest that Ti(III) is severely distorted in the acetylacetonate, Ti(acac)$_3$. This molecule, with $g_{\parallel}$ = 2.000, $g_{\perp}$ = 1.921, is also unusual in that EPR spectra may be observed even at room temperature. A complete theory of Ti(III) in trigonal environments has been published [15], but serious discrepancies with experimental results for several systems were found. The problem probably lies with the fact that the Jahn-Teller effect was ignored, and that attention was centered on the g-values obtained by EPR. Since these are characteristic of the lowest-lying or ground state, the contributions of the (mostly unknown) low-lying states was not considered.

A complex ion which has been studied extensively lately [16] is $[Ti(urea)_6]^{3+}$, as both the iodide and the perchlorate. The compounds are isostructural and belong to a hexagonal space group. The susceptibility was reported over the temperature interval 4–300 K, but there are very few low-temperature points. A crystal field model was used for the data analysis and the optical spectra were fit as well. The data analysis required that the spin-orbit coupling constant be reduced to only 47% of its free ion value, which seems unrealistic.

We have paid special attention to this ion only because one can conclude that the single-ion properties of titanium(III) are not at all simple, and caution must be applied in the study of its compounds.

4.6.2 Vanadium(III), d^2, $^3A_2$

This ion is of more interest for its single-ion properties than for its magnetically-ordered compounds, the latter of which there are hardly any. The zero-field splitting is large, being of the order 5–15 K in the compounds examined to date. The large splitting, which is easily determined from either susceptibilities or specific heats, arises because the octahedral ground state is in fact a ^3F; the $^3A_2(F)$ state becomes the ground state because of axial distortions, but the excited $^3E(F)$ component lies generally only about $1000\ cm^{-1}$ above the $^3A_2$. What is of special interest is that in every case in which the zero-field splitting of V^{3+} has been evaluated, it is always positive (doublet above the singlet) [17]. It has been pointed out [18] that a crystal field calculation shows that $D \simeq (1.25\lambda)^2/v$, where λ is the spin-orbit coupling constant and v is the trigonal field splitting of the $^3T_1$ state into $^3A_2$ and ^3E. When $^3A_2$ is the ground state v is positive and so D is always positive.

One consequence of the large zero-field splittings in vanadium is that the allowed $\Delta m_s = \pm 1$ paramagnetic resonance transition has only been observed in $V^{3+}:Al_2O_3$, by going to very high fields with a pulsed magnetic field; the low-lying orbital levels also give rise to short spin-lattice relaxation times which require that helium temperatures be used for EPR. The forbidden $\Delta m_s = \pm 2$ transition, however, has been observed with several samples, such as with $V^{3+}:[C(NH_2)_3]Al(SO_4)_2 \cdot 6H_2O$ and $V^{3+}:Al_2O_3$. The g-values for trigonally-distorted V^{3+} are typically $g_\parallel \simeq 1.9$ and $g_\perp \simeq 1.7$–1.9.

The magnetization as a function of field, frequency, and temperature has also been measured for several compounds with large zero-field splittings [19, 20]. The isothermal magnetization is given in Eq. (2.12) and plotted in Fig. 2.8.

Two compounds which have been investigated recently are $[C(NH_2)_3]V(SO_4)_2 \cdot 6H_2O$ [20–25] and $Cs_3VCl_6 \cdot 4H_2O$. Magnetization studies [20, 21] on the former compound showed that isothermal conditions were not fulfilled, and that spin-lattice and cross relaxation effects were present at 1–2 K. The ZFS is D/k = 5.45 K, and exchange effects are negligible. The latter compound has a trans-$[VCl_2(H_2O)_4]^+$ coordination sphere [26] yet susceptibility [17] and magnetization [27] measurements in large applied fields have shown that magnetic exchange is still very weak in this compound.

Several recent reports have combined magnetic susceptibility data with spectral data in order to characterize all the electronic states of the vanadium ion. The best-fit parameters [18] for $V(acac)_3$ are $g_\parallel = 1.96$, $g_\perp = 1.78$ and $D = 7.7\ cm^{-1}$ (11 K) and $D = 5.86\ cm^{-1}$ (8.4 K) for $[V(urea)_6][ClO_4]_3$ [16].

4.6.3 Vanadyl, VO^{2+}, d^1

This ion is interesting because of its binuclear nature. Its electronic configuration allows paramagnetic resonance to be observed easily, and many studies have been reported. These include the ion as diluent in a variety of crystals, and even when dissolved in liquids at room temperature. The g values are isotropic at about 1.99, as anticipated, and simple, spin-only magnetism is frequently observed. Molecular structures are usually quite distorted.

4.6.4 Chromium(III), d^3, $^4A_2$

Since chromium(III) is a $\mathscr{S}=\frac{3}{2}$ ion, zero-field splittings are usually found in its compounds. The spin-orbit coupling constant is relatively small, however, so that the zero-field splittings observed to date are quite small. Thus, EPR measurements lead to fairly typical values of only 0.592 cm^{-1} for D in Cr^{3+} : Al(acac)$_3$ and 0.00495 cm^{-1} for Cr^{3+} in [Co(en)$_3$]Cl$_3$ · NaCl · 6 H$_2$O. The g value is almost always isotropic at about 1.98 or 1.99.

The calculated spin-only moment for Cr(III) is 3.88 μ_B. This value is generally observed to be reduced slightly because spin-orbit coupling can cause the excited optical states, which have orbital degeneracy, to mix with the non-degenerate ground state. The relationship is

$$\mu_{eff} = \mu'_{eff}[1 - (4\lambda/10Dq)], \qquad (4.6)$$

where μ'_{eff} is the calculated spin only moment, λ is the spin-orbit coupling constant, and 10Dq is the energy of the lowest excited ($^4T_{1g}$) state. Since $\lambda \simeq 92$ cm^{-1} and $10Dq \simeq 20\,000$ cm^{-1} the correction is small, and effective moments of 3.82 to 3.87 μ_B at room temperature are generally observed for chromium coordination compounds.

There are relatively few measurements at low temperatures on chromium compounds, with the exception of the alums such as KCr(SO$_4$)$_2$ · 12 H$_2$O. These salts have been studied because the magnetic and other non-ideal interactions are generally so weak that the alums provide useful cooling salts for adiabatic demagnetization.

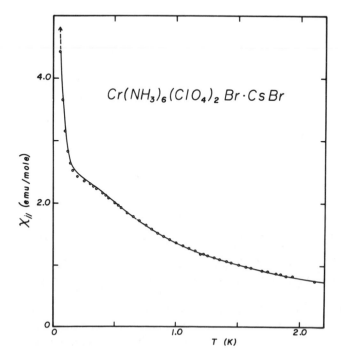

Fig. 4.6. $\chi_\|$ for [Cr(NH$_3$)$_6$] (ClO$_4$)$_2$Br · CsBr. From Ref. [29]

The bimetallic compounds $[Cr(H_2O)(NH_3)_5][Co(CN)_6]$, $[Co(H_2O)(NH_3)_5]$ $[Cr(CN)_6]$ and $[Co(NH_3)_6][Cr(CN)_6]$ have recently been found [28] to obey the Curie-Weiss law in the temperature region 1.2–4.2 K; the g-values are typical at about 1.98, and $\theta = -0.08$ K, 0.02 K, and 0 K, respectively. The cobalt-containing moiety in each case is diamagnetic.

The rhombohedral compound $[Cr(NH_3)_6](ClO_4)_2Br \cdot CsBr$ was recently studied [29] over the temperature interval 40 mK–4.2 K. The compound contains discrete $[Cr(NH_3)_6]^{3+}$ ions which are well-separated, and was attractive for study because the uniaxial symmetry of the crystal readily allowed the anisotropic susceptibilities to be measured. The results are plotted in Fig. 4.6, as $\chi_\parallel$ along with a fit to the relevant equation, Eq. (2.11). What was interesting was that both the sign and the magnitude of the zero-field splitting parameter could be determined, $2D/k = 0.53(2)$ K. It is more common to determine parameters this small by EPR methods, but susceptibility measurements to low-enough temperatures provide an interesting alternative procedure.

4.6.5 Manganese(II), d⁵, ⁶A₁

The spin-free manganese ion is the source of a vast portion of the literature on magnetism. The reasons for this are straightforward. Chemical problems of synthesis are usually minimal. Since it has an odd number of electrons, it is a Kramers ion, and EPR spectra may be observed under a variety of conditions. The g values are isotropic at about 2.0. Zero-field splittings are usually small – of the order of 10^{-2} cm^{-1}, and the ground state well-isolated from the higher energy levels. On the other hand, several systems will be mentioned later which appear to have quite large zero-field splittings, and the large value of the spin sometimes causes magnetic dipole-dipole interactions between the metal ions to be important.

Strong-field or spin-paired manganese has been studied by EPR in such compounds as $K_4[Mn(CN)_6] \cdot 3 H_2O$. Spin-orbit coupling effects are important, g-value anisotropy is found, and the optical spectra are dominated by charge transfer effects. Relatively few magnetic studies pertain to this electronic configuration of the ion.

4.6.6 Iron(III), d⁵, ⁶A₁

The magnetochemistry of iron(III), which is isoelectronic to manganese(II), is relatively straightforward at least for the spin-free compounds. With five unpaired electrons, the ground state is a ⁶S one, and this is well-isolated from the lowest lying excited states. That means again that g is usually quite close to the free-ion value of two, that g is isotropic, and that crystal-field splittings of the ground state are much smaller than a Kelvin. Furthermore, since this is an odd electron (Kramers) ion, resonances are always detected even if the zero-field splittings should be large. Nevertheless, the EPR linewidths are often broad, and this tends to mask any narrow lines which may be present.

Tetrahedral iron(III) is expected to behave similarly, but there are few relevant data. Five-coordinate iron is quite unusual; it is found with some porphyrins and with other molecules of biological importance, but it is best known in the halobisdithiocarbamates. The compounds are unusual because, as a result of the geometry, a $^4A_2$ state becomes the ground state. Thus there are only three unpaired electrons in the system, and the ground state is susceptible to the same zero-field splittings which are found with the other $^4A_2$ state ions, octahedral chromium(III) and tetrahedral cobalt(II). The ZFS has been found to be 7 K in the chlorobisdiethyldithiocarbamate derivative [30]. This compound will be discussed more thoroughly in Chap. 10.

In a strong crystalline field, the ion has but one unpaired electron and a $^2T_2$ ground state. Three orbital states are then low-lying, and spin-orbit coupling effects become very important. Low temperatures are required to detect the EPR, and the g-values deviate greatly from 2. One well-known compound with this electronic configuration is $K_3Fe(CN)_6$. The theory of this electronic state is the same as that for the isoelectronic ruthenium(III) ion (Sect. 8.3).

4.6.7 Chromium(II), d^4, Manganese(III), d^4, and Iron(II), d^6

These ions are grouped together because they are relatively unfamiliar but have essentially similar electronic states.

The problem with these $\mathscr{S} = 2$ ions lies not only with their air-sensitive nature, but especially with the large number of electronic states that they exhibit. Being non-Kramers ions, there are relatively few EPR studies: short spin-lattice relaxation times are found, and large zero-field splittings. For example, D has been reported as $20.6 \, cm^{-1}$ for $Fe^{2+} : ZnSiF_6 \cdot 6H_2O$; $FeSiF_6 \cdot 6H_2O$ was discussed above in Sect. 2.6. The g-values deviate appreciably from 2.

The magnetic moment of $Mn(acac)_3$, 4.8–4.9 μ_B at high temperatures, falls at temperatures below 10 K [18]. It was suggested that a splitting of the orbital degeneracy of the ^5E ground state by $3 \, cm^{-1}$ could account for this behavior.

Mössbauer (4–340 K) and magnetic susceptibility (80–310 K) studies have recently been reported on a number of $[FeL_6](ClO_4)_2$ compounds, where L is a sulfoxide or pyridine N-oxide [31]. Spin-paired ferrous ion is of course diamagnetic; a large number of studies have been reported concerning an equilibrium between the two spin-states [32].

4.6.8 Cobalt(II), d^7

The ground state of this ion will be explored in detail, because cobalt illustrates so many of the concepts important in magnetism, as well as because it provides so many good examples of interesting magnetic phenomena. We begin with six-coordinate, octahedral cobalt(II).

Cobalt(II) with three unpaired electrons, exhibits an important orbital contribution at high temperatures, and a variety of diagnostic rules have been developed to take advantage of this behavior. For example, octahedral complexes typically have a moment of 4.7 to 5.2 μ_B. Tetrahedral complexes usually exhibit smaller moments, in the range of about 4.59 μ_B for $CoCl_4^{2-}$, 4.69 μ_B for $CoBr_4^{2-}$ and 4.77 μ_B for CoI_4^{2-}. The lowest electronic states in octahedral fields are illustrated in Fig. 2.8 where it will be seen that the effects of spin-orbit coupling ($\lambda = -180 \, cm^{-1}$) and crystalline distortions combine to give a spin-doublet ground state, separated by $100 \, cm^{-1}$ or so from the next nearest components. This ground state is the interesting one in cobalt magnetochemistry, so we restrict the remainder of the discussion to low temperatures, at which the population of the other states is small.

The doubly-degenerate level is an effective $\mathscr{S} = \frac{1}{2}$ state, and so unusual features may be anticipated. The theory is essentially due to Abragam and Pryce [33].

Two parameters have been introduced that are useful for the empirical representation of magnetic data. The first of these, a Landé factor usually called α refers to the strong-field ($\alpha = 1$) and weak-field ($\alpha = \frac{3}{2}$) limits and its diminution in value from $\frac{3}{2}$ is a measure of the orbital mixing of $^4T_1(F)$ and $^4T_1(P)$. The lowest electronic level in an axial field with spin-orbit coupling is a Kramers doublet and so cannot be split except by magnetic fields. The orbital contribution of the nearby components of the $^4T_1(F)$ state causes the ground doublet in the weak cubic field limit to have an isotropic $g = 4.33$, a result in agreement with experiments on cobalt in MgO, but large anisotropy in the g value is expected as the crystal field becomes more distorted. The three orthogonal g values are expected to sum in first order to the value of 13 [33].

The second parameter, δ, is an axial crystal field splitting parameter that measures the resolution of the degeneracy of the $^4T_1$ state, and thus is necessarily zero in a cubic crystal. The isotropic g value is then $(\frac{2}{3})(5+\alpha)$, to first order. In the limit of large distortions, δ may take on the values of $+\infty$ or $-\infty$, with the following g values resulting:

$$\delta = +\infty : g_{||} = 2(3+\alpha), \quad g_{\perp} = 0 ,$$
$$\delta = -\infty : g_{||} = 2, \quad g_{\perp} = 4 .$$

For a given α, the two g values are therefore functions of the single parameter δ/λ and so they bear a functional relationship to each other. Abragam and Pryce have presented the general result, but there are more parameters in the theory than can usually be obtained from the available experimental results. With the approximation of isotropic spin-orbit coupling, they derive (to first order) the following equations which provide a useful *estimate* of crystal distortions,

$$g_{||} = 2 + 4(\alpha+2)[\{3/x^2 - 4/(x+2)^2\}/\{1 + 6/x^2 + 8/(x+2)^2\}], \tag{4.7}$$
$$g_{\perp} = 4[\{1 + 2\alpha/(x+2) + 12/x(x+2)\}/\{1 + 6/x^2 + 8/(x+2)^2\}], \tag{4.8}$$

where x is a dummy parameter which for the lowest energy level is positive with limiting values of 2 (cubic field, $\delta = 0$), 0 ($\delta = +\infty$), and ∞ ($\delta = -\infty$). The splitting parameter δ is found as

$$\delta = \alpha\lambda[(x+3)/2 - 3/x - 4(x+2)] . \tag{4.9}$$

The observed g values for octahedral environments with either trigonal or tetragonal fields should therefore all lie on a universal curve. Such a curve is illustrated in Fig. 4.7, and a satisfactory relationship of theory and experiment has been observed. The observation of $g_{||}$ and $g_{\perp}$ allows the solution of Eqs. (4.7) and (4.8) for α and x, and these parameters in turn may be applied to Eq. (4.9).

The situation is far different when the cobalt(II) ion resides in a tetrahedral geometry. A $^4A_2$ ground state results, and with a true spin of $\frac{3}{2}$, the g values are slightly

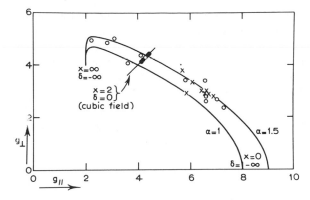

Fig. 4.7 Experimental and theoretical g-values for the configuration $t_2^5 e^2$. Where the site symmetry is rhombic, $g_\perp$ is taken as the mean of g_x and g_y

anisotropic and lie in the range of 2.2 to 2.4. A zero-field splitting of the spin-quartet usually occurs, and it is generally rather large [19, 34]. In Cs_2CoCl_4, the zero field splitting is about 14 K, with $|\pm\frac{1}{2}\rangle$ being the ground state. The system acts as an effective $\mathscr{S}=\frac{1}{2}$ system at low temperatures. When 2D is large and negative, as in the A_3CoX_5 (A=Cs, Rb; X=Cl, Br) series of compounds, then the $|\pm\frac{3}{2}\rangle$ state lies lower. The system again acts as an effective $\mathscr{S}=\frac{1}{2}$ system, but in this case $g'_\| = 3g_\|$ and $g'_\perp = 0$.

A less-common geometry for cobalt(II) is the planar four-coordinate one. Most of the compounds reported to date involve bidentate monoanions such as dimethylglyoximate or dithioacetylacetonate as ligands [35]. They are generally low spin with magnetic moments of 2.2–2.7 μ_B at 300 K; g-values of $g_x = 3.28$, $g_y = g_z = 1.90$ have been reported [35] for bis(pentane-2,4-dithionato)cobalt(II).

4.6.9 Nickel(II), d^8, $^3A_2$

The electronic structure of this ion has been dealt with at length in the previous portions of the book. The reasons for this should be clear: nickel forms complexes with a variety of ligands, as well as the fact that several stereochemistries are common. Restricting the discussion to octahedral complexes, the spin of this ion is large enough to make it a sensitive probe of a variety of phenomena, and yet it is not too high to prevent theoretical analyses. Spin-orbit coupling is important enough to cause the g values to deviate measurably from the free ion values, to values typically in the neighborhood of 2.25. Yet, the g values are commonly found to be isotropic or nearly so, and spin-orbit coupling causes no other problems. The lowest-lying excited states are far enough away (usually at least 8000 cm^{-1}) so as to be unimportant magnetically. Paramagnetic resonance is generally observed without difficulty, frequently at room temperature; on the other hand, nickel(II) is a non-Kramers ion, and the resolution of degeneracies thereby allowed is occasionally observed. Thus, as will be discussed below, the zero-field splitting in $[Ni(C_5H_5NO)_6](ClO_4)_2$, as well as several other salts, is large enough so that paramagnetic resonance absorption is prevented at x-band, even at helium temperatures. This is because the energy of the microwave quantum is smaller than the separation between the $|0\rangle$ and $|\pm1\rangle$ states, the states between which a transition is allowed.

The dominant feature of the magnetochemistry of octahedral nickel(II) is the zero-field splitting of the ground, $^3A_2$, state. The consequences of this phenomenon have been explored above, extensively. Neglecting the rhombic (E) term, the parameter D/k is usually found to be of the order of a few Kelvins. That puts important magnetic anisotropy effects and Schottky anomalies in the specific heat at a convenient temperature region, the easily-accessible helium one. What is of interest is that both signs of D are found, being for example, positive in $Ni(NO_3)_2 \cdot 6H_2O$ [36], $NiSnCl_6 \cdot 6H_2O$ [37], and $[Ni(C_5H_5NO)_6](ClO_4)_2$ [38] and negative in $NiCl_2 \cdot 4H_2O$ [39], $NiCl_2 \cdot 2py$ [40] and $NiZrF_6 \cdot 6H_2O$ [37]. The magnitude of D varies widely, being as small as $+0.58$ K in $NiSnCl_6 \cdot 6H_2O$ and as large as $+6.26$ K in $[Ni(C_5H_5NO)_6](ClO_4)_2$, and even -30 K in the linear chain series $NiX_2 \cdot 2L$, $X=Cl$, Br; $L=$pyridine, pyrazole [40].

Few tetrahedral nickel salts have been investigated extensively at low temperatures; because of spin-orbit coupling, they are expected to become diamagnetic at very low temperatures. Although many compounds have been prepared with planar, four-coordinate nickel, they are diamagnetic at all temperatures.

4.6.10 Copper(II), d⁹, ²E

This ion has one unpaired electron whatever the geometry: octahedral, planar or tetrahedral.

The single-ion magnetic properties of copper(II) are fairly straightforward. Spin-orbit coupling is large, causing the g values to lie in the range 2.0 to 2.3, but because copper has an electronic spin of only $\frac{1}{2}$, there are no zero-field splitting effects. The g values are often slightly anisotropic, being for example 2.223 ($\parallel$) and 2.051 ($\perp$) in $Cu(NH_3)_4SO_4 \cdot H_2O$.

The one problem that does arise with this ion is that it rarely occupies a site of high symmetry; in octahedral complexes, two *trans* ligands are frequently found substantially further from the metal than the remaining four. This has led to many investigations of copper as the Jahn-Teller susceptible ion, par excellence. Dynamic Jahn-Teller effects have also been frequently reported in EPR investigations of copper compounds at low temperatures.

Tetrahedral copper is also well-known, though it is usually quite distorted because of the Jahn-Teller effect. It is found, for example, in Cs_2CuCl_4, a system with an average g-value of 2.20 [41].

4.7 References

1. A more detailed exposition is provided by Figgis B.N., (1966) Introduction to Ligand Fields, Interscience, New York
2. See, for example, Mabbs F.E. and Machin D.J., (1973) Magnetism and Transition Metal Complexes, Chapman and Hall, London
3. König E., (1966) Magnetic Properties of Coordination and Organometallic Coordination Compounds, Landolt-Börnstein, New Series, Vol. II-2, Springer Berlin, Heidelberg, New York, König E. and König G., *idem.*, Vol. II-8, 1976; Vol. II-10, 1979; Vol. II-11, 1981
4. Carlin R.L. and van Duyneveldt A.J., (1977) Magnetic Properties of Transition Metal Compounds, Springer Berlin, Heidelberg, New York, and references therein
5. Cotton F.A. and Wilkinson G., (1980) 4th Ed.: Advanced Inorganic Chemistry, J. Wiley & Sons, New York
6. Figgis B.N. and Lewis J., (1964) in Progress in Inorganic Chemistry, Vol. 6, Ed. by Cotton F.A., Interscience, New York

7. Mitra S., (1972) in Transition Metal Chemistry, Ed. Carlin R.L., Marcel Dekker, N.Y., Vol. 7
8. Mitra S., (1976) in Progress in Inorganic Chemistry, Vol. 22, Ed. Lippard S.J., J. Wiley & Sons, New York
9. Casey A.T. and Mitra S., (1976) in Theory and Applications of Molecular Paramagnetism, Ed. Boudreaux E.A. and Mulay L.N., J. Wiley & Sons, New York, p. 135
10. O'Connor C.J., (1982) in Progress in Inorganic Chemistry, Vol. 29, Ed. Lippard S.J., J. Wiley & Sons, New York, p. 203
11. Cox P.A., Gregson, A.K., Boyd P.D.W. and Murray K.S., in Spec. Period. Report. Electronic Structure and Magnetism of Inorganic Compounds, Vol. 1, 1972; Vol. 2, 1973; Vol. 3, 1974; Vol. 4, 1976; Vol. 5, 1977; Vol. 6, 1980; Vol. 7, 1982
12. McGarvey B.R., (1966) in Transition Metal Chemistry, Vol. 3, Ed. by Carlin R.L., Marcel Dekker, Inc., New York, p. 90
13. This discussion is based upon that by Fackler, Jr. J.P., (1971) Symmetry in Coordination Chemistry, Academic Press, New York
14. Jesion A., Shing Y.H. and Walsh D., Phys. Rev. Lett. **35**, 51 (1975) and references therein
15. Gladney H.M. and Swalen J.D., J. Chem. Phys. **42**, 1999 (1965)
16. Baker J. and Figgis B.N., Aust. J. Chem. **33**, 2377 (1980)
17. Carlin R.L., O'Connor C.J. and Bhatia S.N., Inorg. Chem. **15**, 985 (1976)
18. Gregson A.K., Doddrell D.M. and Healy P.C., Inorg. Chem. **17**, 1216 (1978)
19. Carlin R.L., in "Magneto-Structural Correlations in Exchange Coupled Systems," NATO ASI Series, Edited by Willett R.D., Gatteschi D., and Kahn O., D. Reidel Publ., Dordrecht, 1985
20. Smit J.J., de Jongh L.J., de Klerk D., Carlin R.L. and O'Connor C.J., Physica **B 86–88**, 1147 (1977)
21. Smit J.J., Thesis, Leiden, 1979
22. McElearney J.N., Schwartz R.W., Merchant S., and Carlin R.L., J. Chem. Phys. **55**, 465 (1971)
23. Mahalingam L.M. and Friedberg S.A., Physica **B 86–88**, 1149 (1977)
24. Ashkin J. and Vanderven N.S., Physica **B 95**, 1 (1978)
25. Diederix K.M., Groen J.P., Klaassen T.O., Poulis N.J., and Carlin R.L., Physica **B 97**, 113 (1979)
26. McCarthy P.J., Lauffenburger J.C., Schreiner M.M., and Rohrer D.C., Inorg. Chem. **20**, 1571 (1981)
27. Smit J.J., van Wijk H.J., de Jongh L.J., and Carlin R.L., Chem. Phys. Lett. **62**, 158 (1979)
28. Carlin R.L., Burriel R., Fina J, and Casabo J., Inorg. Chem. **21**, 2905 (1982); Carlin R.L., Burriel R., Pons J., and Casabo J., Inorg. Chem. **22**, 2832 (1983)
29. Chirico R.D. and Carlin R.L., Inorg. Chem. **19**, 3031 (1980)
30. DeFotis G.C., Palacio F., and Carlin R.L., Phys. Rev. B **20**, 2945 (1979)
31. Sams J.R. and Tsin T.B., Inorg. Chem. **14**, 1573 (1975); Sams J.R. and Tsin, T.B., Chem. Phys. **15**, 209 (1976)
32. Goodwin H.A., Coord. Chem. Revs. **18**, 293 (1976); Gütlich P., Structure and Bonding **44**, 83 (1981)
33. Abragam A. and Pryce M.H.L., Proc. Roy. Soc. (London) **A 206**, 173 (1951)
34. Carlin R.L., J. Appl. Phys. **52**, 1993 (1981); Carlin R.L., Science **227**, 1291 (1985)
35. Gregson A.K., Martin R.L. and Mitra S., J.C.S. Dalton **1976**, 1458
36. Myers B.E., Polgar L.G., and Friedberg S.A., Phys. Rev. **B 6**, 3488 (1972)
37. Myers B.E., Polgar L.G., and Friedberg S.A., Phys. Rev. **B 6**, 3488 (1972); Ajiro Y., Friedberg S.A., and Vander Ven N.S., Phys. Rev. **B 12**, 39 (1975); Friedberg S.A., Karnezos M., and Meier D., Proc., 14th Conf. on Low Temperature Physics, Otaniemi, Finland, August, 1975, paper L 59
38. Carlin R.L., O'Connor C.J., and Bhatia S.N., J. Amer. Chem. Soc. **98**, 3523 (1976)
39. McElearney J.N., Losee D.B., Merchant S., and Carlin R.L., Phys. Rev. **B 7**, 3314 (1973)
40. Klaaijsen F.W., Dokoupil Z., and Huiskamp W.J., Physica **79 B**, 457 (1975)
41. Carlin R.L., Burriel R., Palacio F., Carlin R.A., Keij S.F., and Carnegie, Jr., D.W., J. Appl. Phys. **57**, 3351 (1985)

5. Introduction to Magnetic Exchange: Dimers and Clusters

5.1 Introduction

Probably the most interesting aspect of magnetochemistry concerns the interactions between magnetic ions. The remainder of this book is largely devoted to this subject, beginning in this chapter with the simplest example, that which occurs in a dimer. The principles concerning short-range order that evolve here are surprisingly useful for studies on more-extended systems.

The previous discussion has considered primarily the effective Hamiltonian

$$H = g\mu_B H \cdot S + D[S_z^2 - (\tfrac{1}{3})\mathscr{S}(\mathscr{S}+1)] \tag{5.1}$$

which describes only single ion effects. In other words, the properties of a mol of ions followed directly from the energy levels of the constituent ions, the only complication arising from thermal averaging. Now, let two ions interact through an intervening ligand atom as illustrated schematically in Fig. 5.1; this is a simple example of what is called the superexchange mechanism. This interaction, between three atoms linearly arranged in this example, is generally accepted as the most important source of metal-metal interactions or *magnetic exchange* in insulating compounds of the transition metal ions.

The model presented here assumes that the ground state of the system consists of two paramagnetic ions with uncorrelated spins separated by a diamagnetic ligand such as oxide or fluoride ion. That is, all the electrons are paired on the intervening ligand. This is illustrated in Fig. 5.1, where the ground orbital state is a spin-singlet if the metal ion spins are antiparallel (a) or a triplet if parallel (b). The simplest modification of this situation arises from the mixing of small amounts of excited states into the ground state. In particular [1], the partial transfer of electron spin density from, say, a ligand $2p_z$ orbital into a half-filled d_{z^2} orbital on one of the metal ions will yield an excited state which is an orbital singlet as in (c) or a triplet as in (d). This amounts to a

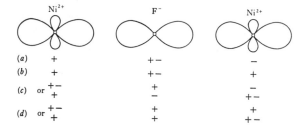

Fig. 5.1. Typical orbitals participating in a linear (180°) superexchange pathway between two Ni^{2+} ions (d_{z^2}) via a F^- ion ($2p_z$) as in $KNiF_3$. From Ref. [1]

configurational mixing of states such as $M^+ F M^{2+}$. The electron on F in (c) and (d) can now couple with the lone electron in a d_{z^2} orbital on an adjoining Ni^{2+}. The preferred coupling is to give a singlet state [(c), called antiferromagnetic] rather than a triplet, (d). This in turn stabilizes the singlet state (a) with respect to the triplet (b). The strength of the interaction will depend on the amount of overlap, and we present here only a model of 180° superexchange. Metal ion orbitals can also overlap with p_π ligand orbitals at a 90° angle, and much recent research effort has gone into finding the factors which influence this superexchange interaction.

This model, which has been applied to a variety of systems [1,2] is somewhat similar to that which is used for explaining spin-spin coupling in NMR spectroscopy [3]. The Hamiltonian in use for metal-metal exchange interaction in magnetic insulators is of the form

$$H = -2\Sigma J_{ij} S_i \cdot S_j, \tag{5.2}$$

where the sum is taken over all pair-wise interactions of spins i and j in a lattice. For the moment, we shall restrict our attention to dimers, and thereby limit the summation to the two atoms 1 and 2 in the dimeric molecule, so that

$$H = -2J S_1 \cdot S_2. \tag{5.3}$$

This is called an isotropic (note the dot product between the two spins, S) or Heisenberg Hamiltonian, and we adopt the convention that negative J refers to antiferromagnetic (spin-paired or singlet ground state in a dimer) interactions, and positive J refers to ferromagnetic (spin-triplet ground state) interactions. The exchange constant J is given in energy units (Kelvins) and measures the strength of the interaction. The reader must be careful in comparing different authors, for a variety of conventions (involving the negative sign and even the factor of 2) are in use. (The symbol J now takes a different meaning from its earlier use as the quantum operator for total angular momentum.)

5.2 Energy Levels and Specific Heats

An antiferromagnetic interaction of the type given in Eq. (5.3), when applied to two ions each of $\mathcal{S} = \frac{1}{2}$ gives a spin-singlet ground state and a spin-triplet 2J in energy above the singlet (Fig. 5.2). Naturally, if the interaction were ferromagnetic, the diagram is simply inverted. For an external field applied along the z-axis of the pair, the complete Hamiltonian will be taken as

$$H = g\mu_B S'_z H_z - 2J S_1 \cdot S_2, \tag{5.4}$$

where S'_z is the operator for the z-component of *total* spin of the pair. The eigenvalues of H are

$$W(\mathcal{S}', m'_s) = g\mu_B m'_s H_z - J[\mathcal{S}'(\mathcal{S}'+1) - 2\mathcal{S}(\mathcal{S}+1)].$$

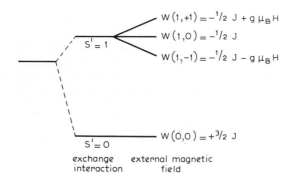

$$W(1,+1) = -\tfrac{1}{2} J + g \mu_B H$$
$$W(1,0) = -\tfrac{1}{2} J$$
$$W(1,-1) = -\tfrac{1}{2} J - g \mu_B H$$

$S' = 1$

$$W(0,0) = +\tfrac{3}{2} J$$

$S' = 0$

exchange   external magnetic
interaction   field

Fig. 5.2. Energy-levels for a pair of $\mathscr{S} = \tfrac{1}{2}$ ions undergoing antiferromagnetic exchange. The $\mathscr{S}' = 1$ level is $-2J$ in energy above the $\mathscr{S}' = 0$ level

For the dimer, $\mathscr{S} = \tfrac{1}{2}$, $\mathscr{S}' = 0$ or 1, and $m'_s = S', S'-1, -S'$. The energy level diagram is as shown in Fig. 5.2. It is assumed for the moment that there is no anisotropy of exchange interaction.

In the limit of zero magnetic field, the partition function is simply

$$Z = 1 + 3\exp(2J/kT)$$

and the derived heat capacity is quickly calculated, using Eq. (3.18) again, as

$$c = \frac{12R(J/kT)^2 \exp(2J/kT)}{[1 + 3\exp(2J/kT)]^2} \tag{5.5}$$

which is of course of exactly the same form as a Schottky specific heat. The reader must be careful in counting the sample; Eq. (5.5) as written refers to a mol of ions, not a mol of dimers. In Fig. 5.3, we illustrate the general behavior of this curve, and in Fig. 5.4, taken from Smart [4], the specific heat behavior of antiferromagnetically coupled dimers of ions of, respectively, spin $\tfrac{1}{2}$, $\tfrac{3}{2}$, and $\tfrac{5}{2}$ is compared. Note that the temperature of the maxima and the low temperature behavior in the latter figure are approximately independent of $\mathscr{S}$.

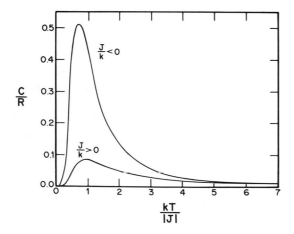

Fig. 5.3. Specific heat of a dimer as a function of $kT/|J|$ for $\mathscr{S} = \tfrac{1}{2}$

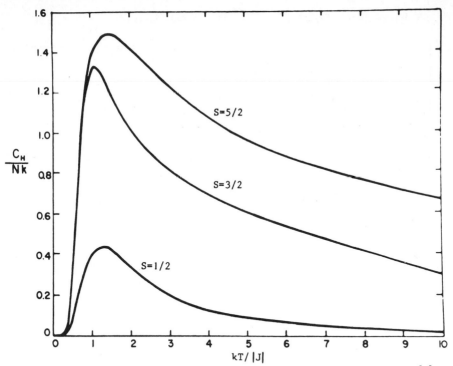

Fig. 5.4. Specific heat vs. $kT/|J|$ for antiferromagnetically coupled $(J<0)$ dimers of $\mathscr{S}=\frac{1}{2},\frac{3}{2},\frac{5}{2}$. From Ref. [4]

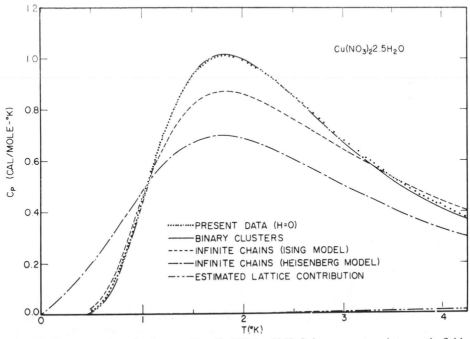

Fig. 5.5. Heat capacity of polycrystalline $Cu(NO_3)_2 \cdot 2\frac{1}{2}H_2O$ in zero external magnetic field. Adapted from Ref. [5]. The data points should be reduced by a factor of 0.963, reducing the apparent good agreement (see Ref. [8] and Chap. 7)

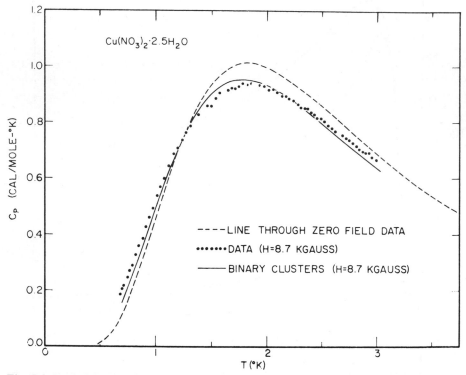

Fig. 5.6. Heat capacity of polycrystalline $Cu(NO_3)_2 \cdot 2\frac{1}{2}H_2O$ in an applied field of $8.7\,kOe$ (0.87 T). From Ref. [5], but see also Ref. [8]

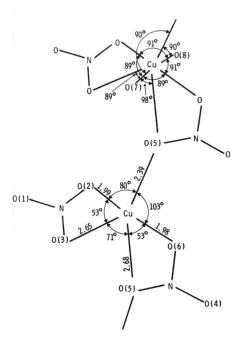

Fig. 5.7. A schematic drawing of the crooked chains which link the copper ions in $Cu(NO_3)_2 \cdot 2\frac{1}{2}H_2O$. The two water oxygen atoms above the lower copper atom have not been included. From Ref. [6]

There seems to be but one example available of a compound that has a specific heat which *appears* to follow Eq. (5.5), and that is $Cu(NO_3) \cdot 2\frac{1}{2} H_2O$ [5]. For most compounds containing exchange-coupled dimers, 2J is so large that the overlap of the magnetic contribution with the lattice contribution is so serious as to prevent their separation and identification. In the case of $Cu(NO_3)_2 \cdot 2\frac{1}{2} H_2O$, however, the singlet-triplet separation is only about 5.2 K [5], with the singlet lower, and the broad maximum (Fig. 5.5) is easily discernible.

The investigation of this system by Friedberg and Raquet was even more interesting because the effect of a magnetic field on the specific heat was also examined. In this case, the far right side of Fig. 5.2 is applicable, and the partition function becomes (in the case of isotropic exchange),

$$Z = 1 + \exp[(2J - g\mu_B H)/kT] + \exp(2J/kT)$$
$$+ \exp[(2J + g\mu_B H)/kT]$$
$$= 1 + \exp(2J/kT)[1 + 2\cosh(g\mu_B H/kT)]. \tag{5.6}$$

Inserting the best-fit parameters to the zero-field data of Fig. 5.5 of g = 2.13 and 2J/k = -5.18 K, a fit to this model at H = 8.7 kOe (0.87 T) is illustrated in Fig. 5.6. The agreement is striking. None of the other models applied to the data, such as infinite chains of atoms, fit either the zero-field or applied-field specific heat results. It was therefore all the more remarkable to find [6] after these investigations that $Cu(NO_3)_2 \cdot 2\frac{1}{2} H_2O$ does *not* contain binary clusters of metal atoms, but is actually chain-like with bridging nitrate groups, as in Fig. 5.7. The chain is not linear but crooked and this appears to be why the short-range or dimer ordering is so important here. The magnetic susceptibility of a dimer, which is discussed in the next section, has a broad maximum at temperatures comparable to the singlet-triplet separation. The susceptibilities of $Cu(NO_3)_2 \cdot 2\frac{1}{2} H_2O$ also fit the theory of the next section with the same exchange constant as derived from the specific heat measurements [7].

Nevertheless, despite the apparently good agreement between experiment and the dimer described here, an even better fit has recently been obtained [8] with an alternating linear chain model. This is because, in part, the data in Fig. 5.5 should be lowered by a factor of 0.963. This will be discussed in Chap. 7.

5.3 Magnetic Susceptibilities

Begin first with the application of Van Vleck's equation to the energy level situation in Fig. 5.2. The isothermal magnetic susceptibility per mol of dimers is readily calculated as:

$$\chi = \frac{\left[\dfrac{2N(g\mu_B)^2}{kT}\right] \exp(J/2kT)}{3\exp(J/2kT) + \exp(-3J/2kT)} = \frac{(2Ng^2\mu_B^2/kT)}{3 + \exp(-2J/kT)},$$

or

$$\chi = (2Ng^2\mu_B^2/3kT)[1 + (\tfrac{1}{3})\exp(-2J/kT)]^{-1}. \tag{5.7}$$

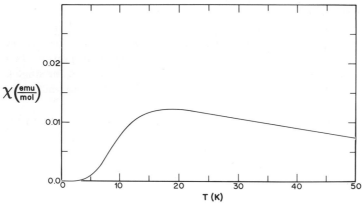

Fig. 5.8. The Bleaney-Bowers equation [Eq. (5.7)], with g=2.2 and 2J/k= − 30 K

(Some authors omit the factor of two in the numerator, in which case the calculated susceptibility refers to a mol of ions, or half a mol of dimers.) Equation (5.7) is plotted in Fig. 5.8 for a typical set of parameters. Note that, for negative (antiferromagnetic) J, χ has a maximum. This may be found by setting $\partial \chi / \partial T = 0$, but the easiest way to find the temperature, T_m, of this maximum is to set $\partial \ln \chi / \partial T = 0$. One finds, with the definitions used here, that the maximum occurs at $J/kT_m \simeq -\frac{4}{5}$. For $J/k \ll T$ (or $T \gg T_m$), the susceptibility, Eq. (5.7), follows a Curie-Weiss law, $\chi = (\frac{3}{4})/(T-\theta)$ with $\theta = J/2k$. This is a special case of the more general connection between a non-zero Curie-Weiss θ and the presence of exchange interaction. (It has already been pointed out that this is not a unique relationship, however.) Note also that $\chi \to 0$ as $-J \to \infty$ or, as we would expect, as the energy state in which the dimer is paramagnetic gets further away (higher in energy) because of a stronger exchange coupling, the susceptibility at a given temperature must decrease. On the other hand, if J should be positive, an $\mathscr{S}=1$ spin-only Curie law susceptibility is obtained as $2J/kT$ becomes large, that is at low temperatures. Indeed, when J is positive, the difference between the susceptibilities calculated according to Eq. (5.7) and the Curie law for spin $\mathscr{S}=\frac{1}{2}$ do not differ greatly until temperatures very low with respect to $2J/k$ are achieved; this is because of the extra factor of 2 which enters when comparing a mol of spins $\mathscr{S}=\frac{1}{2}$ with a mol of dimers (of spin $\mathscr{S}=\frac{1}{2}$ ions). This is illustrated in Fig. 5.9.

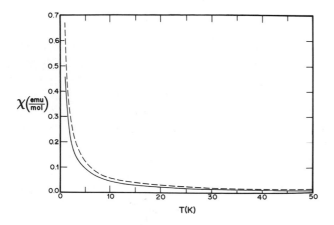

Fig. 5.9. The Bleaney-Bowers equation, compared with the Curie law.
---- Eq. (5.7), g=2.2, J/k=30 K
—— Curie law, g=2.2, $\mathscr{S}=\frac{1}{2}$, per mol

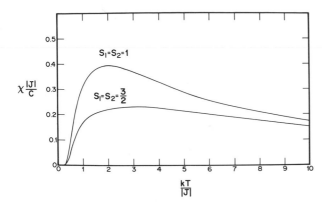

Fig. 5.10. Reduced magnetic susceptibility $\chi|J|/C$ vs. $kT/|J|$ for antiferromagnetically coupled pairs with $\mathscr{S} = 1$ and $\frac{3}{2}$

The susceptibility for a pair of $\mathscr{S} = \frac{1}{2}$ ions antiferromagnetically coupled was illustrated in Fig. 5.8, and a calculation of the reduced susceptibility for $\mathscr{S} = 1$ or $\frac{3}{2}$ pairs is illustrated in Fig. 5.10. While the low temperature behavior in the three cases is the same, the temperature of maximum χ increases with $\mathscr{S}$. A broad, featureless peak is observed because of the continuous population decrease of the various levels as temperature decreases. Data analysis will be most sensitive to the choice of the value of J when $kT/J \approx 1$. When $J > 0$, on the other hand, decreasing temperature decreases the population of the least magnetic levels in the system, so the susceptibility approaches Curie-Weiss behavior. For small values of kT/J (less than about 2), the system is simply a paramagnetic one that obeys the Curie law for an $S = S_1 + S_2 = 3$ system.

Examples to which Eq. (5.7) has been applied are legion [1, 4, 9], the most famous example being copper acetate, $[Cu(OAc)_2 \cdot (H_2O)]_2$. We turn now to a discussion of this system.

5.4 Copper Acetate and Related Compounds

Bleaney and Bowers [10] first brought hydrated copper acetate to the attention of both chemists and physicists in 1952 by their investigation of the EPR spectrum of the pure crystal. They were drawn to the compound because, although most copper salts had relatively straightforward magnetic behavior for $\mathscr{S} = \frac{1}{2}$ systems, it had been reported by Guha [11] that the susceptibility of copper acetate monohydrate passed through a maximum near room temperature and then decreased so rapidly as the temperature fell that it would apparently become zero at about 50 K. No sharp transition had been reported, and the behavior was unlike that found with the usual antiferromagnets (cf. Chap. 6). The EPR spectrum was also unusual in comparison with that of a normal copper salt: at 90 K, a line at X-band (0.3 cm^{-1}) was observed in zero magnetic field. This is inconsistent with the behavior anticipated for an $\mathscr{S} = \frac{1}{2}$ (Kramers) ion. The spectrum was similar but more intense at room temperature, but disappeared at 20 K. The spectra bore certain resemblances to those of nickel(II), which is a $\mathscr{S} = 1$ ion with concomitant zero-field splittings.

The simplest explanation for these results, and the one put forward by Bleaney and Bowers, was that the copper ions must interact antiferromagnetically in pairs. As described in earlier sections of this chapter such a pair-wise exchange interaction yields

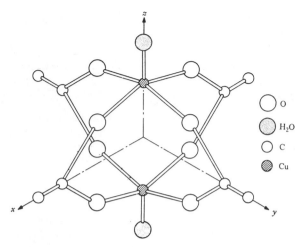

Fig. 5.11. Molecular structure of hydrated copper acetate. From Ref. [1]

a singlet ground state and a triplet excited state. In contrast to the example of $Cu(NO_3)_2 \cdot 2\frac{1}{2} H_2O$, the splitting in copper acetate must be very large in order to cause the reported magnetic behavior, and in fact the singlet-triplet separation should be comparable to thermal energy (kT) at room temperature. Zero susceptibility and the lack of an EPR spectrum at low temperatures, when the triplet state has lost its population, follow immediately. This remarkable hypothesis has since been validated by a variety of experiments.

One result of this success is that Eq. (5.7) is now referred to regularly as the "Bleaney-Bowers equation." The crystal structure of $[Cu(OAc)_2 \cdot (H_2O)]_2$ was reported shortly thereafter by van Niekerk and Schoening [12]. Isolated dimers were indeed found to be present. The structure, illustrated in Fig. 5.11, has also been refined more recently [13, 14]. A pair of copper atoms, separated by 0.2616 nm, is supported by four bridging acetate groups in a D_{4h} array with two water molecules completing the coordination along the Cu–Cu axis. A similar structure is found for the chromous compound [15] but there the exchange interaction is so large that the compound is diamagnetic even at room temperature.

The first EPR results [10] were interpreted in terms of the spin-Hamiltonian

$$H = g\mu_B H \cdot S + D[S_z^2 - (\tfrac{1}{3})\mathscr{S}(\mathscr{S}+1)]$$
$$+ E(S_x^2 - S_y^2) \tag{5.8}$$

with $D = 0.34 \pm 0.03\,\text{cm}^{-1}$, $E = 0.01 \pm 0.005\,\text{cm}^{-1}$ and $g_z = g_{\parallel} = 2.42 \pm 0.03$, $g_x = g_y = g_{\perp} = 2.08 \pm 0.03$. These results apply of course to the excited $\mathscr{S} = 1$ state; the Cu–Cu axis serves as the z-axis. The singlet-triplet separation, or *isotropic exchange interaction*, was estimated, to 20%, as 370 K. More recent results, for both $Cu_2(OAc)_4 \cdot 2 H_2O$ and the zinc-doped monomers [16] $ZnCu(OAc)_4 \cdot 2 H_2O$ (the $[Cu_2(OAc)_4 \cdot 2 H_2O]$ structure will allow replacement of about 0.5% of Cu by Zn) are $g_z = 2.344 \pm 0.005$, $g_x = 2.052 \pm 0.007$, and $g_y = 2.082 \pm 0.007$. The similarity of these values to those found in the more normal or monomeric copper salts supports the suggestion that the copper atoms are, aside from the exchange interaction, subjected to a normal type of crystalline field.

The D and E terms given above have been shown by Abragam and Bleaney [17] to be related to a small anisotropy in the exchange interaction.

There are two energy levels considered for the dimer, one with $\mathscr{S}=1$ which is $-2J$ in energy above the level with $\mathscr{S}=0$. As has been pointed out above, the lower level does not and cannot make any contribution to an EPR spectrum. The excited state is effectively a triplet, as with other $\mathscr{S}=1$ states, and in the presence of isotropic exchange and an external field H_z, the levels are at

$$-J/2+g_z\mu_B H_z,$$

$$-J/2,$$

$$-J/2-g_z\mu_B H_z.$$

With the usual EPR selection rule of $\Delta m_s = \pm 1$, two transitions occur, both at

$$h\nu = g_z\mu_B H_z.$$

This is independent of the value of J, showing that isotropic exchange has no other effect on the spectrum except at temperatures where $kT/J \sim 1$, when the intensity will no longer vary inversely as the absolute temperature because of the triplet-singlet splitting. The exchange constant, J, may be written

$$J = (\tfrac{1}{3})[(J_x - J_x') + (J_y - J_y') + (J_z - J_z')], \tag{5.9}$$

where the primed components denote the anisotropy in the exchange constant; if the exchange were isotropic,

$$J_x' = J_y' = J_z' = 0, \quad \text{and} \quad J = (\tfrac{1}{3})(J_x + J_y + J_z).$$

Furthermore, a constraint on the anisotropic portion is that

$$J_x' + J_y' + J_z' = 0.$$

The zero-field splitting of the $\mathscr{S}=1$ state is then due to the anisotropic contribution, and in fact Abragam and Bleaney show that we may associate the parameters of Eqs. (5.8) and (5.9) as

$$D = 3J_z'/4 = 0.34\,\text{cm}^{-1}$$

and

$$E = \tfrac{1}{4}(J_x' - J_y') = 0.01\,\text{cm}^{-1}.$$

Recalling that the isotropic term was estimated to be of the order of 370 K ($260\,\text{cm}^{-1}$), we see that, relatively, the anisotropic exchange is quite small.

The first careful measurement of the susceptibility of polycrystalline $[\text{Cu(OAc)}_2 \cdot (\text{H}_2\text{O})]_2$ was carried out by Figgis and Martin [18] and the most recent

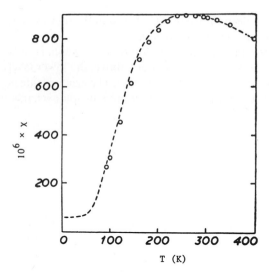

Fig. **5.12.** Experimental and calculated magnetic susceptibilities of $[Cu(OAc)_2 \cdot H_2O]_2$. From Ref. [18]

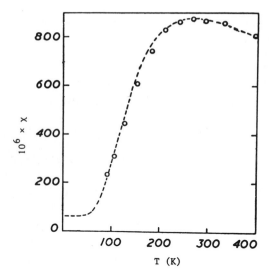

Fig. **5.13.** Experimental and calculated magnetic susceptibilities of anhydrous copper acetate. From Ref. [18]

(single crystal) study [19] is that also of Martin and co-workers. The results are in complete accord with the above discussion and may be fitted by Eq. (5.7), as illustrated in Fig. 5.12, for a more accurate evaluation of J. It was found [18] that $T_{max} = 255$ K, and that $-2J/k = 480$ K when the Hamiltonian of Eq. (5.3) is used. What was even more fascinating, as shown in Fig. 5.13, is that anhydrous copper acetate behaves quite similarly, with $T_m = 270$ K and $-2J/k = 432$ K. The conclusion that anhydrous copper acetate not only retains the gross molecular structure of the hydrate but also behaves antiferromagnetically in much the same fashion is inescapable. Furthermore, replacement of the axial water molecule by, for example, pyridine results in a similar magnetic

[20] and crystallographic [21] situation. In this case, the exchange interaction becomes $-2J/k = 481$ K, and similar results are obtained with substituted pyridines as well as when acetate is replaced by such alkanoates as propionate and butyrate. The hydrated and anhydrous formates, however, are distinctly different, both in structural and magnetic properties [22].

The structure and magnetism of the 2-chloropyridine adducts of both copper trichloroacetate [23] and copper tribromoacetate [24] have been reported. The dimeric molecules have the bridged copper(II) acetate structure, with a long Cu---Cu distance of 0.2766 nm in both molecules (recall that the separation is 0.2616 nm in $[Cu(OAc)_2 \cdot (H_2O)]_2$). Naturally, the water molecules in Fig. 5.11 are replaced by 2-chloropyridine, and the methyl groups on the acetate bridges by the larger tri-halomethyl, $-CX_3$, groups. The magnetic susceptibility exhibits the familiar maximum (at about 200 K for the trichloro compound, and 160 K for the tribromo derivative) and the data for the $-CCl_3$ compound are fit by the Bleaney-Bowers equation with $g = 2.26$ and $2J = -217$ cm^{-1} (-312 K). The exchange constant drops to -180 cm^{-1} (-260 K) for the $-CBr_3$ material [24]. This is the smallest exchange constant yet reported for any dimeric copper(II) acetate derivative, and appears to be due more to the presence of the $-CBr_3$ group rather than the nature of the axial ligand.

The single crystal measurements [19] of magnetic anisotropy of $[Cu(OAc)_2 \cdot (H_2O)]_2$ are important for two reasons. First, it was found that the exchange interaction is isotropic (within the sensitivity limits of susceptibility measurements) since all the data could be fit with but one value of $2J$. Secondly, contrary to several suggestions in the literature, both the g-values and $2J$ were found to be temperature independent over the temperature interval 80 to 300 K. The best values of the magnetic parameters found in this work are $g_{\parallel} = 2.34$, $g_{\perp} = 2.07$, and $-2J = 286$ cm^{-1} (412 K), where "parallel" and "perpendicular" refer to the Cu–Cu axis.

Finally, dimeric copper acetate hydrate has also been subjected to neutron inelastic scattering spectroscopy [25, 26]. This experiment is powerful because it can distinguish between magnetic excitations ("magnons") shared by an ensemble of magnetic centers in a lattice and the localized excitations anticipated in a molecule such as $[Cu(OAc)_2 \cdot (H_2O)]_2$. The method has the advantage of being a direct spectroscopic observation, and showed the singlet-triplet splitting (actually in $Cu_2(CD_3COO)_4 \cdot 2D_2O$, but the deuteration was shown to be of no magnetic significance) to be 298 ± 4 cm^{-1} (429 K) and to be temperature independent between 10 and 300 K. This appears to be the final word on the value of $-2J$ for hydrated copper acetate.

The great success of this magnetic model for copper acetate and its analogues has led to its application as an indicator of structure to almost any copper compound that has a subnormal magnetic moment at room temperature. In many cases, the structure has been correctly deduced [27]. But, as was the case with hydrated copper nitrate, magnetic properties turn out to be a more fallible indicator of crystal structure than does X-ray crystallography [9]. For example, the aniline adducts of copper acetate behave magnetically like the other dimers described above [20], though the singlet-triplet energy was reduced substantially so that the temperature of maximum susceptibility was not measured directly. The parameter $-2J/k$ was estimated as only about 150 K for both the m- and p-toluidine adducts of copper butyrate. It has since been shown [28] that the p-toluidine adduct of copper acetate is not binuclear, but instead polymeric chains are formed. The powder susceptibility of $Cu(OAc)_2 \cdot 2p$-toluidine $\cdot 3 H_2O$ was fitted [28] as an (anisotropic) Ising linear chain (Chap. 7) rather than to the Bleaney and Bowers relationship, but, as has been pointed out [29], copper is usually a Heisenberg or isotropic ion. A more complete set of measurements is required before one can be satisfied that this system is well-understood. Furthermore [30], the p-toluidine adduct of copper(II) propionate, which is composed of one-dimensional polymeric chains, exhibits a magnetic susceptibility (over a limited range of temperatures) which fits the Bleaney-Bowers scheme quite well.

The structure and physical properties of polynuclear metal carboxylates have been reviewed [31].

A further question remains to be discussed and that is, what is the mechanism of spin-spin or antiferromagnetic coupling in the copper acetate compounds? Many models have been presented [1].

The first bonding scheme, presented in the earliest [18] magnetic study, proposed that there was a direct bond formed between the two metal atoms. After all, the copper atoms are only 0.2616 nm apart in $[Cu(OAc)_2 \cdot (H_2O)]_2$. It was proposed that a δ-bond was formed by a lateral overlap of two $d_{x^2-y^2}$ orbitals, and a variety of spectral data appear to be consistent with this proposal [1]. Superexchange interaction through the carboxylate groups has been widely suggested as the more important contributor, however, and a great deal of indirect information tends to favor this scheme. For example, though the structure of copper propionate p-toluidine is not the usual binuclear one, the susceptibility behaves similarly to that of the binuclear molecules, and yet the Cu–Cu separation is increased to a range of 0.3197 to 0.3341 nm. This distance, and the arrangement of the chains of atoms, make a direct overlap seem unlikely. Furthermore, the quinoline adduct of copper(II) trifluoroacetate is binuclear with the molecular structure common to the copper acetate dimers [32]. The Cu–Cu distance is 0.2886 nm, a full 0.0272 nm longer than the corresponding distance in $[Cu(OAc)_2 \cdot (H_2O)]_2$. The susceptibility obeys the Bleaney-Bowers relationship over the range 80–300 K with g=2.27 and $2J = -310 \mathrm{cm}^{-1}$ ($2J/k = -446 \mathrm{K}$). The large difference in Cu–Cu separation between the magnetically similar acetate and trifluoroacetate adduct demonstrates that the metal-metal distance in these dimers is not an important factor in determining the strength of the Cu–Cu interaction. This in turn lends further weight to the importance of the superexchange mechanism.

The results described above on the dimeric 2-chloropyridine adduct of copper trichloroacetate [23] and tribromoacetate [24], especially concerning the influence of the $-CX_3$ groups, would tend to argue in favor of a superexchange mechanism for exchange interaction in these molecules, rather than a direct metal-metal interaction. Indeed, it was found that the ordering of the exchange constants is identical to the ordering of the group polarizabilities of the substitutuents on the acetate group (R=$-H$, $-CF_3$, $-CH_3$, $-CCl_3$ or $-CBr_3$). On the other hand, the mean Cu–O–C–O–Cu bridging pathway through the triatomic bridges of 0.6374 nm ($-CCl_3$ compound) is somewhat shorter that most other observed values. This is contrary to a proposal [33] that a shorter bridging pathway is associated with a larger Cu–Cu interaction; for $[(CH_3)_4N]_2[Cu(HCO_2)_2(SCN)_2]_2$, and $[(CH_3)_4N]_2[Cu(CH_3CO_2)_2(SCN)_2]_2$, both of which have the copper acetate type structure, it was found that 2J was not dependent on the copper-copper internuclear separation [34].

5.5 Some Other Dimers

So many dimeric systems have been investigated that it is impossible to review all of them here. The physical principles involved in the study of most of the systems have been elaborated upon above, so that only a selection of chemically-interesting studies will be mentioned. Some further systems have been described in reviews [1,9].

In Fig. 5.10, we plotted the reduced magnetic susceptibility calculated as a function of reduced temperature for dimeric metal complexes as a function of spin. Broad maxima which shift to higher temperature as the spin increases, are found when the exchange constant is negative (antiferromagnetic). Maxima are not obtained when the exchange interaction is positive, and therefore, a very careful fitting of theory to experiment is required in order to prove that positive exchange coupling is in fact occurring. Broad maxima in the specific heats are required in both situations [4,9].

The susceptibility of dimers in which both constituent metal ions are spin $\mathscr{S} = \frac{1}{2}$ is, in principle, the most straightforward to analyze. For the most part, the relevant ions are therefore Ti(III) and Cu(II); several vanadyl complexes have also been investigated.

The bis(cyclopentadienyl)titanium(III) halides form a series [35] of binuclear molecules of the type $[(\eta^5-C_5H_5)_2Ti-X_2-Ti-(\eta^5-C_5H_5)_2]$, with two halide bridging atoms between the metal atoms. The series is interesting because X may be any one of the four halide ions. Bromide, being more polarizible than chloride, frequently allows a larger superexchange interaction. However, bromide is also larger than chloride and tends to separate the metal atoms further. The balance of these two factors, as well as other imponderables, helps to determine the relative strength of the exchange in a given situation. The magnetic susceptibilities indeed follow Eq. (5.7), the singlet-triplet equation, with $-2J$ values of the order of 62 cm^{-1} (F), 220 cm^{-1} (Cl), 276 cm^{-1} (Br), and 168–179 cm^{-1} (I). The relative behavior of the values of the exchange constants of the chloride and bromide derivatives follows a common, but not universal, trend, but in the absence of enough detailed structural data (especially concerning the dimensions and angles of the Ti$_2$X$_2$ cores), the origin of the trends in the Ti–Ti interaction must remain uncertain. The crystal structure of $[(C_5H_5)_2TiCl]_2$ has been reported [35], but not that for any of the other compounds. The crystal structures of $[(CH_3C_5H_4)_2TiCl]_2$ and $[(CH_3C_5H_4)_2TiBr]_2$ have been analyzed [36], but even though the molecules have similar geometries, they do not belong to the same space group. The exchange constants $(-2J)$ in $[(C_5H_5)_2TiBr]_2$ and $[(CH_3C_5H_4)_2TiBr]_2$ are found to have the same value of 276 cm^{-1} (387 K), while the analogous chloride complexes are found to have singlet-triplet separations of, respectively, 220 and 320 cm^{-1} (317 and 461 K). There simply aren't enough detailed data on enough related systems where only one parameter changes to allow unambiguous comparisons to be made. Though it has been suggested [35] that direct exchange (metal-metal bonding) may contribute to the exchange interaction in these compounds, so many other factors are changing as well that such a suggestion must be considered as tentative at best at this time. (A change in crystal space group can change not only such factors as hydrogen bonding and crystal lattice energies but also such variables as molecular dimensions. Changes in any of these can have unpredictable changes on magnetic exchange interactions.) This statement will be justified below in the examination of the dimers of other metals, for which there are more data available.

An unusual example of 1,3-magnetic exchange has recently been reported [36, 37]. The compounds are of the type $[(\eta^5-C_5H_5)_2Ti]_2ZnCl_4$, which contain collinear metal atoms in units of the sort

$$\text{>Ti} \diagup_{Cl}^{Cl} \text{Zn} \diagup_{Cl}^{Cl} \text{Ti<}$$

with a Ti–Zn distance of 0.3420 nm, a Zn–Cl–Ti angle of 89.9°, a Cl–Ti–Cl angle of 82.1° and a Ti–Zn–Ti angle of 173.4°. The zinc atoms are tetrahedral. The Ti–Ti distance is 0.6828 nm. Similar compounds with Cl replaced by Br, and Zn replaced even by Be were also reported. Despite the intervening diamagnetic MX_4 unit, the compounds follow the Bleaney-Bowers relation with $2J/k$ of the order -10 to -20 K. Substitution on the ring has little effect on the magnitude of the exchange interaction, nor does the replacement of Zn by Be. The beryllium compound shows that d orbitals are not a requirement of the central metal atom in order to have superexchange.

The structural and magnetic properties of copper(II) dimers have probably been investigated more extensively than those of any other metal. This is partly because of the facile synthesis of a variety of compounds and partly because the exchange is often so strong as to put the observed magnetic anomalies in a convenient temperature region for many experimentalists. The theoretical treatment is generally straightforward, depending only on the use of the Bleaney-Bowers equation, and small changes in geometry among related complexes allow one to search for trends in structural/magnetic correlations. Hodgson, in 1975 [9] reviewed in this vein 100 or more copper dimers, Hatfield mentions other examples [38], and the recent literature contains a large number of studies. Several of the more interesting examples from the recent literature will be discussed.

One interesting series of molecules is based on the ligand pyridine N-oxide, which can serve as a bridge between two copper ions. This structure is illustrated in Fig. 5.14. The copper atoms are further apart (by 0.061 nm) than they are in $[Cu(OAc)_2 \cdot (H_2O)]_2$, yet the exchange has increased in strength. The singlet-triplet separation is some $600 \, cm^{-1}$ (860 K), indicating that the pyridine N-oxide provides a very efficient superexchange path. A number of substituted pyridine N-oxide derivatives which retain the same structure have also been examined [38].

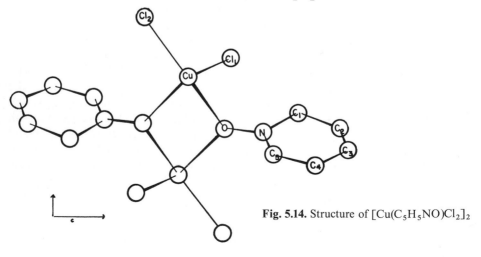

Fig. 5.14. Structure of $[Cu(C_5H_5NO)Cl_2]_2$

A series of dihydroxo-bridged Cu(II) dimers have been examined [39]. The relevant magnetic moiety is planar

and it has been suggested that there is a linear relationship between the Cu–O–Cu angle and the strength of the exchange interaction. When the angle becomes small enough, even the sign of the exchange apparently changes, and the triplet state becomes the ground state. However, the situation is more complicated because it has recently been shown [40, 41] that the strength of the exchange is also a function of the dihedral angle between the two CuO_2 planes. As this angle decreases from 180°, the exchange interaction becomes weaker. Furthermore, the linear correlation referred to above is valid neither for single hydroxo-bridged Cu(II) ions [42] nor for dihydroxo-bridged Cr(III) ions.

An interesting series of molecules has been reported [43] where the link between the copper ions is only provided by hydrogen bonds. A representative molecule is illustrated.

The oxygen-oxygen separation is very short (0.231 nm) and the copper atoms are separated by about 0.500 nm. The exchange constant of $-2J = 95\,cm^{-1}$ (137 K) is relatively strong for such an usual linkage. Several similar compounds were also examined, and it was found that there is no direct correlation of exchange constants with either oxygen-oxygen distances or copper-copper distances.

Several rubeanates have been studied [44]; the structure is sketched:

Although the Cu–Cu distance in the binuclear unit is as large as 0.561 nm, the singlet-triplet separation is 594 cm^{-1} (855 K). An isomer of this compound, in which the major structural change involves only the positions of the water molecules and sulfate ligands [45], has the metals separated by 0.5648 nm. The compound exhibits a very large exchange constant of -523 cm^{-1} (-753 K). Clearly, an efficient superexchange path is what determines the strength of magnetic exchange, not the metal-metal separation.

A variety of compounds has been synthesized with a view towards determining how well the ligand bridging the copper ions transmits the superexchange interaction. As one example [46], consider two copper ions bridged by an aromatic ring such as pyrazine, $N(CHCH)_2N$, versus a similar but saturated molecule called Dabco, $N(CH_2CH_2)_3N$. In the first case, the exchange interaction is only $2J = -6.4$ cm^{-1} (-9.2 K), but the compound with the Dabco ligand exhibits an exchange constant smaller than 1 K. Presumably the delocalized π-system of pyrazine is an important factor here in promoting exchange interaction. Interestingly, when the metals are bridged by the extended bridging ligand

$$H_2N \!-\!\!\left\langle \bigcirc \right\rangle\!\!-\!\!\left\langle \bigcirc \right\rangle\!\!-\! NH_2 \, ,$$

the magnetic interaction is -7.4 cm^{-1} (-10.7 K), which is comparable to that found for the pyrazine compound.

Finally, a variety of studies have been reported on the exchange interaction between two copper ions as transmitted by chloride or bromide ions. These generally provide a less effective path than many of the compounds described above, but there are a number of factors at work here which have not yet been sorted out. Thus, the coordination geometry of copper is rarely a constant factor in a comparison of a series of compounds. Four-coordinate copper often binds weakly to a fifth nearby ligand, adding another complication. An example is provided by $\{CuCl_2[(CH_2)_4SO]_2\}_2$, where the ligand is tetramethylene sulfoxide [47]. The copper ion in each *trans*-$CuCl_2[(CH_2)_4SO]_2$ monomer has a significant tetrahedral distortion from planarity; long Cu–Cl interactions between monomer units create discrete dichloro-bridged dimers. The copper-chlorine-copper bridging geometry is asymmetric with one long bond (0.3020 nm) and one short bond (0.2270 nm); the Cu–Cl–Cu bridging angle is 88.5°. The exchange constant, measured by magnetic susceptibility, is $2J/k = -24$ K, but a comparison with similar molecules shows that there is no obvious structural-magnetic correlation.

Again, the dichloro-bridged molecule $[CuLCl_2]_2$, where L is a N,N,N'-triethyl ethylenediamine, $(C_2H_5)_2NCH_2CH_2NH(C_2H_5)$, has recently been studied [48]. The geometry at copper is distorted trigonal bipyramidal with the halide ligands occupying apical and equatorial positions of each copper ion. The apical Cu–Cl bridging distance is 0.2284 nm while the equatorial Cu–Cl bridging separation is 0.2728 nm. The $Cu(Cl)_2Cu$ moiety is planar but quite asymmetric. The exchange interaction in this case is negligibly small. The difficulties which arise in establishing a magneto-structural correlation for copper salts have been discussed by Willett [49, 50].

One of the most vexing problems concerning the magnetochemistry of copper dimers is the question of the existence of ferromagnetic exchange. If 2J were positive, the

$\mathscr{S}=1$ state of Fig. 5.2 would be the lower one, and several such cases have been reported in the literature. However, these reports have been questioned.

Recall the calculations reported in Fig. 5.8, in which the Bleaney-Bowers relationship is plotted for the (antiferromagnetic) case of $g = 2.2$ and $2J/k = -30$ K. The susceptibility is small in magnitude and goes through the characteristic maximum (at about 19 K, in this case). Turn again to Fig. 5.9, in which the same calculation is carried out, but with $2J = 30$ K, i.e., a ferromagnetic interaction. The susceptibility of course is much larger, but the curve is featureless. That is, the susceptibility increases regularly with decreasing temperature, and the Bleaney-Bowers curve ($g = 2.2$, $2J = 30$ K) for a mol of spin $\mathscr{S} = \frac{1}{2}$ ions (or $N/2$ dimers, where N is Avogadro's number) is indistinguishable on this scale from the Curie law susceptibility for $N/2$ ions of spin $\mathscr{S} = 1$ with $g = 2.2$. Also plotted is the Curie law for a mol of $\mathscr{S} = \frac{1}{2}$ ions with $g = 2.2$; this curve is indistinguishable from the other above ~ 20 K, and deviates strongly only below about 4 K. In other words, for a ferromagnetic $2J$ of say 30 K, susceptibility measurements of very high accuracy and at low temperatures are required before one can safely distinguish this case from a simple Curie law behavior of paramagnetic ions.

The compound bis(N, N-diethyldithiocarbamato) copper(II), $Cu(S_2CNEt_2)_2$ (hereafter, $Cu(dtc)_2$), has been investigated frequently. The interest in the molecule lies with the fact that it has a binuclear structure (Fig. 5.15a) and that several measurements

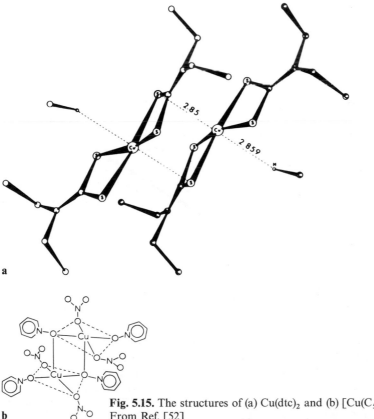

a

2.85

2.859

b

Fig. 5.15. The structures of (a) $Cu(dtc)_2$ and (b) $[Cu(C_5H_5NO)_2(NO_3)_2]$. From Ref. [52]

[51,52] have been interpreted in terms of a strong ferromagnetic intradimer interaction. This analysis depended upon susceptibility data taken above 4.2 K and the use of a modified Bleaney-Bowers equation

$$\chi = [Ng^2\mu_B^2/3k(T-\theta)][1 + (\tfrac{1}{3})\exp(-2J/kT)]^{-1}$$

for 1 mol of interacting $\mathscr{S} = \tfrac{1}{2}$ ions. The parameters reported as fitting the data are $\langle g \rangle = 2.041$, $2J/k = 34.5$ K, and $\theta = -1.37$ K. The negative Curie-Weiss constant was interpreted in terms of an antiferromagnetic interdimer interaction.

If such large ferromagnetic interactions were important in $Cu(dtc)_2$, the result should be apparent in susceptibility data taken to low temperatures [53].

The inverse zero-field ac susceptibility of $Cu(dtc)_2$ at low temperatures is displayed in Fig. 5.16. Above 2 K, χ obeys the Curie-Weiss law with $C = 0.397$ emu-K/mol and $\theta = 0.25$ K. This Curie constant corresponds to the reasonable value of $\langle g \rangle = 2.06$ for a mol of $\mathscr{S} = \tfrac{1}{2}$ ions. Below 2 K, a deviation is observed, with χ^{-1} becoming larger.

Several other susceptibility measurements were also reported [54], such as the ac susceptibility as a function of applied magnetic field and the high-frequency (adiabatic) susceptibility. All the data are consistent with the ferromagnetic intrapair interaction being of the order of only $J/k = 0.96$ K, with an antiferromagnetic interpair interaction of $J'/k = -0.007$ K. These results, which have been verified independently [55], proved that the intrapair exchange is relatively weak, despite the close proximity of the metal ions to each other.

A similar story applies to $[Cu(C_5H_5NO)_2(NO_3)_2]_2$, whose structure appears in Fig. 5.15b. The ligand C_5H_5NO is pyridine N-oxide, and early measurements [52, 56]

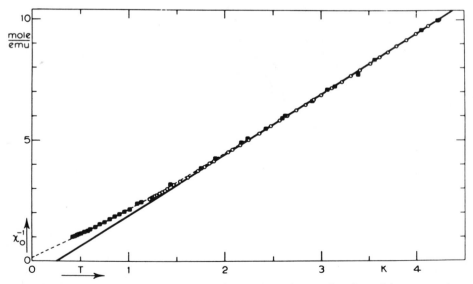

Fig. 5.16. Inverse zero-field susceptibility of $Cu(S_2CNEt_2)_2$ as a function of the temperature. Measurements on the "raw material" are given as black squares, while the results of a sample consisting of powdered single crystals are shown as the open circles. The solid line represents the Curie-Weiss fit with $C = 0.397$ emu-K/mol and $\theta = 0.25$ K. From Ref. [53]

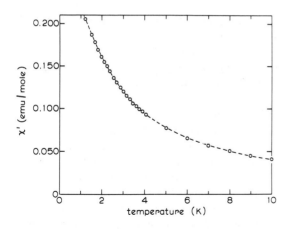

Fig. 5.17. Zero-field susceptibility of polycrystalline [Cu(C$_5$H$_5$NO)$_2$(NO$_3$)$_2$]. From Ref. [54]

suggested to the authors that this dimeric molecule also had a ferromagnetic ground state, with the singlet lying 10 cm^{-1} ($\sim$14 K) higher in energy.

The susceptibility of [Cu(C$_5$H$_5$NO)$_2$(NO$_3$)$_2$] at zero static field, χ, is displayed in Fig. 5.17 over the limited temperature interval between 1.2 K and 10 K, and was reported [57] to 80 K. The earlier data [56] agree quite well at temperatures below 30 K, but deviate at the higher temperatures. In the first case [56] Curie-Weiss behavior was observed only above 11 K and the reported Weiss constant was +2 K. In the case of the newer data [57] Curie-Weiss behavior was observed over practically the whole measured temperature interval, with deviations only beginning to appear at the lowest temperatures. The fitted parameters are C=0.441 emu-K/mol and $\theta = -0.8$ K and thus g=2.17 for $\mathscr{S} = \frac{1}{2}$.

For an independent check of the absolute χ results the magnetization M of the pyridine N-oxide compound was determined at several temperatures as a function of the static magnetic field. A typical example for $T=4.21$ K is displayed in Fig. 5.18 for a 474.5 mg sample, along with a similar measurement for 52.5 mg of manganese ammonium Tutton salt, Mn(NH$_4$)$_2$(SO$_4$)$_2 \cdot 6$H$_2$O. The latter is known (Chap. 1) as one of the best examples of a Curie law paramagnet, with $\mathscr{S} = \frac{5}{2}$ and Curie constant C=4.38 emu-K/mol. The ratio of the slopes of the two curves in Fig. 5.18 is 1.21(1) which corresponds to a ratio of 11.4(1) for the molar magnetization of the compounds. Consequently for [Cu(C$_5$H$_5$NO)$_2$(NO$_3$)$_2$], χ=M/H=89.9(8) $\times 10^{-3}$ emu/(mol-Cu) at 4.21 K, in excellent agreement with the *ac* susceptibility result of χ=90.7(6)

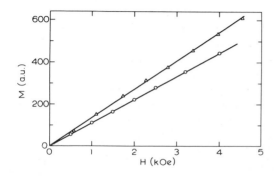

Fig. 5.18. Relative magnetizations of manganese Tutton salt ($\triangle$) and of [Cu(C$_5$H$_5$NO)$_2$(NO$_3$)$_2$] (o). From Ref. [54]

$\times 10^{-3}$ emu/(mol-Cu) at 4.20 K. The isothermal susceptibility as a function of applied field was also reported.

All the data presented [57] require that the compound $[Cu(C_5H_5NO)_2(NO_3)_2]$ behaves as an $\mathscr{S} = \frac{1}{2}$ paramagnet down to liquid helium temperatures. Any exchange interaction that may be present in the compound is only beginning to manifest itself at about 1.2 K.

If the compound were to consist of ferromagnetically aligned pairs, with an $\mathscr{S} = 1$ ground state and with an $\mathscr{S} = 0$ state at about 14 K higher in energy, that would necessarily have been evident in these results. One may calculate, according to the Boltzmann factor, that the relative population of the triplet state would be 96% at 4.2 K and 100% at 2 K. None of the data [57] admit an analysis in terms of $\mathscr{S} = 1$.

It is surprising that the interactions are so weak in this compound, for the structural results indeed show the existence of dimers. The bridge between the metal atoms is an oxygen atom from pyridine N-oxide, and the Cu-O-Cu angle is close to a right angle. So, either this geometry is unfavorable in general for the transmission of ferromagnetic superexchange interaction or else pyridine N-oxide simply provides a very weak superexchange path in this compound. The latter is certainly true in a number of pyridine N-oxide complexes which have been studied recently [58] but in those cases a M–O---O–M superexchange path is operative.

Numerous efforts are still in progress to prepare and characterize copper dimers with spin $\mathscr{S} = 1$ ground states. A curious situation arises when azide, N_3^-, is the bridging ligand, for it can bind in two ways [59, 60]. When the binuclear unit is of the form

$$Cu \underset{N-N-N}{\overset{N-N-N}{<\quad>}} Cu,$$

antiferromagnetic interaction is found. However, a compound with the different kind of binding,

$$Cu \underset{N}{\overset{O}{<\quad>}} Cu$$

has been suggested to exhibit a strong intradimeric ferromagnetic interaction.

Let us turn now to systems of higher spin. Though there are fewer examples of these in the literature than there are for copper, a large amount of work has been done for several ions, especially nickel, chromium, and iron. Though the paramagnetic properties of spin $\mathscr{S} = 1$ vanadium(III) are well-understood, there are relatively few examples of V(III)-containing dimers. One example consists of two seven-coordinate vanadium atoms bridged by alkoxy groups [61]. For the isotropic Hamiltonian $H = -2J\mathbf{S}_1 \cdot \mathbf{S}_2$ applied to two $\mathscr{S} = 1$ ions, a diamagnetic ground state is obtained when J is negative (antiferromagnetic). The equation resulting for the susceptibility per mol of dimers is

$$\chi = \frac{2Ng^2\mu_B^2}{kT} \frac{e^{2J/kT} + 5e^{6J/kT}}{1 + 3e^{2J/kT} + 5e^{6J/kT}}. \tag{5.10}$$

This equation has been illustrated in Fig. 5.10; for the vanadium compound, $-2J = 17.1 \text{ cm}^{-1}$, or 24.6 K. Alkoxy-bridged dimers appear to be coupled only weakly compared to oxo-bridged metal ions.

The major $\mathcal{S} = 1$ ion used in the study of dimers is nickel(II), and the most interesting examples (albeit the most difficult to analyze) are those in which the zero-field splitting is comparable in magnitude to the exchange interaction. The best known (i.e., most thoroughly studied) examples of such nickel compounds are the dihalide-bridged materials, $[Ni(en)_2X_2]Y_2$, where en is ethylenediamine and X and Y are both chloride or bromide [62–66], but the case of bridging thiocyanate $(X = NCS^-)$ with $Y = I^-$ has also been examined [62]. The system $X = Cl^-$, $Y = ClO_4^-$ or $B(C_6H_5)_4^-$ has also been reported [64,65]. Finally, these systems are important as the double-halide bridge which is present is characteristic of many more-extended chemical and magnetic chain systems (Chap. 7).

There have been two separate problems involved in the study of these molecules, neither of them well-resolved as yet. The first concerns the exact evaluation of the magnetic parameters which describe the several systems, and the second has concerned the rationalization of the values of these parameters. Chemists, after all, are interested not only in the phenomenological description of a system but also in a fundamental interpretation at the molecular level.

Rather than apply Eq. (5.10) to nickel dimers, it has been more common to apply an equation in which the zero-field splitting of each Ni is also included. Because of the anisotropy that zero-field splittings introduce, there are then three susceptibility equations, one each for the x, y, and z directions. These are too lengthy to reproduce here, but are given by Ginsberg [62].

The compounds $[Ni_2(en)_4Cl_2]Cl_2$ and $[Ni_2(en)_4Br_2]Br_2$ are di-(μ-halo) bridged dimers in which a ferromagnetic coupling appears to take place via an approximately $96°$ Ni–X–Ni $(X = Cl)$ interaction. Susceptibilities of powdered samples were first [62]

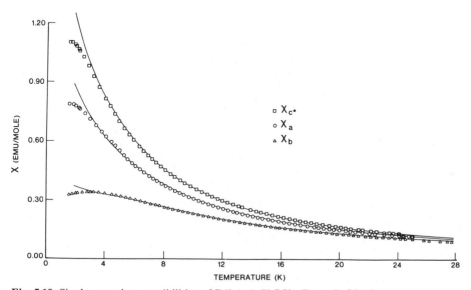

Fig. 5.19. Single crystal susceptibilities of $[Ni_2(en)_4Cl_2]Cl_2$. From Ref. [63]

measured over a wide temperature interval (1.5–300 K) and magnetic field range (1–15.3 kOe, 0.1–1.53 T) and compared to calculations carried out similarly to those described above. The effects of interdimer interaction were examined in a molecular field approximation (see Chap. 6), and the effects of the likely large zero-field splittings were also examined. Unfortunately, zero-field splitting is qualitatively similar to an antiferromagnetic interdimer interaction in its effect on the dimer susceptibility. These systems are good examples of the situation where so many factors are at work that single crystal susceptibilities (the calculated susceptibilities for the model are quite anisotropic) are required for a final analysis of the situation.

The experimentally determined [63] single crystal susceptibilities of $[Ni_2(en)_4Cl_2]Cl_2$ are illustrated in Fig. 5.19, along with the best fits to the data. The model used included the zero-field splittings of the nickel ions, an exchange parameter J, defined as usual by the energy term $H = -2JS_1 \cdot S_2$, and a molecular field correction for interdimer magnetic interactions. The resulting parameters are reported in Table 5.1, Column A.

Table 5.1. Magnetic Parameters for $[Ni_2(en)_4Cl_2]Cl_2$.

	A^a	B^b	C^b
$\langle g \rangle$	2.25 ± 0.02	2.14	2.12
D/k, K	$-14 \quad \pm 1$	-9.4	16
J/k, K	$5.0 \quad \pm 0.5$	14.2	15.0
zJ'/k, K	$-0.3 \quad \pm 0.1$	-0.24	-0.03

[a]   Best fitted values, Ref. [63].
[b]   The two best fits of the powder data as reported in Ref. [62].

It is clear from these results that there is a relatively large single-ion zero-field splitting, with doublet lying lower. Earlier measurements, for which the analysis was ambiguous (Columns B and C) could not resolve the sign of D, but a satisfactory fit to the data presented in Ref. [63] could not be obtained with a positive value for D. The intradimer exchange constant is smaller than that reported earlier but the sign remains ferromagnetic. The value for $\langle g \rangle$ obtained here is typical of that usually found for nickel, and there is a small antiferromagnetic interdimer interaction. In fact, the data may be seen to level off at the lowest temperatures (1.5 K), suggesting that long-range order (Chap. 6) may occur at slightly lower temperatures.

One problem with this comparison of powder and single crystal results is that there appears to be a crystal phase transition in the temperature region 10–27 K, in which the material changes from monoclinic (high temperatures) to triclinic. One other report on the measurement of powder susceptibilities [64] therefore restricted the data analysis to temperatures above 27 K, and this resulted in the parameters $J/k = 9.6$ K and $D/k = -5.2$ K. (The reader of this section must pay particular attention to the way that the different parameters are defined in the original sources. There is no agreement!)

A neutron scattering investigation of $[Ni_2(en)_4Br_2]Br_2$ suggested [66] that this very similar compound is described by the parameters $J/k = 5.1$ K and $D/k = -9.8$ K; a small correction for intermolecular interactions was also required.

Kahn and his collaborators [64, 65] have attempted to rationalize the value of J as

the geometry of the Ni $\diagdown$ Cl $\diagup$ Ni unit varies with counter ion Y=Cl$^-$, ClO$_4^-$,

and B(C$_6$H$_5$)$_4^-$. While the Ni–Cl–Ni angle varies only slightly through the series (96.6, 95.4, 95.6°, respectively), the greatest change is in the Ni–Ni distance (0.3743, 0.3678, 0.3636 nm, respectively). But it is difficult to interpret the significance of this trend. Furthermore, these workers use their own values of J (9.6, 12.8, 13.7 K, respectively) and we have already seen that other authors have determined different values. Further, the exchange constants are of necessity determined by the temperature dependence of the susceptibility at low temperatures, while the crystal structures have been determined at room temperature. So it is difficult to argue too strongly in favor of any rationale of a very small change of the values of parameters which are not determined unambiguously.

Irrespective of these points, there seems to be little argument that the exchange coupling in these molecules is indeed ferromagnetic, or positive, in sign and that the zero-field splitting is of magnitude comparable to 2J and negative in sign.

It is interesting to note how closely the magnetic parameters reported for a [Ni(Cl)$_2$Ni]$^{2+}$ dimer unit compare with those reported [67] for the [Ni(Cl)$_2$Ni]$_\infty$ linear chains (Chap. 10) which occur, for example, in NiCl$_2 \cdot$ 2py. The intrachain exchange parameter in this compound likewise is ferromagnetic in sign, $J/k = 5.35$ K, and D/k is even larger than that reported for the dimers, -27 K. It is indeed satisfying that the magnetic properties for this structural unit appear to be more or less independent of the number of the units joined together, which is consistent with the chemical and geometric features being generally quite similar.

The remaining dimer in this series is [Ni$_2$(en)$_4$(SCN)$_2$]I$_2$, which has a di(μ-thiocyanato) structure. The powder susceptibility again fits a model with a molecular ground state of total spin $\mathscr{S} = 2$ (i.e., ferromagnetic interaction), with an intracluster exchange constant of 8.6 K. This is an important result because the Ni atoms are far apart (0.58 nm) and because they are connected by two rather long three-atom bridges. The occurrence of exchange coupling of the same order of magnitude in both [Ni$_2$(en)$_4$X$_2$]X$_2$ and [Ni$_2$(en)$_4$(SCN)$_2$]I$_2$, in spite of the great difference in Ni–Ni distance (which is about 0.37 nm in the chloride complex), demonstrates the relative unimportance of the metal-metal distance in determining the strength of exchange interactions, so long as there exist appropriate pathways for exchange coupling through bridging ligands.

Another metal ion which forms many dimers is chromium(III). The well-known kinetic inertness of Cr(III) and the central position that this ion has played in the long history of the study of coordination compounds has allowed the synthesis of a number of well-characterized molecules with closely similar structures. Thus, there has once again been a search for a magnetic/structural chemical correlation. The geometry around each chromium is always quasi-octahedral, and the ion has spin $\mathscr{S} = \frac{3}{2}$. Applying the spin Hamiltonian $\boldsymbol{H} = -2J\boldsymbol{S_1} \cdot \boldsymbol{S_2}$ in the usual fashion, the ground state for antiferromagnetic coupling has spin $\mathscr{S} = 0$ and is non-magnetic. Chromium then differs from copper in having a large number of excited states in its exchange manifold; the lowest excited state is at $|J/k|$, followed by others at $|2J/k|$ and $|3J/k|$. Thus, even for a relatively small $|J/k|$, the effects of exchange will be observable at relatively high temperatures. Zero-field splittings, so important for $\mathscr{S}=1$ nickel dimers, are typically less than a Kelvin for chromium(III), and therefore have been ignored in the data analyses of chromium dimers. The susceptibility equation assuming isotropic exchange for a mol of dimers is [9]

$$\chi = \frac{2Ng^2\mu_B^2}{kT}\left\{\frac{\exp(2J/kT)+5\exp(6J/kT)+14\exp(12J/kT)}{1+3\exp(2J/kT)+5\exp(6J/kT)+7\exp(12J/kT)}\right\}. \quad (5.11)$$

This equation is plotted in Fig. 5.10. Occasionally a small term called biquadratic exchange is also introduced. This is of the form $H'=-j(S_1\cdot S_2)^2$, and modifies the above equation somewhat. The parameter j is typically found as being about 5% of the magnitude of J, and is best looked upon only as another fitting parameter.

There are several magneto-structural relationships observed with dihydroxo-bridged Cr(III) dimers but no universal correlation has been found as yet. The magnitude of the exchange parameter depends upon the Cr–O–Cr bridging angle [68], the Cr–O bond length [69] and the dihedral angle between the bridge planes [70]. A number of data have been summarized [71, 72] and applied to a model which has been constructed to include the different structural parameters. Three fitting parameters were required. Any such analysis suffers for the fact that the structural parameters are obtained at room temperature while the exchange constants depend on measurements at low temperatures. The sensitivity of the magnetic parameters is quite large, as is illustrated by the following discussion.

The pH-dependent equilibria that μ-hydroxy complexes of Cr(III) undergo allow [73] the synthesis of the related molecules

where L is the bidentate ligand,

The bridging Cr–O–Cr angles at the oxo and hydroxo atoms in *2* are 100.6 and 95.0°, respectively, while the dihydroxo-bridged series of complexes has angles of 97.6–103.4°, depending on the anion. Again depending on the sterochemistry of (optically active) L and on the nature of the counterion, *1* is found to have $-2J = 31$ to $37\,cm^{-1}$ (45–53 K). The susceptibility of *2* goes through a very broad maximum at about 50–100 K, which may be fit by Eq. (5.11) with $-2J = 46\,cm^{-1}$ (66 K). The data for *3* exhibit a broader maximum, at around 170 K and could not readily be analyzed. [Equation (5.11) is relatively insensitive to small changes in J when J is large.] The best fit gave $-2J \simeq 83\,cm^{-1}$ (120 K).

The sensitivity of the exchange constant to the $Cr(OH)_2Cr$ environment is illustrated by the series of compounds $[(en)_2Cr(OH)_2Cr(en)_2]X_4 \cdot 2H_2O$ and $[(NH_3)_4Cr(OH)_2Cr(NH_3)_4]X_4 \cdot 4H_2O$, where X is chloride or bromide [74]. The coordination spheres are quite similar, yet the ethylenediamine salts have $-2J \simeq 15\,cm^{-1}$ (22 K) while the NH_3 salts have $-2J \simeq 1\,cm^{-1}$ (1.4 K). The hydrogen-bonding differs between the two sets of salts, the hydrogen atom in the (en) salts lying more or less in the Cr–O–Cr plane. In the (NH_3) salts, the hydrogen atom is displaced out of this plane by an angle of at least 30° but more likely 50–60°. This variation appears to be the most important parameter in determining the relative exchange constants. Again, changing the NH_3 salt to the dithionate compound

$$[(NH_3)_4Cr(OH)_2Cr(NH_3)_4](S_2O_6)_2 \cdot 4H_2O$$

changes [75] $-2J$ to $6.2\,cm^{-1}$ (8.9 K). The difficulties involved in finding a magneto-structural correlation are illustrated by the fact that the exchange constant for

$$[(NH_3)_4Cr(OH)_2(NH_3)_4]Cl_4 \cdot 4H_2O$$

is $J = -0.63\,cm^{-1}$ when obtained from spectral analysis [74] and $-2.48\,cm^{-1}$ (a factor of 4 larger) when obtained from susceptibility data [75].

Many iron salts have been studied because of the biochemical importance of iron. The physical principles are similar to those described earlier, but the manifold of states is larger for a dimer because the spin is $\frac{5}{2}$. We mention three examples from the recent literature.

Iron(III) bridged by oxalates as in

has been prepared [76] and shown to interact with $-2J = 7.2\,cm^{-1}$ (10.4 K). This is a typical value for binuclear iron(III). Similarly, when iron is bridged by two hydroxy or methoxy groups in a Schiff base complex $-2J$ is found [77] to be 15 or $16\,cm^{-1}$ (22–23 K). This value changes but a small amount as the coordination sphere changes. When L is N,N′-ethylenebis(salicylamide) in dihydroxybridged $[FeL(OH)]_2$ [78], $-2J = 13.8\,cm^{-1}$ (20 K).

All of the above discussion has focussed on the particular metal ion involved, both chemically and magnetically. Reedijk [79] has actively sought to prepare and study dimers (and clusters) of the several metal ions which are bridged by fluorine atoms. These are of interest because of the simplicity of the bridging atom, and the inability of fluoride to engage in π-bonding. The synthetic problems have been greater, however.

The decomposition of tetrafluoroborate salts has proved to be a useful general synthetic method. A typical reaction is

$$[Co(H_2O)_6](BF_4)_2 + 4\,DMPz \rightarrow [CoF_2(DMPz)_2] + 2\,BF_3 \cdot DMPz$$

where the solvent is ethanol and DMPz is 3,5-dimethylpyrazole. This particular example has a $\text{Co} \overset{\displaystyle F}{\underset{\displaystyle F}{<>}} \text{Co}$ core, and is an infinite polymer. An intermediate of stoichiometry $[CoF(DMPz)_3]_2(BF_4)_2$ can be isolated from the above reaction, and this material is dimeric with bridging fluorides. The exchange constant in the dimer is -1 K or smaller, which reflects the constrained geometry of the $Co-F_2-Co$ core, as well as the poor superexchange path furnished by fluoride [80]. Several copper compounds with CuF_2Cu cores [81] likewise have essentially zero exchange interaction.

There has been a number of studies recently which have used orbital models to try to explain and predict superexchange interaction in dimers [82–89]. One of the questions asked, for example, is what are the conditions for the exchange constant to be ferromagnetic in sign? Or, how should an exchange constant vary as the Cu–OH–Cu angle varies in a hydroxy-bridged complex? To date empirical correlations seem to be more successful than a priori prediction.

One of the more interesting results has concerned complexes of a Schiff base in which a Cu(II) ion is bound, respectively, to $\mathscr{S} = \frac{3}{2}$ Cr(III) or $\mathscr{S} = \frac{5}{2}$ Fe(III) by two bridging oxygen atoms. It is reported that the CuFe compound exhibits an antiferromagnetic interaction while the CuCr compound exhibits a ferromagnetic interaction. A detailed orbital picture has been presented to account for this difference in behavior [90].

One of the criticisms of these models [89] is that the angular dependence of the coupling constant on the angle M–L–M is not necessarily associated with the directional properties of the atomic p and d orbitals. Model calculations show that the experimental trends can be explained on the basis of spherically symmetric atomic orbitals.

5.6 EPR Measurements

One of the principal applications of EPR to the study of transition metal ions is the determination of spin-Hamiltonian parameters [17]. When metal ions are put into diamagnetic hosts in small concentration, this is one of the most accurate ways of determining g-values and zero-field splittings that are relatively small. Exchange and other effects often limit the amount of information that can be obtained by EPR on concentrated materials, and so a study [91] of the $Cr_2Cl_9^{3-}$ ion as the Cs^+, Et_4N^+, and Pr_4N^+ salts is of considerable interest and illustrates a procedure of broad application.

The complex ion is formed by the sharing of a face between two adjacent $CrCl_6$-octahedra. Thus, the metal atoms are bridged by three chloride ions, and are 0.312 nm apart in the Cs salt. The early powder susceptibility measurements [92] go down in temperature only to 80 K and so those data are not very sensitive to exchange, which was estimated to be $2J/k = -10$ K for the usual Hamiltonian, $H = -2JS_1 \cdot S_2$. Under such an interaction which is antiferromagnetic in sign, two chromium(III) ions of $\mathscr{S} = \frac{3}{2}$ as described above form a manifold of four levels of total spin 0 (at $\frac{15}{2}J$), 1 (at $\frac{11}{2}J$), 2 (at $\frac{3}{2}J$), and 3 (at $-\frac{9}{2}J$), where the terms in parentheses refer to the energies of the different levels with respect to the free ion levels. With negative J, the $\mathscr{S} = 0$ level lies lowest, the other levels are expected to have a Boltzmann population, and so a non-Curie susceptibility that goes to zero at $T \to 0$ is predicted. Each of the several levels is expected to give an EPR spectrum characteristic of the total spin of the particular level, as in the case of copper acetate, with a zero-field splitting that is once again due to anisotropy in the exchange and dipole-dipole forces. For the Cs^+ salt, only the $\mathscr{S} = 1$ level was observed, while the $\mathscr{S} = 2$ state was the only one observed for $(Et_4N)_3[Cr_2Cl_9]$. Spectra were observed from all the manifolds of different total-spin with the tetra-n-propylammonium salt, which allowed a large deviation from the Landé interval rule to be observed. In particular, in addition to the usual bilinear term in $S_1 \cdot S_2$, it was found necessary for the data analysis to add smaller terms in $(S_1 \cdot S_2)^2$ and even $(S_1 \cdot S_2)^3$. These latter isotropic terms may arise not so much from superexchange interaction but from the effects of such phenomena as exchange striction, the interaction of exchange with the elastic constants of the crystalline material. The interaction constants were estimated by the temperature variation of the intensity, and $2J/k = -16$ to -20 K was found for the three compounds, with very little exchange coupling between the pairs. Susceptibility data are consistent with this interpretation [93].

5.7 Clusters

The methods described above are of course applicable to any discrete cluster of magnetic atoms. Trimers and tetramers are well-known [1, 9, 79, 94] and we are limited only by the ease of chemical synthesis and the increasing mathematical complexity of the problem as the symmetry of the cluster decreases and the interactions increase in both number and kind. Thus, consider the metal atom triad of Fig. 5.20, with three exchange constants. The metal atoms may or may not be alike, zero-field splittings may contribute if the spin is greater than $\frac{1}{2}$, and the exchange constants may or may not be the same. The more parameters that are introduced, the more detailed must be the experimental data in order to separate the different factors contributing to the magnetic properties.

An interesting example is offered by linear trimeric bis(acetylacetonato)nickel(II), $[Ni(acac)_2]_3$, which is schematically illustrated in Fig. 5.21. The crystal structure has

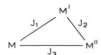

Fig. 5.20. Three metal atoms with inequivalent magnetic exchange

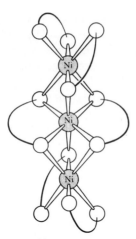

Fig. 5.21. Molecular structure of $[Ni(acac)_2]_3$. From Ref. [96]

been analyzed in detail [95]. For the isolated trimer, the exchange Hamiltonian may be written [96] as

$$H = -2J[S_1 \cdot S_2 + S_2 \cdot S_3] - 2J'(S_1 \cdot S_3),$$ (5.12)

where J is the exchange constant between adjacent nickel atoms (1,2 and 2,3), and J' is the exchange term between the two terminal nickel atoms (1,3) within the trimer. The spin of the trimeric molecule as well as the order of the energy level manifold depends on the relative values of J and J', as well as the sign of J. For example, for positive J, the ground state is always paramagnetic, but the spin depends on the value of J'/J. For negative J, the ground state is diamagnetic only in the case of $0.5 < J'/J < 2$; otherwise it is paramagnetic. The susceptibility [96] and magnetization [97] of polycrystalline samples of the compound have been measured, and the data analysis was found to depend sensitively upon whether or not zero-field splittings were included. The molecular ground state has a spin $\mathscr{S} = 3$. A ferromagnetic interaction of $-2J = 25\,\text{cm}^{-1}$ (36 K) between neighboring nickel ions was determined, as well as an antiferromagnetic interaction of $-2J = 9\,\text{cm}^{-1}$ (13 K) between the terminal nickel(II) ions. These parameters depend on the determination of a zero-field splitting of the $\mathscr{S} = 3$ trimer ground state of $D = -1.3\,\text{cm}^{-1}$, which corresponds to a single-ion zero-field splitting of $-2.17\,\text{cm}^{-1}$ (-3.1 K). The antiferromagnetic exchange between the terminal ions is remarkable for its strength since it involves the interaction of the two ions via four ligand atoms.

The compound $[(\eta^5\text{-}C_5H_5)_2TiCl]_2MnCl_2 \cdot 2\,THF$ (THF = tetrahydrofuran) has also been investigated [37]. This material contains a collinear trimetallic molecule. A

Ti — Cl — Mn — Cl — Ti core is found, and the central manganese atom is six-coordinate, having a tetrahydrofuran molecule coordinated to it both above and below the plane. The entire $Ti\text{-}Cl_2\text{-}Mn\text{-}Cl_2\text{-}Ti$ unit is planar. The molar magnetic moment decreases with decreasing temperature, implying the presence of an exchange

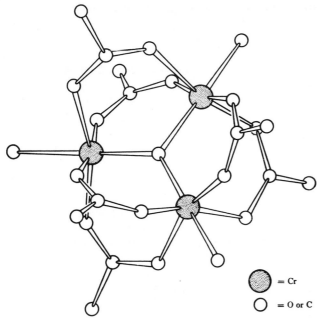

$= Cr$

$= O$ or C

Fig. 5.22. Molecular structure of the trimeric cation, $[Cr_3O(CH_3CO_2)_6(H_2O)_3]^+$. From Ref. [1]

interaction $[-2J \simeq 8\,cm^{-1}\ (11.5\,K)]$ between the manganese and titanium atoms, but the nature of the 1,3-titanium-titanium coupling (if any) could not be elucidated.

An interesting series of clusters contains the trimeric ion, $[M_3O(CH_3CO_2)_6(H_2O)_3]^+$. Compounds where M is Cr or Fe have been studied most extensively, and the mixed Cr_2Fe compound has also been examined. Some of the problems raised here are quite illuminating, for they illustrate the difficulty involved in studying large clusters.

The molecular structure [98] of $[Cr_3O(CH_3CO_2)_6(H_2O)_3]^+$ is illustrated in Fig. 5.22. Each Cr^{3+} ion is nearly octahedrally coordinated by oxygen, and the three metal atoms are arranged, at ambient temperature, in an equilateral triangle about a central oxide anion. Early measurements of the specific heat and susceptibility [99] suggested that an antiferromagnetic interaction occurred among the three ions, and with a Hamiltonian of the form

$$H = -2J[S_1 \cdot S_2 + S_2 \cdot S_3 + \alpha S_1 \cdot S_3] \tag{5.13}$$

a fair fit to the data of a number of investigations was obtained with $J/k = -15\,K$ and α between 1 and 1.25. A value of α different from 1 signifies that the magnetic interactions are more like that of an isosceles triangle than the structural equilateral triangle. Since the ground state of the system of three $\mathscr{S} = \frac{3}{2}$ ions corresponds to a total net spin of $\frac{1}{2}$, and is therefore paramagnetic, intercluster coupling is expected to occur at sufficiently low temperatures. Susceptibility measurements provide no evidence for this down to 0.38 K, however. Attempts to add higher-order interactions [100] to fit data better to the equilateral triangle model have proved to be invalid [101].

The recent discovery [101] of a crystallographic phase transition for

$$[Cr_3O(CH_3COO)_6(H_2O)_3]Cl \cdot 6 H_2O$$

at about 210 K serves to make all the previous data analyses suspect. The specific heat was measured over a wide temperature interval (1.5–280 K), but the estimation of the lattice specific heat below 20 K, where the major intracluster interaction occurs, depended on a fit to the total specific heat between 30 and 100 K, which is a questionable procedure. Straightforward application of both the equilateral and isosceles triangle models of intracluster magnetic exchange did not give satisfactory fits to the magnetic specific heat, but the observation of the phase transition led the authors to propose a new model for this system. Though a structural equilateral triangle obtains at room temperature, it was assumed that the symmetry was slightly distorted through the phase transition processes, but the unit cell of the low temperature phase preserves four formula units in two sets of equivalent pairs. In other words, *two* sets of isosceles triangles were assumed, and an excellent fit to the magnetic heat was obtained with the following parameters for the spin-Hamiltonian

$$H = -2J_0(S_1 \cdot S_2 + S_2 \cdot S_3 + S_1 \cdot S_3) - 2J_1 S_2 \cdot S_3. \tag{5.14}$$

Set 1:   Set 2:

$2J_0/k = -30\,K$ $2J_0/k = -30\,K$

$2J_1/k = -4.5\,K$ $2J_1/k = +1.5\,K$.

These results are not inconsistent with any other available data, and the presence of two sets of units is in fact confirmed [102, 103] by the analysis of the emission and optical spectra. These spectral results also show that intercluster interactions are negligibly small [104].

The tetranuclear ion, $[Cr_4(OH)_6(en)_6]^{6+}$ offers another interesting example of the difficulties associated with the study of clusters. In this case, the early magnetic studies were interpreted on the basis of the incorrect structural model. It was discovered that Pfeiffer's cation, $[Cr_4(OH)_6(en)_6]^{6+}$, as it was found in $[Cr_4(OH)_6(en)_6](N_3)_6 \cdot 4 H_2O$, is a Cr_4 planar rhomboid [105],

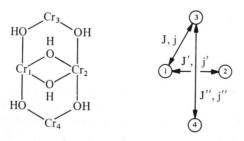

with Cr–Cr distances: Cr(1)–Cr(2) = 0.293 nm, Cr(1)–Cr(3) = 0.361 nm, and Cr(3)–Cr(4) = 0.655 nm. The structure is illustrated in Fig. 5.23. Using a Hamiltonian of the form

$$H = -2J[S_1 \cdot S_3 + S_2 \cdot S_3 + S_1 \cdot S_4 + S_2 \cdot S_4] - 2J'S_1 \cdot S_2 \tag{5.15}$$

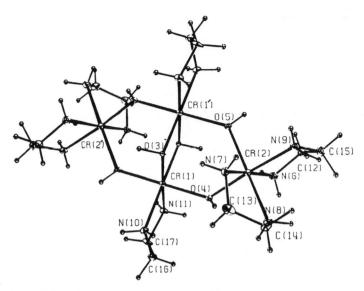

Fig. 5.23. A view of the $[Cr_4(OH)_6(en)_6]^{6+}$ cation. From Ref. [105]

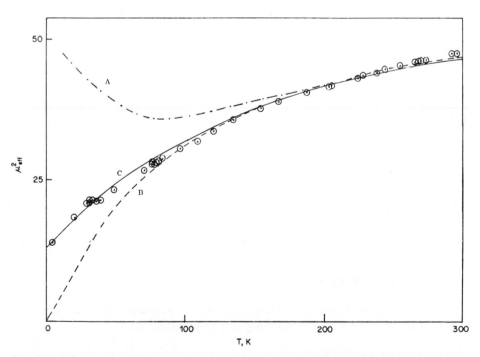

Fig. 5.24. Effective magnetic moment squared (calculated per tetranuclear ion) vs. temperature, calculated for a trigonal planar model (curve A, $J/k = -20\,K$), tetrahedral model (curve B, $J/k = -10\,K$), and planar-rhomboid model (curve C, $J/k = -10.5\,K$, $J_{12}/k = -20\,K$). Experimental data are indicated by the circles. From Ref. [106]

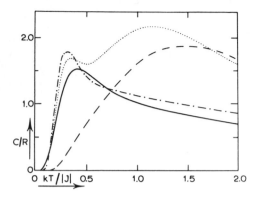

Fig. 5.25. The magnetic heat capacities associated with various models; ―――: trigonal planar; ―·―·―: tetrahedral; ―――: planar-rhomboid, $J'/J = 0.5$; ·····: planar-rhomboid, $J'/J = 1$. From Ref. [107]

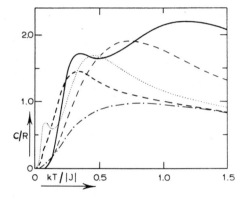

Fig. 5.26. The variation of the magnetic heat capacity arising from a planar-rhomboid model with the ratio J'/J; ―――: $J'/J = 1$; ―――: $J'/J = 0.75$; ·····: $J'/J = 0.55$; ―――: $J'/J = 0.45$; ―·―·―: $J'/J = 0.25$. From Ref. [107]

a fit to the powder susceptibility over a wide temperature interval was obtained [106], as illustrated in Fig. 5.24, with parameters $2J/k$ of about -20 K and $2J'/k$ of about -40 K. The large distance between Cr(3) and Cr(4) was assumed to make any significant spin-spin interaction between these atoms unlikely.

Subsequently, the magnetic specific heat of $[(Cr_4(OH)_6(en)_6] (SO_4)_3 \cdot 10 H_2O$ was obtained [107]. Two broad peaks were observed, at about 2.3 and 20 K. In Fig. 5.25, we illustrate the specific heat behavior calculated [107] for several tetrameric models, and at least one broad peak is always obtained; in Fig. 5.26, the specific heat for the planar rhomboid model, Eq. (5.15), is illustrated for a variety of ratios J/J'. The magnetic specific heat of the sulfate salt resembles these curves, but a final fit to the data was not obtained until the new interaction

$$H = -2J''(S_3 \cdot S_4)$$

was also included. The sensitivity of the specific heat of this model to the new parameter J'' is illustrated in Fig. 5.27. The resulting best fit parameters are

$$2J/k = -22.8 \text{ K}$$

$$2J'/k = -42.6 \text{ K}$$

$$2J''/k = -\ \ 7.6 \text{ K}.$$

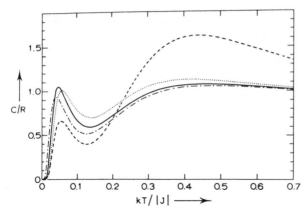

Fig. 5.27. The variation of the magnetic specific heat due to a planar-rhomboid model with the ratio J''/J; $---$: $J''/J=0$; $\cdots$: $J''/J=0.17$; ——: $J''/J=0.19$; $-\cdot-\cdot-$: $J''/J=0.20$. From Ref. [107]

All parameters are antiferromagnetic in sign, and the major interactions are comparable in magnitude to those reported by Gray and co-workers. Sorai and Seki [107] also showed that the specific heat is more sensitive at low temperatures to the effect of the ratio J''/J' than is the susceptibility. Clearly, a complete magnetic study should always involve both isothermal susceptibility and heat capacity measurements, as well as complete structural information. But bulk measurements on systems with so many parameters to evaluate often do not present enough structure to allow an unambiguous determination, as we shall now see.

Güdel and co-workers [108–110] have studied the related molecule $[Cr_4(OD)_6(ND_3)_{12}]Cl_6 \cdot 4\,D_2O$, called rhodoso chloride. The compound was deuterated in order to allow inelastic neutron scattering experiments to be carried out on it. The bilinear exchange terms listed above were included in the data analysis, along with the biquadratic terms

$$-j\{(S_1 \cdot S_3)^2 + (S_1 \cdot S_4)^2 + (S_2 \cdot S_3)^2 + (S_2 \cdot S_4)^2\}$$
$$-j'(S_1 \cdot S_2)^2 - j''(S_3 \cdot S_4)^2 .$$

The susceptibility of polycrystalline material is simply not sensitive to all these parameters, but they could be evaluated from the neutron measurements. The final best values are [110]

$$-2J = 17.4\,\mathrm{cm}^{-1}\ (25\,\mathrm{K})$$

$$-2J' = 25.2\,\mathrm{cm}^{-1}\ (36\,\mathrm{K})$$

$$-2J'' = 1.7\,\mathrm{cm}^{-1}\ (2.4\,\mathrm{K})$$

$$j = 0.1\,\mathrm{cm}^{-1}\ (0.1\,\mathrm{K})$$

$$j' = 1.6\,\mathrm{cm}^{-1}\ (2.3\,\mathrm{K})$$

$$j'' = \text{set to zero} .$$

All the interactions are antiferromagnetic and nearest-neighbor exchange is one order of magnitude larger than next-nearest-neighbor exchange. Exchange-striction may be responsible for the biquadratic terms. Though the parameters are close to those as determined for the Pfeiffer complex, small changes in the J/J' ratio affect the level ordering, so that Pfeiffer's salt, as the azide, appears to have a spin-triplet ground state, while the rhodoso chloride has a spin singlet lowest.

Another tetranuclear chromium complex which has been studied [111] is $[Cr\{(OH)_2Cr(en)_2\}_3](S_2O_6) \cdot 8\,H_2O$. The structure, schematically, is

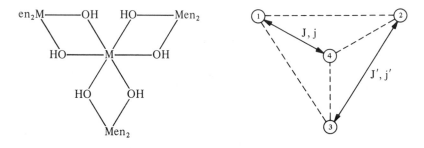

with the coupling scheme as indicated. The spin Hamiltonian is

$$H = -2J[S_1 \cdot S_4 + S_2 \cdot S_4 + S_3 \cdot S_4]$$
$$- 2J'[S_1 \cdot S_2 + S_2 \cdot S_3 + S_1 \cdot S_3]$$

plus biquadratic terms. The square of the magnetic moment for four uncoupled chromium(III) ions with g = 2 is calculated as $60\mu_B^2$, but the experimental result, even at 300 K, is reduced below $50\mu_B^2$. This is clear evidence for antiferromagnetic exchange coupling. As the temperature is decreased, μ^2 decreases, goes through a minimum near 50 K, and then increases to about $43\mu_B^2$ at the lowest temperatures. These results require that a total spin $\mathscr{S} = 3$ level lie lowest in energy. The parameter $-2J$ was found to be about $17\,\mathrm{cm}^{-1}$ (24 K) with the next-nearest neighbor exchange (J') of about 10% of that value. The data are not sensitive enough to discriminate between zero and non-zero values for the biquadratic exchange terms. We may conclude that dihydroxo-bridged chromium(III) compounds exhibit exchange constants of similar order of magnitude.

A cluster complex of continuing interest is of the formula $Cu_4OX_6L_4$ where X is chloride or bromide and L is, variously, chloride, bromide, pyridine, or oxygen donor such as a phosphine oxide, R_3PO, or amine N-oxide, R_3NO [112–116]. The structure, illustrated in Fig. 5.28, consists of a regular tetrahedron of copper ions, at the center of which is an oxide ion. Six halogen ions bridge adjacent copper ions of the tetrahedron. Each copper has trigonal bipyramidal coordination, distorted to varying degrees depending on the axial ligand L. Though the exchange interaction between the copper atoms is substantial, it is impossible to distinguish the relative importance of the Cu–O–Cu and Cu–X–Cu exchange pathways. Several orbital and exchange models have been proposed [112–115] to fit the available magnetic susceptibility data, the problem of concern being that the magnetic moment $\mu_{\mathrm{eff}}(T)$ decreases monotonically with decreasing temperature for some of the compounds, while it passes through a

$Cu_4OL_4X_6$

● O
○ Cu
● X
◑ L

Fig. 5.28. Molecular structure of the tetranuclear copper(II) cluster complex $[Cu_4OX_6L_4]$

maximum at low temperatures for others. It is not possible to explain the maximum in $\mu_{eff}(T)$ in terms of intracluster isotropic exchange with equivalent coupling constants. Intercluster exchange interactions appear to be negligibly small.

Recently, a model has been proposed [116] that allows the available data for 8 compounds to be fit. It allows for the magnetic effects of distortion from the mean, approximately tetrahedral, configuration of the $Cu_4OX_6L_4$ molecules. It assumes that the exchange constant $J(d)$ between two copper atoms in the cluster may be written as

$$J(d) = J(0) + d(\partial J/\partial d)_0 ,$$

where d is a distortion vector and $J(0)$ is the exchange constant in the absence of distortion. It is assumed that the distortion is small but that $(\partial J/\partial d)_0$ is large and isotropic. To the extent that this relationship is valid, magnetic exchange becomes a cause of distortion from tetrahedral symmetry. The model was further extended to modify this static picture by introducing dynamic distortions that interconvert equivalent configurations, or fluxionality. Fluxionality is expected in tetrahedral systems such as these since any distortion that lowers the symmetry occurs along a degenerate vibrational normal mode. Three possibilities arise: 1) slow fluxionality, in which the energy barrier between different configurations is so high that the system will be effectively nonfluxional; 2) fast fluxionality, with a low barrier, and 3) temperature-dependent fluxionality. All three cases were found among the eight compounds.

5.8 The Ising Model

The Hamiltonian we have used so far for magnetic exchange is, as mentioned earlier, an isotropic one and is often referred to as the Heisenberg model. We may rewrite

$$H = -2JS_1 \cdot S_2$$

explicitly as

$$H = -2J[\gamma(S_{1x}S_{2x} + S_{1y}S_{2y}) + S_{1z}S_{2z}] \qquad (5.16)$$

and observe that the case with $\gamma = 1$, the Heisenberg model, corresponds to using all the spin-components of the vectors S_1 and S_2. A very anisotropic expression, called the Ising model, is obtained when γ is set equal to zero. This may appear as a very artificial situation, and yet as we shall see, the Ising model is exceedingly important in the theory of magnetism as well as in other physical many-body problems. This is true, in part, because solutions of the Ising Hamiltonian are far more readily obtained than those of the Heisenberg Hamiltonian. The history and application of the Ising model have been reviewed [117].

The Ising model is the simplest many-particle model that exhibits a phase transition. However, for an extended three dimensional lattice, the calculations are so complicated that no exact calculations have yet been done; the phase transition does not occur in one-dimension with either the Ising or Heisenberg models, as we shall see in Chap. 7. In a two-dimensional lattice, the Ising model does exhibit a phase transition, and several physical properties may be calculated. Although it may not appear physically realistic, it will be useful to introduce the Ising model here and to illustrate the calculations by calculating the susceptibility of an isolated dimer.

The essence of the Ising model is that spins have only two orientations, either up or down with respect to some axis. We consider only the operator S_z, with eigenvalues $+\frac{1}{2}$ and $-\frac{1}{2}$ and so a dimer will exhibit but four states: $+\frac{1}{2}, +\frac{1}{2}; +\frac{1}{2}, -\frac{1}{2}; -\frac{1}{2}, +\frac{1}{2};$ and $-\frac{1}{2}, -\frac{1}{2}$. For the pair, we choose as the Hamiltonian

$$H = -2JS_{z1}S_{z2} + g\mu_B H_z(S_{z1} + S_{z2}) \qquad (5.17)$$

which has eigenvalues which may be obtained by inspection,

$$-(J/2 + g\mu_B H_z)$$

$$J/2 \text{ (twice)}$$

$$-(J/2 - g\mu_B H_z)$$

and we may write the partition function as

$$Z_a = e^{-\langle H \rangle / kT}$$
$$= e^{(J/2 + g\mu_B H_z)/kT} + 2e^{-J/2kT} + e^{(J/2 - g\mu_B H_z)/kT}$$
$$= 2e^{J/2kT}[e^{-J/kT} + \cosh(g\mu_B H_z/kT)]. \qquad (5.18)$$

The molar magnetization M is defined as

$$M = NkT(\partial \ln Z_a / \partial H_z)_T \qquad (5.19)$$

and thus straightforward calculation leads to

$$M = \frac{Ng\mu_B \sinh(g\mu_B H_z/kT)}{e^{-J/kT} + \cosh(g\mu_B H_z/kT)}.$$

Since

$$\chi = (\partial M / \partial H_z)_T$$

we find

$$\chi = \frac{2Ng^2\mu_B^2}{kT} \left[\frac{1 + e^{-J/kT} \cosh(g\mu_B H_z/kT)}{[e^{-J/kT} + \cosh(g\mu_B H_z/kT)]^2} \right]$$

or in the limit of zero-field

$$\chi_o = \frac{Ng^2\mu_B^2}{kT} (1 + e^{-J/kT})^{-1}. \tag{5.20}$$

Note that this solution is actually only $\chi_{\parallel}$, the susceptibility parallel to the z-axis, and refers to a mol of dimers; an additional factor of 2 in the denominator is required in order to refer to a mol of magnetic ions. The behavior of Eq. (5.20) is compared with that of Eq. (5.7) in Fig. 5.29. A general discussion has been published [118].

The specific heat per mol of ions of the two-spin Ising system at zero field is easily calculated from Eq. (3.18) and Eq. (5.18), above, as

$$c = R(J/kT)^2 \operatorname{sech}^2(J/2kT) \tag{5.21}$$

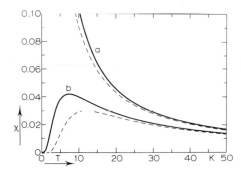

Fig. 5.29. Magnetic susceptibility of dimers of $\mathscr{S} = \frac{1}{2}$ ions coupled in the Heisenberg (dashed curves) and Ising (drawn curves) approximations [Eqs. (5.7) and (5.20)]. $|J/k| = 10$ K with a: $J > 0$ and b: $J < 0$

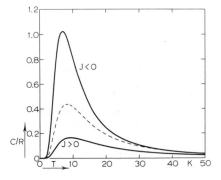

Fig. 5.30. Specific heat of dimers according to the Heisenberg (drawn curve) and Ising (dashed curve) approximations [Eqs. (5.5) and (5.21)], $|J/k| = 10$ K

and Eqs. (5.5) and (5.21) are compared in Fig. 5.30. To date, there are no relevant data to compare this theory with experiment, although this model has been applied to the tetrameric compound cobalt acetylacetonate [119].

5.9 References

1. Martin R.L., (1968) New Pathways in Inorganic Chemistry, Ebsworth E.A.V., Maddock A.G., and Sharpe A.G., Eds., Cambridge University Press, London, Chapter 9
2. Anderson P.W., (1963) Magnetism, (Rado G.T. and Suhl H., Eds), Vol. 1, p. 25, Academic Press, Inc., New York; Stevens K.W.H., Phys. Reports (Phys. Lett. C) **24C**, #1, February, 1976
3. Carrington A. and McLachlan A.D., (1967) Introduction to Magnetic Resonance, Harper and Row, New York
4. Smart J.S., (1965) Magnetism, (Rado G.T. and Suhl H., Eds), Vol. III, p. 63, Academic Press, Inc., New York
5. Friedberg S.A. and Raquet C.A., J. Appl. Phys. **39**, 1132 (1968)
6. Morosin B., Acta Cryst. **B26**, 1203 (1970)
7. Berger L., Friedberg S.A., and Schriempf J.T., Phys. Rev. **132**, 1057 (1963)
8. Bonner J.C., Friedberg S.A., Kobayashi H., Meier D.L., and Blöte H.W.J., Phys. Rev. B **27**, 248 (1983)
9. Ginsberg A.P., Inorg. Chem. Acta Reviews **5**, 45 (1971); Hodgson D.J., Prog. Inorg. Chem. **19**, 173 (1975); Doedens R.J., Prog. Inorg. Chem. **21**, 209 (1976)
10. Bleaney B. and Bowers K.D., Proc. Roy. Soc. (London) **A214**, 451 (1952)
11. Guha B.C., Proc. Roy. Soc. (London) **A206**, 353 (1951)
12. van Niekerk J.N. and Schoening F.R.L., Acta Cryst. **6**, 227 (1953)
13. de Meester P., Fletcher S.A., and Skapsky A.C., J. Chem. Soc. – Dalton **1973**, 2575
14. Brown G.M. and Chidambaram R., Acta Cryst. **B29**, 2393 (1973)
15. Cotton F.A., De Boer B.G., La Prade M.D., Pipal J.R., and Ucko D.A., Acta Cryst. **B27**, 1664 (1971)
16. Kokoszka G.F., Allen, Jr. H.C., and Gordon G., J. Chem. Phys. **42**, 3693 (1965)
17. Abragam A. and Bleaney B., (1970) Electron Paramagnetic Resonance of Transition Ions, Oxford University Press, Oxford
18. Figgis B.N. and Martin R.L., J. Chem. Soc. **1956**, 3837
19. Gregson A.K., Martin R.L., and Mitra S., Proc. Roy. Soc. (London) **A320**, 473 (1971)
20. Kokot E. and Martin R.L., Inorg. Chem. **3**, 1303 (1964); Dubicki L., Harris C.M., Kokot E., and Martin R.L., Inorg. Chem. **5**, 93 (1966)
21. Barclay G.A. and Kennard C.H.L., J. Chem. Soc. **1961**, 5244
22. Gregson A.K. and Mitra S., J. Chem. Phys. **51**, 5226 (1969); de Jongh L.J. and Miedema A.R., Adv. Phys. **23**, 1 (1974)
23. Moreland J.A. and Doedens R.J., Inorg. Chem. **17**, 674 (1978)
24. Porter L.C. and Doedens R.J., Inorg. Chem. **23**, 997 (1984)
25. Güdel H.U., Stebler A., and Furrer A., Inorg. Chem. **18**, 1021 (1979)
26. Güdel H.U., Stebler A., and Furrer A., J. Phys. C **13**, 3817 (1980)
27. Whyman R. and Hatfield W.E., Transition Metal Chemistry **5**, 47 (1969)
28. Komson R.C., McPhail A.T., Mabbs F.E., and Porter J.K., J. Chem. Soc. **A1971**, 3447
29. Carlin R.L. and van Duyneveldt A.J., (1977) Magnetic Properties of Transition Metal Compounds, Springer, Heidelberg, Berlin, New York
30. Yawney D.B.W., Moreland J.A., and Doedens R.J., J. Am. Chem. Soc. **95**, 1164 (1973)
31. Catterick J. and Thornton P., Adv. Inorg. Chem. Radiochem. **20**, 291 (1977)
32. Moreland J.A. and Doedens R.J., J. Am. Chem. Soc. **97**, 508 (1974)
33. Goodgame D.M.L., Hill N.J., Marsham D.F., Skapski A.C., Smart M.L., and Trouton P.G.H., Chem. Commun. 629 (1969)

34. See Charlot M.F., Verdaguer M., Journaux Y., de Loth P., and Daudey J.P., Inorg. Chem. **23**, 3802 (1984); Julve M., Verdaguer M., Gleizes A., Philoche-Levisalles M., and Kahn O., Inorg. Chem. **23**, 3808 (1984) and references therein for recent theoretical work on this problem

35. Coutts R.S.P., Martin R.L., and Wailes P.C., Aust. J. Chem. **26**, 2101 (1973); Jungst R., Sekutowski D., Davis J., Luly M., and Stucky G., Inorg. Chem. **16**, 1645 (1977)

36. Jungst R., Sekutowsky D., and Stucky G., J. Am. Chem. Soc. **96**, 8108 (1974)

37. Sekutowsky D., Jungst R., and Stucky G.D., Inorg. Chem. **17**, 1848 (1978)

38. Hatfield W.E., (1976) in: Theory and Applications of Molecular Paramagnetism (Boudreaux E.A. and Mulay L.M., Eds.,) J. Wiley and Sons, New York, p. 349

39. Crawford V.H., Richardson H.W., Wasson J.R., Hodgson D.J., and Hatfield W.E., Inorg. Chem. **15**, 2107 (1976)

40. Charlot M.F., Jeannin S., Jeannin Y., Kahn O., Lucrece-Abaul J., and Martin-Frere J., Inorg. Chem. **18**, 1675 (1979)

41. Charlot M.F., Kahn O., Jeannin S., and Jeannin Y., Inorg. Chem. **19**, 1410 (1980)

42. Haddad M.S., Wilson S.R., Hodgson D.J., and Hendrickson D.N., J. Am. Chem. Soc. **103**, 384 (1981)

43. Bertrand J.A., Fujita E., and VanDerveer D.G., Inorg. Chem. **19**, 2022 (1980)

44. Girerd J.J., Jeannin S., Jeannin Y., and Kahn O., Inorg. Chem. **17**, 3034 (1978)

45. Chauvel C., Girerd J.J., Jeannin Y., Kahn O., and Lavigne G., Inorg. Chem. **18**, 3015 (1979)

46. Haddad M.S., Hendrickson D.N., Cannady J.P., Drago R.S., and Bieksza D.S., J. Am. Chem. Soc. **101**, 898 (1979)

47. Swank D.N., Needham G.F., and Willett R.D., Inorg. Chem. **18**, 761 (1979)

48. Marsh W.E., Patel K.C., Hatfield W.E., and Hodgson D.J., Inorg. Chem. **22**, 511 (1983)

49. Fletcher R., Hansen J.J., Livermore J., and Willett R.D., Inorg. Chem. **22**, 330 (1983)

50. Willet R.D.: in Magneto-Structural Correlations in Exchange Coupled Systems, Edited by Willett R.D., Gatteschi D., and Kahn O., NATO ASI Series, D. Reidel Publ. Co., Dordrecht, 1985

51. Villa J.F. and Hatfield W.E., Inorg. Chem. **10**, 2038 (1971)

52. McGregor K.T., Hodgson D.J., and Hatfield W.E., Inorg. Chem. **12**, 731 (1973)

53. van Duyneveldt A.J., van Santen J.A., and Carlin R.L., Chem. Phys. Lett. **38**, 585 (1976)

54. van Santen J.A., van Duyneveldt A.J., and Carlin R.L., Inorg. Chem. **19**, 2152 (1980)

55. Boyd P.D.W., Mitra S., Raston C.L., Rowbottom G.L., and White A.H., J. Chem. Soc. – Dalton 13 (1981)

56. McGregor K.T., Barnes J.A., and Hatfield W.E., J. Amer. Chem. Soc. **95**, 7993 (1973)

57. Carlin R.L., Burriel R., Cornelisse R.M., and van Duyneveldt A.J., Inorg. Chem. **22**, 831 (1983); a rebuttal is provided by Hatfield W.E., Inorg. Chem. **22**, 833 (1983)

58. Carlin R.L. and de Jongh L.J., to be published

59. Kahn O., Sikorav S., Gouteron J., Jeannin S., Jeannin Y., Inorg. Chem. **22**, 2877 (1983); Sikorav S., Bkouche-Waksman I., and Kahn O., Inorg. Chem. **23**, 490 (1984)

60. Kahn O., in: Magneto-Structural Correlations in Exchange Coupled Systems, Edited by Willett R.D., Gatteschi D., and Kahn O., NATO ASI Series, D. Reidel Publ. Co., Dordrecht, 1985

61. Shepherd R.E., Hatfield W.E., Ghosh D., Stout C.D., Kristine F.J., and Ruble J.R., J. Am. Chem. Soc. **103**, 5511 (1981)

62. Ginsberg A.P., Martin R.L., Brookes R.W., and Sherwood R.C., Inorg. Chem. **11**, 2884 (1972)

63. Joung K.O., O'Connor C.J., Sinn E., and Carlin R.L., Inorg. Chem. **18**, 804 (1979)

64. Journaux Y. and Kahn O., J. Chem. Soc. Dalton **1979**, 1575

65. Bkouche-Waksman I., Journaux Y., and Kahn O., Trans. Met. Chem. (Weinheim) **6**, 176 (1981)

66. Stebler A., Güdel H.U., Furrer A., and Kjems J.K., Inorg. Chem. **21**, 380 (1980)

67. Klaaijsen F.W., Dokoupil Z., and Huiskamp W.J., Physica B **79**, 547 (1975)

68. Cline S.J., Kallesøe, Pedersen E., and Hodgson D.J., Inorg. Chem. **18**, 796 (1979)

69. Hatfield W.E., MacDougall J.J., and Shepherd R.E., Inorg. Chem. **20**, 4216 (1981)

70. Cline S.J., Glerup J., Hodgson D.J., Jensen G.S., and Pedersen E., Inorg. Chem. **20**, 2229 (1981)
71. Glerup G., Hodgson D.J., and Pedersen E., Acta Chem. Scand. **A37**, 161 (1983)
72. Glerup J., Hodgson D.J.: in Magneto-Structural Correlations in Exchange Coupled Systems, Edited by Willett, R.D., Gatteschi, D., and Kahn, O., NATO ASI Series, D. Reidel Publ. Co., Dordrecht, 1985
73. Michelsen K., Pedersen E., Wilson S.R., and Hodgson D.J., Inorg. Chem. Acta **63**, 141 (1982)
74. Decurtins S. and Güdel H.U., Inorg. Chem. **21**, 3598 (1982)
75. Cline S.J., Hodgson D.J., Kallesøe S., Larsen S., and Pedersen E., Inorg. Chem. **22**, 637 (1983)
76. Julve M. and Kahn O., Inorg. Chim. Acta **76**, L39 (1983)
77. Chiari B., Piovesana O., Tarantelli T., and Zanazzi P.F., Inorg. Chem. **22**, 2781 (1983); See also: Chiari B., Piovesana O., Tarantelli T., and Zanazzi P.F., Inorg. Chem. **23**, 3398 (1984)
78. Borer L., Thalken L., Ceccarelli C., Glick M., Zhang J.H., and Reiff W.M., Inorg. Chem. **22**, 1719 (1983)
79. Reedijk J. and ten Hoedt R.W.M., Recueil, J. Roy. Neth. Chem. Soc. **101**, 49 (1982)
80. Smit J.J., Nap G.M., de Jongh L.J., van Ooijen J.A.C., and Reedijk J., Physica **97B**, 365 (1979)
81. Rietmeijer F.J., de Graaff R.A.G., and Reedijk J., Inorg. Chem. **23**, 151 (1984)
82. Hay P.J., Thibeault J.C., and Hoffmann R., J. Am. Chem. Soc. **97**, 4884 (1975)
83. Kahn O. and Charlot M.F., Nouv. J. Chim. **4**, 567 (1980)
84. Kahn O., Inorg. Chim. Acta **62**, 3 (1982)
85. Kahn O., Galy J., Tola P., and Coudanne H., J. Am. Chem. Soc. **100**, 3931 (1978)
86. Kahn O., in: Magneto-Structural Correlations in Exchange Coupled Systems, Edited by Willett, R.D., Gatteschi, D., and Kahn, O., NATO ASI Series, D. Reidel Publ. Co., Dordrecht, 1985
87. Bouchez P., Block R., and Jansen L., Chem. Phys. Lett. **65**, 212 (1979)
88. van Kalkeren G., Schmidt W.W., and Block R., Physica B **97**, 315 (1979)
89. Bominaar E.L. and Block R., Physica B **121**, 109 (1983)
90. Journaux Y., Kahn O., Zarembowitch J., Galy J., and Jaud J., J. Am. Chem. Soc. **105**, 7585 (1983)
91. Beswick J.R. and Dugdale D.E., J. Phys. C (Solid State) **6**, 3326 (1973); Benson P.C. and Dugdale D.E., J. Phys. C (Solid State) **8**, 3872 (1975)
92. Earnshaw A. and Lewis J., J. Chem. Soc. **1961**, 396
93. Kahn O. and Briat B., Chem. Phys. Lett. **32**, 376 (1975)
94. Sinn E., Coordin. Chem. Rev. **5**, 313 (1970)
95. Hursthouse M.B., Laffey M.A., Moore P.T., New D.B., Raithby P.R., and Thornton P., J. Chem. Soc. Dalton **1982**, 307
96. Ginsberg A.P., Martin R.L., and Sherwood R.C., Inorg. Chem. **7**, 932 (1968)
97. Boyd P.D.W. and Martin R.L., J. Chem. Soc. Dalton **1979**, 92
98. Chang S.C. and Jeffrey G.A., Acta Cryst. **B26**, 673 (1970)
99. Earnshaw A., Figgis B.N., and Lewis J., J. Chem. Soc. (A) **1966**, 1656 and references therein
100. Uryû N. and Friedberg S.A., Phys. Rev. **140**, A1803 (1965)
101. Sorai M., Tachiki M., Suga H., and Seki S., J. Phys. Soc. Japan **30**, 750 (1971)
102. Ferguson J. and Güdel H.U., Chem. Phys. Lett. **17**, 547 (1972); Schenk K.J. and Güdel H.U., Inorg. Chem. **21**, 2253 (1982)
103. Dubicki L., Ferguson J., and Williamson B., Inorg. Chem. **22**, 3220 (1983)
104. See also the recent discussion by Jones D.H., Sams J.R., and Thompson R.C., J. Chem. Phys. **81**, 440 (1984); but also see: Güdel, H.U., J. Chem. Phys. **82**, 2510 (1985)
105. Flood M.T., Marsh R.E., and Gray H.B., J. Am. Chem. Soc. **91**, 193 (1969)
106. Flood M.T., Barraclough C.G., and Gray H.B., Inorg. Chem. **8**, 1855 (1969)
107. Sorai M. and Seki S., J. Phys. Soc. Japan **32**, 382 (1972)
108. Güdel H.U., Hauser U., and Furrer A., Inorg. Chem. **18**, 2730 (1979)
109. Güdel H.U., Furrer A., and Murani A., J. Mag. Mag. Mat. **15–18**, 383 (1980)

110. Güdel H.U., in: Magneto-Structural Correlations in Exchange Coupled Systems, Edited by Willett, R.D., Gatteschi, D., and Kahn, O., NATO ASI Series, D. Reidel Publ. Co., Dordrecht, 1985
111. Güdel H.U. and Hauser U., Inorg. Chem. **19**, 1325 (1980)
112. Lines M.E., Ginsberg A.P., Martin R.L., and Sherwood R.C., J. Chem. Phys. **57**, 1 (1972)
113. Wong H., tom Dieck H., O'Connor C.J., and Sinn E., J. Chem. Soc. Dalton **1980**, 786
114. Dickinson R.C., Baker, Jr. W.A., Black T.D., and Rubins R.S., J. Chem. Phys. **79**, 2609 (1983); Black T.D., Rubins R.S., De D.K., Dickinson R.C., and Baker, Jr. W.A., J. Chem. Phys. **80**, 4620 (1984)
115. Jones D.H., Sams J.R., and Thompson R.C., Inorg. Chem. **22**, 1399 (1983)
116. Jones D.H., Sams J.R., and Thompson R.C., J. Chem. Phys. **79**, 3877 (1983)
117. McCoy B.M. and Wu T.T., (1973) The Two Dimensional Ising Model, Harvard University Press, Cambridge, Mass., (USA)
118. Nakatsuka S., Osaki K., and Uryû N., Inorg. Chem. **21**, 4332 (1982)
119. Bonner J.C., Kobayashi H., Tsujikawa I., Nakamura Y., and Friedberg S.A., J. Chem. Phys. **63**, 19 (1975)

6. Long Range Order.
Ferromagnetism and Antiferromagnetism

6.1 Introduction

We now extend the concept of exchange to include interactions through a three-dimensional (3D) crystalline lattice. The interactions themselves are short-ranged, and probably are important only for the first through fourth nearest neighbors, but the effects are observed over large distances in a sample. Transitions to long-range order from the paramagnetic state are characterized both by characteristic specific heat anomalies and by susceptibility behavior quite different from what has been described above. In other words, the transition from the paramagnetic state to a magnetically ordered state with long-range correlations between the magnetic moments is in fact a phase transition.

If the moments on a given lattice are all aligned *spontaneously* in the same direction, then the ordered state is a ferromagnetic one. No external field is required for this ordering, and in general an external field will destroy a ferromagnetic phase. The spontaneous ordering of spins persists below a certain critical (Curie) temperature, usually called T_c, and the susceptibility obeys a Curie-Weiss law at temperatures well above T_c, where the spins act as a paramagnetic system. One of the interesting facts here is that some of the phenomena associated with the phase transition, such as the specific heat anomaly, are spread over exceedingly small temperature intervals.

Qualitatively, this is what is happening: At high temperatures, the paramagnetic spins are uncorrelated, which means that their relative spin orientations are completely random. That is, a paramagnet is the magnetic analogue of an ideal gas. Just as intermolecular interactions become more important in a gas as the temperature is lowered towards the boiling point or the pressure is increased, magnetic interactions become more important as kT decreases and becomes comparable to the exchange constant J. In a classical model, the spontaneous magnetization in a ferromagnet would be zero above T_c and then increases below T_c to the maximum possible or saturation magnetization at $T=0$.

In reality, short-range correlations among the moments begin to accumulate even above T_c, and this is called short-range order. The magnetization increases from zero as the temperature drops below T_c, but does not reach the saturation value until well below T_c; this will be illustrated below. Thus the increase in order is a continuous function of T (at $T < T_c$) and perfect alignment is achieved only at 0 K.

A ferromagnetically aligned material actually breaks up into microscopic domains; in each domain, all the spins have the same alignment, but each domain is differently oriented than its neighbors. This decreases the free energy of the system. The subject of domains is discussed more thoroughly by Morrish [1].

6.2 Molecular Field Theory of Ferromagnetism [1]

The magnitude of the spontaneous magnetization M of a ferromagnet is very large. The molecular field theory (MFT) assumes then that there is some sort of internal magnetic field H_m which orients the spins, while of course thermal agitation opposes this effect. The temperature T_c is that temperature above which the spontaneous magnetization loses the battle. An estimate of a typical H_m may be obtained by equating the two energies at T_c,

$$g \mathscr{S} \mu_B H_m = k T_c ,$$

and for metallic iron, $T_c \simeq 1000 \, \mathrm{K}$, $g \simeq 2$ and $\mathscr{S} \simeq 1$, and so

$$H_m \simeq 10^{-20}/(2 \times 10^{-23}) \simeq 500 \, \mathrm{T} ,$$

which is larger than any laboratory field. The value of H_m calculated is also larger than any dipole field, and requires the quantum concept of exchange in order to explain it.

Weiss introduced the MFT concept by assuming the existence of an internal field H_m which is proportional to the magnetization,

$$H_m = \lambda M , \tag{6.1}$$

where λ is called the Weiss field constant. Restricting the discussion to temperatures *above* T_c for the moment, assume that the Curie law holds, but now the total magnetic field, H_T, acting on the sample is the sum of H_m and the external field, H_{ext}.

$$H_T = H_{ext} + H_m \tag{6.2}$$

so

$$M/H_T = M/(H_{ext} + \lambda M) = C/T .$$

Multiplying through and rearranging,

$$M[1 - (\lambda C/T)] = C H_{ext}/T \tag{6.3}$$

or

$$\chi = M/H_{ext} = C/(T - \lambda C), \tag{6.4}$$

which is an exact result. When $H_{ext} = 0$, M is not zero at T_c (because of H_m) and so Eq. (6.3) yields under these conditions, $T_c = \lambda C$ and

$$\chi = C/(T - T_c) \tag{6.5}$$

which is the Curie-Weiss Law. Recall that Eq. (6.5) applies only above T_c, and that it is usually written as

$$\chi = C/(T - \theta) \tag{6.6}$$

which suggests that $\theta/T_c = 1$. While Eqs. (6.5) and (6.6) illustrate the close relationship between a Curie-Weiss θ and ferromagnetic exchange, few real materials obey the relationship between θ and T_c exactly.

Now let us turn to the spontaneous magnetization in the ferromagnetic region, $T < T_c$. Recall [Eq. (A1.1)] that (using $\mathscr{J}$ again as the total angular momentum quantum number)

$$M = Ng\mu_B \mathscr{J} B_J(\eta)$$

with

$$\eta = g\mu_B H_T/kT,$$

and writing

$$H_T = H_{ext} + \lambda M,$$

then

$$\eta = g\mu_B(H_{ext} + \lambda M)/kT. \tag{6.7}$$

Let us examine the spontaneous behavior by setting $H_{ext} = 0$. Now, $\eta \to \infty$ as $T \to 0$, and recall that

$$B_J(\eta \to \infty) = 1$$

so that

$$M(T \to 0) = Ng\mu_B \mathscr{J},$$

the maximum M possible, and taking ratios,

$$M(T)/M(T=0) = B_J(\eta). \tag{6.8}$$

But, we also have from Eq. (6.7) that

$$\eta = g\mu_B[\lambda M(T)]/kT$$

when $H_{ext} = 0$, or

$$M(T) = \eta kT/\lambda g\mu_B$$

which gives another expression for the ratio of $M(T)$ and $M(0)$,

$$M(T)/M(0) = \eta kT/Ng^2\mu_B^2 \mathscr{J}\lambda. \tag{6.9}$$

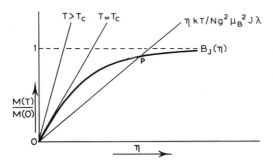

Fig. 6.1. Graphical method for the determination of the spontaneous magnetization at a temperature T

Equations (6.8) and (6.9) provide two independent relationships for $M(T)/M(0)$, and so there must be two solutions at a given temperature. A trivial solution is obtained by setting the ratio equal to zero. The other solution is usually found graphically by plotting both ratios vs. η and finding the intersection. This method is sketched in Fig. 6.1. Notice that for temperatures above T_c, the only intersection occurs at the origin which means that the spontaneous magnetization vanishes. This is consistent with the model. The curve for $T=T_c$ corresponds to a curve tangent to the Brillouin function at the origin, which is a critical temperature, while, at $T<T_c$, the intersections are at $T=0$ and the point P.

Earlier we derived the relation for the Brillouin function as a function of the total angular momentum, $\mathscr{J}$

$$B_J(\eta)=[(\mathscr{J}+1)/3]\eta$$

for $\eta \ll 1$, and so the initial slope of $B_J(\eta)$ vs. η is $(\mathscr{J}+1)/3$. From Eq. (6.9) the initial slope is $kT/Ng^2\mu_B^2\mathscr{J}\lambda$ and so, equating the two initial slopes at $T=T_c$,

$$(\mathscr{J}+1)/3=kT_c/Ng^2\mu_B^2\mathscr{J}\lambda,$$

we see that

$$T_c=Ng^2\mu_B^2\mathscr{J}(\mathscr{J}+1)\lambda/3k \tag{6.10}$$

which predicts an increase in transition temperature with increasing total angular momentum and molecular field constant. Furthermore, substituting into Eq. (6.9), we have

$$M(T)/M(0)=(\eta kT/Ng^2\mu_B^2\mathscr{J})\,[Ng^2\mu_B^2\mathscr{J}(\mathscr{J}+1)]/3kT_c$$
$$=[(\mathscr{J}+1)/3]\,(T/T_c)\eta \tag{6.11}$$

which is a universal curve which should be applicable to all ferromagnets [1], and is illustrated in Fig. 6.2 for several values of $\mathscr{J}$.

Although the exact shape of this curve is somewhat in error [2], the general trend agrees with experiment, and in particular shows that the picture of totally aligned spins which are often drawn are really applicable only as $T\rightarrow 0$.

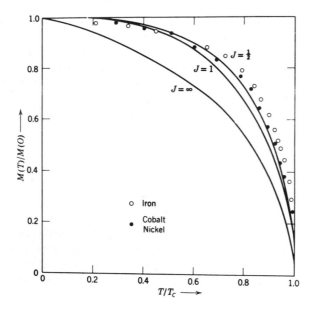

Fig. 6.2. The spontaneous magnetization as a function of temperature. The curves are obtained from theory; the points represent experimental data

6.3 Thermal Effects

An ordered substance – whether it be a ferromagnet or an antiferromagnet – has complete spin order at (and only at) 0 K. As the temperature increases, increasing thermal energy competes with the exchange energy, causing a decrease in the magnetic order, or increasing the disorder. Or, to put it another way, since the Third Law requires that the entropy of a substance be zero at 0 K, as the temperature increases spin correlations will decrease and the entropy of the magnetic system increases, and so there must be a magnetic contribution to the specific heat. This follows from the well-known equation

$$\Delta S_{\mathrm{M}}(T) = \int_0^T c_{\mathrm{p}} \mathrm{d}(\ln T), \tag{6.12}$$

where the symbol S_{M} will be used for the magnetic entropy. Alternatively, look at the system as it is cooled down. As the spin system becomes more correlated, the entropy must decrease. The result is as shown in Fig. 6.3, the specific heat rises smoothly as T_{c} is approached from below and drops discontinuously, as it must by MFT, at $T/T_{\mathrm{c}} = 1$.

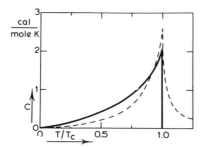

Fig. 6.3. The molecular-field variation of the magnetic specific heat with temperature (full line), compared with the measured values for nickel (broken line), to which the molecular-field results are scaled for the best fit. From Ref. [4]

Since the MFT takes no account of any short range order correlations above T_c, there can be no entropy acquired above T_c. A comparison of the magnetic specific heat of nickel with the MFT curve shows that, though the broad features are similar, the experimental data rise more sharply (for $T < T_c$) and fall more slowly (for $T > T_c$). The magnetic specific heat of nickel above T_c is due primarily to short-range order effects, presaging the onset of long-range order, and these effects are completely neglected by the classical theory. These results are typical of all long-range order magnetic transitions.

6.4 Molecular Field Theory of Antiferromagnetism [1]

The exchange energy is very sensitive to the spacing of the paramagnetic ions, as well as a number of other factors such as the nature of the superexchange path, and most magnetic insulators (i.e., the transition metal compounds which are the subject of this book) do not order ferromagnetically. Rather, neighboring spins are more frequently found to adopt an anti-parallel or antiferromagnetic arrangement. The paramagnetic-antiferromagnetic transition likewise is a cooperative one, accompanied by a characteristic long-range ordering temperature which is usually called the Neel temperature, T_N. However, we adopt the convention of de Jongh and Miedema [2] and use T_c as the abbreviation for a critical temperature, whether the transition be a ferromagnetic or antiferromagnetic one. This is in part acknowledgement that much of the theory in use is independent of the sign of the exchange as well as because of the fact that, as we shall see, short-range ferromagnetic interactions are sometimes quite important for substances which undergo long-range antiferromagnetic ordering.

We may consider the typical antiferromagnet as consisting of two interpenetrating sublattices, with each sublattice uniformly magnetized with spins aligned parallel, but with the spins on one sublattice antiparallel to those on the other. A simple cubic lattice of this type is illustrated in Fig. 6.4, taken from Ref. [4]. This model implies that the important quantity is not the magnetization of the sample as a whole, for this should vanish at low temperatures in a perfect antiferromagnet. Rather it is the magnetization of each sublattice which is important, and this should behave in a fashion similar to that already described for a ferromagnet. The sublattice magnetizations cannot be

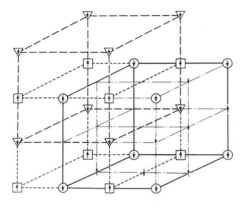

Fig. 6.4. Antiferromagnetism in a simple cubic lattice. The spins of the ions at the corners of the small cubes are arranged so that they form a series of interpenetrating cubic lattices with double the cell size. Three of these large cubes are shown, in heavy outline (ions denoted by circles), dashed line (ions denoted by triangles), dotted line (ions denoted by squares). The ions at the corners of any one large cube have all their spins parallel. From Ref. [4]

measured by a macroscopic technique such as magnetic susceptibility, but require a microscopic procedure. This would include neutron scattering and nuclear resonance experiments [1–4].

The MFT is used as above, but it is assumed that ions on one sublattice (A) interact only with ions on the B sublattice, and vice versa. Then, the field H_A acting on the A sublattice is

$$H_A = H_{ext} + H_{int\,B} = H_{ext} - \lambda M_B$$

while similarly,

$$H_B = H_{ext} - \lambda M_A .$$

The negative signs are used because antiferromagnetic exchange effects tend to destroy alignment parallel to the field. Above T_c, the Curie Law is assumed, and one writes

$$M_A = \tfrac{1}{2}C(H_{ext} - \lambda M_B)/T$$

$$M_B = \tfrac{1}{2}C(H_{ext} - \lambda M_A)/T$$

and the total magnetization is $M = M_A + M_B$. Following procedures [1] similar to those used above, a modified Curie or Curie-Weiss law is derived,

$$\chi = C/(T + \theta'),$$

where θ' is $\lambda C/2$ in this case. Note the sign before the θ' constant, which is opposite to that found for ferromagnets. Again, MFT says that $T_c/\theta' \simeq 1$, which is generally not true experimentally. Note also that any interaction with next-nearest or other neighbors has been ignored.

It is probably worth reminding the reader that the common empirical usage of the Curie-Weiss law writes it as

$$\chi = C/(T - \theta) \tag{6.13}$$

and a plot of χ^{-1} vs. T yields a positive θ for systems with dominant ferromagnetic interactions, and a negative one for antiferromagnetic interactions.

On the other hand, it has been pointed out [5] that writing the Curie-Weiss law as

$$(\chi T)^{-1} = C^{-1}(1 - \theta/T) \tag{6.14}$$

is a more convenient way of determining the Curie constant. A plot of $(\chi T)^{-1}$ vs. T^{-1} is a straight line (when $T \gg \theta$) and yields directly the reciprocal of the Curie constant when we let $T^{-1} \to 0$. (This assumes that corrections have been made for both diamagnetism and temperature independent paramagnetism.) Furthermore, this procedure is a more sensitive indicator of the nature of the exchange interactions.

A list of a representative sample of antiferromagnets will be found in Table 6.1; more extensive tables are found in Refs. [6–8].

Table 6.1. Some antiferromagnetic substances.

Substance	T_c, K	Substance	T_c, K
$NiCl_2$	52	$[Ni(en)_3](NO_3)_2$	1.25
$NiCl_2 \cdot 2H_2O$	7.258	$K_2Cu(SO_4)_2 \cdot 6H_2O$	0.029
$NiCl_2 \cdot 4H_2O$	2.99	$Ni[(NH_2)_2CS]_6Br_2$	2.25
$NiCl_2 \cdot 6H_2O$	5.34	$Co[(NH_2)_2CS]_4Cl_2$	0.92
$MnCl_2$	1.96	$Cs_3Cu_2Cl_7 \cdot 2H_2O$	1.62
$MnCl_2 \cdot 2H_2O$	6.90	$RbFeCl_3 \cdot 2H_2O$	11.96
$MnCl_2 \cdot 4H_2O$	1.62	$CeCl_3$	0.345
$MnCl_2 \cdot 4D_2O$	1.59	$Cs_2[FeCl_5(H_2O)]$	6.43
$CoCl_2$	25	$Rb_2[FeCl_5(H_2O)]$	10.0
CoF_2	37.7	$K_2[FeCl_5(H_2O)]$	14.06
$CoCl_2 \cdot 2H_2O$	17.5	$Cs_2[FeBr_5(H_2O)]$	14.2
$CoCl_2 \cdot 6H_2O$	2.29	$Rb_2[FeBr_5(H_2O)]$	22.9
$CuCl_2 \cdot 2H_2O$	4.35	$Cs_2[RuCl_5(H_2O)]$	0.58
$Rb_2MnCl_4 \cdot 2H_2O$	2.24	$[Ru(NH_3)_5Cl]Cl_2$	0.525
$Rb_2NiCl_4 \cdot 2H_2O$	4.65	$[Cu(en)_3]SO_4$	0.109
$Cs_2MnCl_4 \cdot 2H_2O$	1.86	$Rb_2MnBr_4 \cdot 2H_2O$	3.33
$MnBr_2 \cdot 4H_2O$	2.125	$CsCoCl_3 \cdot 2H_2O$	3.38
$CoBr_2 \cdot 6H_2O$	3.150	$RbCoCl_3 \cdot 2H_2O$	2.975
$CoBr_2 \cdot 6D_2O$	3.225	$NiBr_2 \cdot 6H_2O$	8.30
MnO	117	$FeCl_2 \cdot 4H_2O$	1.1
MnF_2	67.4	$K_3Fe(CN)_6$	0.129
$CsMnCl_3 \cdot 2H_2O$	4.89	$\alpha\text{-}RbMnCl_3 \cdot 2H_2O$	4.89
$CsMnBr_3 \cdot 2H_2O$	5.75	Cs_3MnCl_5	0.601
$Cs_2MnBr_4 \cdot 2H_2O$	2.82	$KMnCl_3 \cdot 2H_2O$	2.74
$CoCs_3Cl_5$	0.282	ReK_2Br_6	15.3
$CoCs_3Br_5$	0.523	ReK_2Cl_6	12.3
$CoRb_3Cl_5$	1.14	$NiBr_2 \cdot 4tu$	2.28
$CoCs_2Cl_4$	0.222	$NiI_2 \cdot 6tu$	1.77
$RbMnF_3$	83.0	$CuSO_4 \cdot 5H_2O$	0.029
$CsNiCl_3$	4.5	$CuSeO_4 \cdot 5H_2O$	0.046
$KNiF_3$	246	$[Cu(NH_3)_4]SO_4 \cdot H_2O$	0.42
Cr_2O_3	308	$NiCl_2 \cdot 6NH_3$	1.45
$CrCl_3$	16.8	$NiI_2 \cdot 6NH_3$	0.31
$CrBr_3$	32	$Co_3La_2(NO_3)_{12} \cdot 24H_2O$	0.181
$CrCs_2Cl_5 \cdot 4H_2O$	0.185	$Ni_3La_2(NO_3)_{12} \cdot 24H_2O$	0.393
$Cr(CH_3NH_3)(SO_4)_2 \cdot 12H_2O$	0.02	$Mn_3La_2(NO_3)_{12} \cdot 24H_2O$	0.230
$Co(NH_4)_2(SO_4)_2 \cdot 6H_2O$	0.084	$Cu_3La_2(NO_3)_{12} \cdot 24H_2O$	0.089
$Ir(NH_4)_2Cl_6$	2.15	$Mn(NH_4)_2(SO_4)_2 \cdot 6H_2O$	0.17
IrK_2Cl_6	3.08	$Gd_2(SO_4)_3 \cdot 8H_2O$	0.182
$IrRb_2Cl_6$	1.85	$GdCl_3 \cdot 6H_2O$	0.185

As with ferromagnetism, the onset of antiferromagnetism causes a sharp anomaly in the magnetic specific heat. It is frequently λ-shaped and is found to occur over a small temperature interval. A typical set of data, in this case for $NiCl_2 \cdot 6H_2O$ [9], is illustrated in Fig. 6.5. The observation of such a λ-shaped curve is adequate proof that a phase transition has occurred, but of itself cannot distinguish whether the transition is to an antiferromagnetic or ferromagnetic state, or whether the transition is in fact

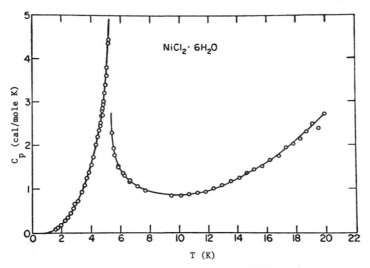

Fig. 6.5. Specific heat of $NiCl_2 \cdot 6H_2O$. From Ref. [9]

magnetic in origin. Further experiments, such as susceptibility measurements, as well as the specific heat measured in an external magnetic field, are required before the correct nature of the phase transition may become known. The reader will also notice the important lattice specific heat contribution in these data.

A general discussion of specific heats of magnetically ordered materials has been provided by Cracknell and Tooke [10].

Now, an antiferromagnetic (AF) substance will generally follow the Curie-Weiss law above T_c. If it is a cubic crystal and if g-value anisotropy and zero-field splittings are unimportant, the susceptibility χ is expected to be isotropic. As T_c is approached from above, single-ion effects as well as short range order effects may begin to cause some anisotropy, and below T_c a distinctive anisotropy is required by theory and found experimentally. Thus, the following discussion concerns only measurements on oriented single crystals, and should not be confused with crystalline field anisotropy effects.

Imagine a perfect antiferromagnetically aligned crystal at $0\,K$. There is some crystallographic direction, to be discovered only by experiment and called the preferred or easy axis, parallel to which the spins are aligned. The easy axis may or may not coincide with a crystallographic axis. A small external magnetic field, applied parallel to the easy axis, can cause no torque on the spins, and since the spins on the oppositely-aligned sublattices cancel each other's magnetization, then $\chi_{\parallel}=0$ at $T=0$. Note that "parallel" in this case denotes only the relative orientation of the axis of the aligned spins and the external field. As the temperature rises, the spin alignment is upset as usual by thermal agitation, the external field tends to cause some torque, and an increasing $\chi_{\parallel}$ is observed.

This molecular field result is illustrated in Fig. 6.7 where it will be seen that this theory assigns T_c to that temperature at which χ has a maximum value. (This is different from the result which will be discussed below.)

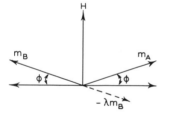

Fig. 6.6. The magnetization of an antiferromagnet when the field is applied at right angles to the spin orientation. Each sublattice rotates through a small angle ϕ, yielding a net magnetic moment

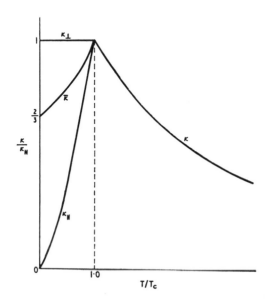

Fig. 6.7. The temperature-dependence of susceptibility from a molecular-field model of an antiferromagnetic crystal. The single-crystal and powder values are indicated. From Ref. [4]

A field applied perpendicular to the easy axis tends to cause spins to line up, but (Fig. 6.6) the net resulting couple on each pair of dipoles should be zero. The net result of this effect is that $\chi_\perp$ remains approximately constant below T_c, especially in the molecular field theory (Fig. 6.7). A typical set of data [3] is illustrated in Fig. 6.8, where the described anisotropy is quite apparent. Notice also that the susceptibility of a powder is described as

$$\chi_{\text{powder}} = \langle \chi \rangle = (\chi_\parallel + 2\chi_\perp)/3 \tag{6.15}$$

and so $\langle \chi \rangle$ at 0 K is $\frac{2}{3}$ its value at the temperature at which it is a maximum. Equation (6.15) also requires that $\langle g^2 \rangle = (g_\parallel^2 + 2g_\perp^2)/3$.

The MFT also yields [11] two other equations of some interest:

$$\theta = 2\mathscr{S}(\mathscr{S}+1)zJ/3k \tag{6.16}$$

and, in the limit of zero field for the usual easy-axis antiferromagnet

$$\chi_\perp(T=0) = Ng^2\mu_B^2/4zJ . \tag{6.17}$$

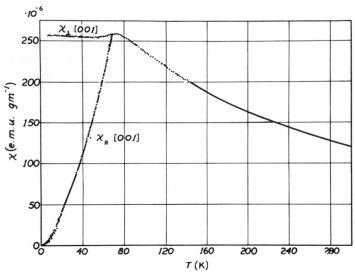

Fig. 6.8. The susceptibility of MnF$_2$ parallel and perpendicular to the [001] axis of the crystal (data of S. Foner). From Ref. [3]

In Eq. (6.16), θ is the Curie-Weiss constant, $\mathscr{S}$ is the spin, J/k the exchange constant in Kelvins, and z the magnetic coordination number of a lattice site. The parameters in Eq. (6.17) have been defined, and the equation applies to $\chi_\perp$ at $T=0\,\mathrm{K}$. Both relationships show the intimate correlation between the observable parameters and the exchange constant.

It is important to distinguish the ordering temperature, T_c, from the temperature at which a maximum occurs in the susceptibilities. They are not the same, contrary to the molecular field result illustrated in Fig. 6.7. Fisher [12] showed that the temperature variation of the specific heat $c(T)$ of an antiferromagnet is essentially the same as that of the temperature derivative of the susceptibility. He established the relation

$$c(T) = A(\partial/\partial T)\,[T\chi_\parallel(T)], \tag{6.18}$$

where the constant of proportionality A is a slowly varying function of temperature. This expression implies that any specific heat anomaly will be associated with a similar anomaly in $\partial(T\chi_\parallel)/\partial T$. Thus the specific heat singularity (called a λ-type anomaly) normally observed at the antiferromagnetic transition temperature T_c (see below) is associated with a positively infinite *gradient* in the parallel susceptibility at T_c. The maximum in $\chi_\parallel$ which is observed experimentally in the transition region must lie somewhat *above* the actual ordering temperature, and the assignment of T_c to T_{max} by the MFT is due, once again, to the exclusion of short-range order effects. The exact form of Eq. (6.18) has in fact been substantiated by careful measurements [13] on CoCl$_2 \cdot 6\,\mathrm{H_2O}$ and on MnF$_2$ [14], and agrees with all other known measurements.

This may be an appropriate place to point out that there are some qualitative correlations between magnetic dilution and T_c that are interesting. Anhydrous metal compounds order at higher temperatures, generally, than do the hydrates; the metal atoms are usually closer together in compounds such as $NiCl_2$ than in the several hydrates, and water usually does not furnish as good superexchange paths as do halogen atoms. The first four entries in Table 6.1 are in accord with the suggestion, though the lack of an exact trend is clear. The ordering temperature will clearly increase with increasing exchange interaction, but there are no simple models available which suggest why, for example, T_c for $NiCl_2 \cdot 6H_2O$ should be higher than that for $NiCl_2 \cdot 4H_2O$. The determining factor is the superexchange interaction, and this depends on such factors as the ligands separating the metal atoms, the distances involved, as well as the angles of the metal-ligand-metal exchange path. The single-ion anisotropy is also a contributing factor.

Anhydrous $MnCl_2$ has an unusually low transition temperature for such a magnetically concentrated material.

The molar entropy change associated with any long-range spin ordering is always $\Delta S_M = R \ln(2\mathscr{S} + 1)$, where we may ignore Schottky terms for the moment. Once the magnetic specific heat is known, it may be integrated as in Eq. (6.12) in order to obtain the magnetic entropy. The full $R \ln(2\mathscr{S} + 1)$ of entropy is never acquired between $0 < T < T_c$ because, although long-range order persists below T_c, short-range order effects always contribute above T_c. Calculations [11] for $\mathscr{S} = \frac{1}{2}$ face-centered cubic lattices in either the Ising or Heisenberg model, for example, show that respectively 14.7 and 38.2% of the total entropy must be obtained above T_c. As an example, the entropy change [9] for magnetic ordering in $NiCl_2 \cdot 6H_2O$ is illustrated in Fig. 6.9; only 60% of the entropy is acquired below T_c.

As was mentioned in Chap. 3, the magnetic ordering specific heat contribution follows a T^{-2} high temperature behavior. In fact [11], the relationship is

$$cT^2/R = 2\mathscr{S}^2(\mathscr{S} + 1)^2 zJ^2/3k^2 \tag{6.19}$$

which again allows an estimation of the exchange parameter, J/k.

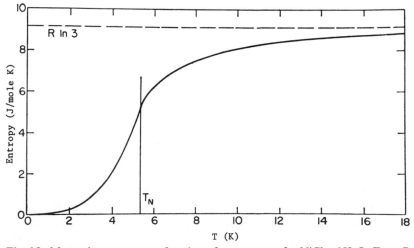

Fig. 6.9. Magnetic entropy as a function of temperature for $NiCl_2 \cdot 6H_2O$. From Ref. [9]

6.5 Ising, XY, and Heisenberg Models [2, 11]

The molecular field theory offers a remarkably good approximation to many of the properties of three-dimensionally ordered substances, the principle problem being that it neglects short-range order. As short-range correlations become more important, say as the dimensionality is reduced from three to two or one, molecular field theory as expected becomes a worse approximation of the truth. This will be explored in the next chapter.

In order to discuss exchange at the atomic level, one must introduce quantum-mechanical ideas, and as has been alluded to already, we require the Hamiltonian

$$H = -2J\Sigma S_i \cdot S_j. \tag{6.20}$$

The mathematical complexities of finding an exact solution of this Hamiltonian for a three-dimensional lattice which will allow one to predict, for example, the shape of the specific heat curve near the phase transition, are enormous, and have to date prevented such a calculation. On the other hand, the careful blending of experiment with numerical calculations has surely provided most of the properties we require [2]. For example, the specific heats of *four* isomorphic copper salts, all ferromagnets, are plotted on a universal curve in Fig. 6.10. It is clear [2, 15] that the common curve is a good approximation of the body-centered cubic (b.c.c.) Heisenberg ferromagnet, spin $\mathscr{S} = \frac{1}{2}$, whose interactions are primarily nearest neighbor.

The first approximation usually introduced in finding solutions of Eq. (6.20) is to limit the distance of the interactions, most commonly to nearest neighbors. The

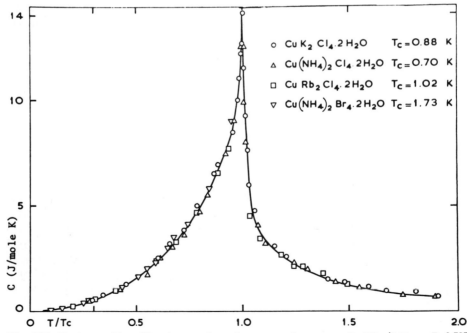

Fig. 6.10. Heat capacities of four isomorphous ferromagnetic copper salts ($\mathscr{S} = \frac{1}{2}$). From Ref. [2]

dimensionality of the lattice is also of special importance, as will also be discussed in the next chapter; in particular, the behavior of the various thermodynamic quantities changes more between changes of the lattice dimensionality (1, 2, or 3) than they do between different structures (say, simple cubic, face-centered cubic, body-centered cubic) of the same dimensionality. The last choice to be made, and again a significant one, is the choice of magnetic model or approximation to be investigated. There are three limiting cases that have been extensively explored. Two of these, the Ising and Heisenberg models, have been introduced already; the third, the XY model, is also of some importance. Expanding Eq. (6.20),

$$\boldsymbol{H} = -2\Sigma\{J_{xy}(S_{ix}S_{jx} + S_{iy}S_{jy}) + J_z S_{iz}S_{jz}\} \tag{6.21}$$

the Heisenberg model lets $J_{xy} = J_z$. When J_{xy} is zero, the Ising model obtains, and the XY model is found when J_z is set equal to zero. There are experimental realizations of each of these limiting cases, as well as of a number of intermediate situations.

Isotropic interactions are required in order to apply the Heisenberg model, and this suggests first of all that the metal ions should reside in sites of high symmetry. In veiw of all the possible sources of anisotropy – thermal contraction on cooling, inequivalent ligands, low symmetry lattice, etc. – it is surprising that there are a number of systems which do exhibit Heisenberg behavior. Clearly, single ion anisotropy, whether it be caused by zero-field splittings or g-value anisotropy, must be small, and this is why the S-state ions Mn^{2+}, Fe^{3+}, Gd^{3+}, and Eu^{2+}, offer the most likely sources of Heisenberg systems. Recall that g-value anisotropy is related to spin-orbit coupling, which is zero, to first order, for these ions. Not every compound of each of these ions will necessarily provide a realization of the Heisenberg model. For example, $MnBr_2 \cdot 4H_2O$ is much more anisotropic than the isostructural $MnCl_2 \cdot 4H_2O$. Copper(II) offers the next-best examples of Heisenberg systems that have been reported because the orbital contribution is largely quenched. Being an $\mathscr{S} = \frac{1}{2}$ system, there are no zero-field splittings to complicate the situation, and the g-value anisotropy is typically not large. For example, $g_\parallel = 2.38$ and $g_\perp = 2.06$ for $K_2CuCl_4 \cdot 2H_2O$. Trivalent chromium and nickel(II) are also potential Heisenberg ions, though only in those compounds where the zero-field splittings are small compared to the exchange interactions.

Ising ions require large anisotropy, and since the magnetic moment varies with the g-value, it is possible to take g-value anisotropy as a reliable guide to finding Ising ions. The large anisotropy in g-values of octahedral cobalt(II) was mentioned in Chap. 2, and this ion provides some of the best examples of Ising system. Similarly, tetrahedral cobalt(II) can be highly anisotropic when the zero-field splitting is large, and forms a number of Ising systems. Thus, Cs_3CoCl_5 has $2D/k = -12.4\,K$, and with $T_c = 0.52\,K$, this compound follows the Ising model [16]. On the other hand, if the zero-field splitting of tetrahedral cobalt(II) is small compared to the exchange interactions, then it should be a Heisenberg ion! If, however, the zero-field splitting is large and positive, then tetrahedral cobalt(II) may by an XY-type ion. This will be discussed in more detail in Sect. 6.13.

Because of zero-field splittings, Dy^{3+} is also frequently an Ising ion [2]. The presence of anisotropy is more important than the cause of it in a particular sample.

Table 6.2. Critical entropy parameters of theoretical 3D Ising models, $\mathscr{S}=\frac{1}{2}$, and their experimental approximants. The numbers in parentheses refer to the number of nearest magnetic neighbors.

Compound or model	$J/k(K)$	$T_c(K)$	T_c/θ	S_c/R	$(S_\infty - S_c)/R$	$(S_\infty - S_c)/S_c$
Ising, diamond (4)			0.6760	0.511	0.182	0.356
Ising, s.c. (6)			0.7518	0.5579	0.1352	0.2424
Ising, b.c.c. (8)			0.7942	0.5820	0.1111	0.1909
Ising, f.c.c. (12)			0.8163	0.5902	0.1029	0.1744
MFT (∞)			1.000	0.693	0.000	0.000
Rb_3CoCl_5	-0.511	1.14	0.74	0.563	0.137	0.24
Cs_3CoCl_5	-0.222	0.52	0.79	0.593	0.106	0.18
$[Fe(C_5H_5NO)_6](ClO_4)_2$	-0.32	0.719	–	0.55	0.19	0.34
$DyPO_4$	-2.50	3.390	0.678	0.505	0.185	0.37
$Dy_3Al_5O_{12}$	-1.85	2.54	0.68	0.489	0.204	0.42
$DyAlO_3$	$\simeq -2$	3.52	0.62	0.521	0.172	0.33

The way to characterize the magnetic model system to which a compound belongs is to compare experimental data with the calculated behavior. For example, since some short-range order above T_c presages even three-dimensional ordering, it is interesting to observe that the amount of that short-range order depends on the nature of the lattice, as well as the magnetic model system. Thus, the entropy change $\Delta S_M/R = \ln(2\mathscr{S}+1)$ may be 0.693 for $\mathscr{S}=\frac{1}{2}$, but only $0.5579/0.693 \times 100 = 80.5\%$ of this entropy change is acquired below T_c for an Ising simple-cubic lattice. This amount is called the critical entropy. Table 6.2, largely taken from the extensive article [2] of de Jongh and Miedema, reports some of the critical properties of 3D Ising models and compounds. The total entropy to be acquired by magnetic ordering for $\mathscr{S}=\frac{1}{2}$ system is $S_\infty = 0.693R$, and S_c indicates the amount of entropy acquired below T_c. One fact immediately apparent from the calculations is that some 15–25% of the entropy must appear as short-range entropy, above T_c. Increasing the spin $\mathscr{S}$ of a system increases the total entropy change of the system from $T=0$ to $T=\infty$. Nearly all this increase in entropy takes place in the region below the critical temperature.

The isomorphous compounds Cs_3CoCl_5 and Rb_3CoCl_5 provide two of the best examples of three-dimensional Ising systems [2, 16, 17]. The structure [18], shown in Fig. 6.11, consists of isolated tetrahedral CoX_4^{2-} units, along with extra cesium and chloride ions. All the magnetic ions are equivalent, and although the crystal is tetragonal, each Co ion has six nearest Co neighbors in a predominantly simple-cubic environment. As has been pointed out above, $2D/k$ is large and negative in the cesium compound, and appears [17] to be negative and even larger in magnitude in the rubidium compound. Thus, only the $|S_z\rangle = |\pm\frac{3}{2}\rangle$ states are thermally populated and need be considered, $g_\perp = 0$, and the two systems meet the requirements of being Ising lattices with effective $\mathscr{S}=\frac{1}{2}$. The ordering temperatures are 0.52 K (Cs) and 1.14 K (Rb), and the specific heats of both systems are plotted in Fig. 6.12 on a reduced scale. (The Schottky contribution lies at much higher temperatures, and the lattice contribution is essentially zero at these temperatures.) The curve is a calculated one, based on series expansion techniques, for a 3D Ising lattice, and perfect agreement is obtained for the

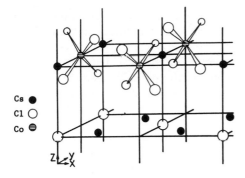

Cs ●
Cl ○
Co ◒

Fig. 6.11. Crystal structure of Cs_3CoCl_5. From Ref. [18]

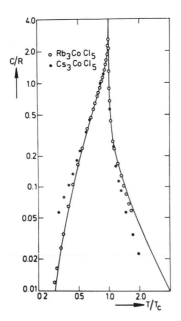

Fig. 6.12. Heat capacities of $CoRb_3Cl_5$ and $CoCs_3Cl_5$ compared with the theoretical prediction for the simple cubic Ising model. From Ref. [2]

rubidium system. The lower ordering temperature of the cesium compound causes dipole-dipole interactions (Sect. 6.6) to be more important, which probably causes the small discrepancies; another way of putting this is that in the cesium compound, a larger effective number of neighbors must be assumed, which suggests that it may be closer to a b.c.c. Ising system.

Another three-dimensional Ising system which has recently been studied is $[Fe(C_5H_5NO)_6](ClO_4)_2$, where the ligand is pyridine N-oxide. Because of large crystal field splittings, the ^5D state is split such that there is a doublet ground state with the lowest lying excited state some 154 K higher in energy. The ground state doublet is very anisotropic, with $g_\parallel \simeq 9.0$ and $g_\perp \simeq 0.6$, and so Ising behavior was anticipated. This was substantiated by specific heat [19] and susceptibility measurements [20]. Indeed, the system is so anisotropic ($\chi_\parallel$ so much greater than $\chi_\perp$) that the susceptibility of the polycrystalline powder could be taken as characteristic of $\chi_\parallel$ alone.

Let us turn now to the Heisenberg model, in Table 6.3, which is also taken from de Jongh and Miedema [2]. Calculations for the Heisenberg model are more difficult than those for the Ising model, and thus only $\mathscr{S} = \frac{1}{2}$ and $\mathscr{S} = \infty$ are listed, the latter case corresponding to a classical spin. The experimental examples listed are those whose anisotropy amounts to 1% or less. Notice immediately that S_c/R is smaller for the Heisenberg model than for the Ising model, which says that short-range ordering is more important in the Heisenberg system. For a given spin $\mathscr{S}$, the tail of the specific heat curve above T_c is nearly three times as large as for the Ising model. The ratio T_c/θ also indicates this. Furthermore, short-range order effects are also enhanced by the lowering of the spin, $\mathscr{S}$. Incidentally, $\mathscr{S} = \infty$ corresponds to the classical limit for spin and is often approximated by spin $\mathscr{S} = \frac{5}{2}$ or $\frac{7}{2}$ systems.

From the experimentalists' point of view, the problem with the Heisenberg model is to find materials that are sufficiently isotropic to warrant application of the available theory. A cubic crystal structure, for example, is obviously a preferable prerequisite, yet a material that is of high symmetry at room temperature, which is where most crystallographic work is done, may undergo a crystallographic phase transition on going to the temperature where the magnetic effects become observable. Single ion anisotropies must of course be as small as possible, and in this regard it is of interest that $KNiF_3$ is a good example of a Heisenberg system. With a spin $\mathscr{S} = 1$, zero-field splittings may be anticipated, but in this particular case the salt assumes a cubic perovskite structure and exhibits very small anisotropy. The high transition temperature, 246 K, on the other hand, causes difficulty in the accurate determination of such quantities as the magnetic specific heat. The compound $RbMnF_3$ is isomorphous and also isotropic but then the high spin ($\frac{5}{2}$) makes difficult the theoretical calculations for comparison with experimental data.

The specific heat of four Heisenberg ferromagnets has already been referred to and shown in Fig. 6.10. The compounds have tetragonal unit cells, but the c/a ratios are close to 1. The difficulty in applying the Heisenberg model here lies with the problem that nearest-neighbor interactions alone do not fit the data; rather, second neighbors also have to be included. In fact, recent work [21] suggests that as many as seventeen equivalently interacting magnetic neighbors need to be considered.

The compound $CuCl_2 \cdot 2H_2O$ serves as a good example of the 3D Heisenberg $\mathscr{S} = \frac{1}{2}$ antiferromagnet, and is of further interest because so many other copper compounds seem to be ferromagnetic. The crystal has a chain-like structure of copper atoms bridged by two chlorine atoms, and this has led some investigators to try to interpret the properties of the compound as if it were a magnetic chain as well (cf. Chap. 7). For example, 33% of the spin entropy appears above T_c. De Jongh and Miedema [2] have argued, however, that pronounced short-range order is characteristic of a 3D Heisenberg magnet, and that theory predicts in this case as much as 38% of the entropy should be obtained above T_c.

There are relatively few 3D XY antiferromagnets known, and they have all been found in the past few years. They contain octahedral cobalt, have effective spin $\mathscr{S} = \frac{1}{2}$, and all order below 1 K. The molecules are $[Co(C_5H_5NO)_6]X_2$, with $X = ClO_4^-$, BF_4^-, NO_3^-, and I^- [22–26] and $CoX_2 \cdot 2$ pyrazine [27, 28], with $X = Cl^-$ or Br^-. The ligand C_5H_5NO is pyridine N-oxide. As discussed earlier (Sect. 2.5), octahedral cobalt exhibits unusual magnetic anisotropy. When $g_\perp \gg g_\parallel$, then XY magnetic ordering may be found. A uniaxial crystal structure enhances the possibility of exhibiting XY-like

Table 6.3. Critical entropy parameters of theoretical 3D Heisenberg models. The nearest-neighbor $\mathscr{S}=\tfrac{1}{2}$ and $\mathscr{S}=\infty$ models are listed. The values refer to ferromagnets. In case of T_c/θ the values for antiferromagnets ($\mathscr{S}=\tfrac{1}{2}$) have been added (minus sign). For references to the experimental data see the text or Ref. [2].

Model or Compound	$\mathscr{S}$	T_c(K)	J/k(K)	T_c/θ	S_c/R	$(S_\infty - S_c)/R$	$(S_\infty - S_c)/S_c$
Heisenberg, s.c. (z=6)	$\tfrac{1}{2}$			0.56 (+)	0.43	0.26	0.60
				0.64 (−)			
Heisenberg, b.c.c. (z=8)	$\tfrac{1}{2}$			0.63 (+)	0.45	0.24	0.53
				0.70 (−)			
Heisenberg, f.c.c. (z=12)	$\tfrac{1}{2}$			0.67 (+)	0.46	0.23	0.50
				0.72 (−)			
Heisenberg, s.c. (z=6)	∞			0.72 (+)		0.42	
Heisenberg, b.c.c. (z=8)	∞			0.77 (+)		0.34	
Heisenberg, f.c.c. (z=12)	∞			0.79 (+)		0.31	
Ferromagnets							
$(NH_4)_2CuCl_4 \cdot 2H_2O$	$\tfrac{1}{2}$	0.701	0.23	0.77			
$K_2CuCl_4 \cdot 2H_2O$	$\tfrac{1}{2}$	0.877	0.30	0.74			
$Rb_2CuCl_4 \cdot 2H_2O$	$\tfrac{1}{2}$	1.02			0.47	0.22	0.46
$(NH_4)_2CuBr_4 \cdot 2H_2O$	$\tfrac{1}{2}$	1.83	0.60	0.74			
$Rb_2CuBr_4 \cdot 2H_2O$	$\tfrac{1}{2}$	1.87	0.63	0.74			
EuO	$\tfrac{7}{2}$	69		0.80			
EuS	$\tfrac{7}{2}$	16.4		0.81			
Antiferromagnets							
$CuCl_2 \cdot 2H_2O$	$\tfrac{1}{2}$	4.36	?	?	0.43	0.23	0.54
NdGaG	$\tfrac{1}{2}$	0.516	~ -0.34	~ 0.76	0.46	0.26	0.58
SmGaG	$\tfrac{1}{2}$	0.967	~ -0.60	~ 0.81	0.42	0.27	0.65
$KNiF_3$	1	246	−44	0.72			
$RbMnF_3$	$\tfrac{5}{2}$	83.0	−3.40	0.70			
MnF_2	$\tfrac{5}{2}$	67.33	−1.76	0.79	1.53	0.26	0.17

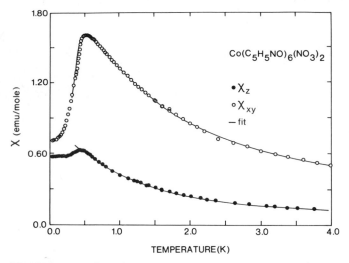

Fig. 6.13. χ_{xy} and χ_z for $[Co(C_5H_5NO)_6](NO_3)_2$, an $\mathscr{S}=\frac{1}{2}$, three-dimensional XY antiferromagnet. From Ref. [26]

character; the pyridine N-oxide compounds listed above are rhombohedral, and the pyrazine compounds are tetragonal. Lowering the symmetry to monoclinic, for example, tends to destroy the equivalence between the X- and Y-directions. This was observed when pyridine N-oxide was replaced as a ligand by γ-picoline N-oxide; the compound $[Co(\gamma\text{-picoline N-oxide})_6](ClO_4)_2$ has a low symmetry coordination sphere and behaves as an Ising system [29, 30].

The large anisotropy of the XY model may be illustrated by an examination of the susceptibilities of $[Co(C_5H_5NO)_6](NO_3)_2$ [26]. These are illustrated in Fig. 6.13, where it may be seen that $\chi_{xy} > \chi_z$ throughout. The susceptibility χ_{xy} is defined with the measuring field in the xy plane perpendicular to the z-axis of the crystal. No anisotropy within the xy-plane was observable. Thus the XY model may be said to represent an easy plane of magnetization.

The critical parameters for an XY system are listed in Table 6.4, along with the available experimental data.

Table 6.4. Critical entropy parameters of the XY model.[a]

	T_c, K	S_c/R	$(S_\infty - S_c)/R$
s.c. XY	–	0.44	0.25
$[Co(C_5H_5NO)_6](BF_4)_2$	0.357	0.45	0.25
$[Co(C_5H_5NO)_6](ClO_4)_2$	0.428	0.47	0.24
$[Co(C_5H_5NO)_6](NO_3)_2$	0.458	0.44	0.25
$[Co(C_5H_5NO)_6]I_2$	0.500	0.42	0.26

[a]   Taken from Ref. [25].

6.6 Dipole-Dipole Interactions

We have so far only considered exchange as the source of interactions between metal ions. While this is undoubtedly the main phenomenon of interest, magnetic dipole-dipole interactions can be important in some instances, particularly at very low temperatures. Weiss showed long ago that dipole-dipole forces cannot cause ordering at ambient temperatures.

Writing a magnetic moment as

$$\mu = g\mu_B S$$

then the classical Hamiltonian for dipole-dipole interaction is of the form

$$H_{d-d} = \sum_{\substack{i,j \\ i \neq j}} [\mu_i \cdot \mu_j / r_{ij}^3 - 3(\mu_i \cdot r_{ij})(\mu_j \cdot r_{ij})/r_{ij}^5], \tag{6.22}$$

where r_{ij} is the vector distance between the magnetic atoms i and j. Since the spin of an ion enters Eq. (6.22) as the square, this effect will be larger, the larger the spin. Notice that solution of this Hamiltonian is an exact, classical type calculation, contrary to what has been expressed above in Eq. (6.20). Therefore, given the crystal structure of a material, good estimates of the strength of the interaction can be made. Well-known methods [31] allow the calculation of the magnetic heat capacity. It is important to observe that λ-like behavior of the specific heat is obtained, irrespective of the atomic source of the ordering. The calculation converges rapidly, for the magnetic specific heat varies as r_{ij}^{-6}.

The high-temperature tail of the λ anomaly in the heat capacity again takes the form

$$c/R = b/T^2,$$

where the constant b may be calculated from the above Hamiltonian. Accurate measurements [32] on CMN have yielded a value of b of $(5.99 \pm 0.02) \times 10^{-6} \, K^2$, while the theoretical prediction is $(6.6 \pm 0.1) \times 10^{-6} \, K^2$. The close agreement gives strong support to the dipole-dipole origin of the specific heat, although there is available at present no explanation for a heat capacity less than that attributable to dipole-dipole interaction between the magnetic ions. The crystal structure [33] of CMN shows that it consists of $[Ce(NO_3)_6]_2[Mg(H_2O)_6]_3 \cdot 6\,H_2O$, with the ceriums surrounded by 12 nitrate oxygens at 0.264 nm. These separate the ceriums – the only magnetic ions present – quite nicely, for the closest ceriums are three at 0.856 nm and three more at 0.859 nm. There appears to be no effective superexchange path. More recently T_c was found at 1.9 mK, and the formation of ferromagnetic domains was suggested [34, 35]. CMN is discussed in more detail in Chap. 9.

The f-electrons of the rare earth ions are well-shielded by the outer, filled shells, and therefore are generally not available for chemical bonding. This is the reason that superexchange is generally thought to be weak compared to the dipole-dipole interactions for the salts of these ions.

Magnetic dipole-dipole interaction appears to be the main contributor to the magnetic properties of most rare earth compounds such as $Er(C_2H_5SO_4)_3 \cdot 9\,H_2O$,

$DyCl_3 \cdot 6\,H_2O$, $Dy(C_2H_5SO_4)_3 \cdot 9\,H_2O$ and $ErCl_3 \cdot 6\,H_2O$ [35], and it also affects the EPR spectra of many metal ion systems. This is because the moment or spin enters the calculation as the square, and many of the lanthanides have larger spin quantum numbers than are found with most iron-series ions. While exchange interaction narrows EPR lines, dipole-dipole interactions broaden them, and this is the reason diamagnetic diluents are usually used for recording EPR spectra. Dipole-dipole interactions set up local fields, typically of the order of 5 or 10 mT. With many magnetic neighbors, these fields tend to add randomly at a reference site, and may be considered as a field additional to the external field. Thus, the effective magnetic field varies from site to site, which in turn causes broadening.

The magnetic properties of rare earth compounds are described in Chap. 9.

Lastly, some anisotropy is often observed in the susceptibilities of manganese(II) compounds. The effect is observed at temperatures too high to be caused by zero-field splittings, and anisotropic exchange effects are not characteristic of this ion. Rather, dipole-dipole interactions are usually assigned [36] as the source of the anisotropy, and this is important because of the high spin ($\frac{5}{2}$) of this ion.

6.7 Exchange Effects on Paramagnetic Susceptibilities

A recurring problem concerns the effect of magnetic exchange on the properties of paramagnetic substances. That is, as the critical temperature is approached from above, short-range order accumulates and begins to influence, for example, the paramagnetic susceptibilities. It is relatively easy to account for this effect in the molecular field approximation [37, 38], and this provides a procedure of broad applicability.

Consider a nickel ion, to which Eqs. (2.4), (2.9) would be applied to fit the susceptibilities as influenced only by a zero-field splitting. To include an exchange effect, a molecular exchange field is introduced. This field is given by

$$H_i' = \frac{2zJ}{Ng_i^2\mu_B^2}\chi_i'H_i',$$

with $i = \parallel$ or $\perp$ and where χ_i' is the exchange-influenced susceptibility actually measured and where the external field H_i and the resulting exchange field are in the i direction. For convenience, we assume axial symmetry and isotropic molecular g values. Then, with this additional exchange field existing when there is a measuring field, the measured magnetization in the i direction is given by

$$M = \chi_i(H_i + H_i').$$

But then, since by definition the measured susceptibility is given by

$$\chi_i' = \lim_{H_i \to 0} M_i/H_i$$

the exchange-corrected susceptibility is given by

$$\chi_i' = \frac{\chi_i}{1 - (2zJ/Ng_i^2\mu_B^2)\chi_i}, \qquad i = \parallel, \perp. \tag{6.23}$$

Fits to experimental data thereby allow an evaluation of zJ/k.

More generally, this procedure can be used whenever a theoretical expression for χ is available; if the symmetry is lower than axial, the i directions can refer to any particular set of axes in conjunction with the direction cosines of the molecules with respect to those crystal axes.

6.8 Superexchange

Some of the ideas concerning superexchange mechanisms were introduced in Sect. 5.1 while, in a sense, this whole book discusses the subject. Direct exchange between neighboring atoms is not of importance with transition metal complexes. The idea that magnetic exchange interaction in transition metal complexes proceeds most efficiently by means of cation-ligand-cation complexes is implicit throughout the discussion, but the problem of calculating theoretically the magnitude of the superexchange interaction for a given cation-ligand-cation configuration to a reasonable accuracy is a difficult task. A variety of empirical rules have been developed, and a discussion of these as well as of the theory of superexchange is beyond the purposes of this book. A recent reference to articles and reviews is available [39].

It is of interest, however, to point out recent progress in one direction, and that is an empirical correlation of exchange constants for a $180°$ superexchange path M^{2+}–F–M^{2+} in a variety of 3d metal (M $=$ Mn, Co, or Ni) fluorine compounds [39]. Two series of compounds were investigated, AMF_3 and A_2MF_4, where A $=$ K, Rb, or Tl, and M $=$ Mn, Co, or Ni. Taking exchange constants from the literature, plots of these were made as a function of metal-ligand separation. For the series of compounds for which data are available, it was found that $|J/k|$ has an R^{-n} dependence, where R is the separation between the metal ions, and n is approximately 12. Thus, exchange interactions are remarkably sensitive to the separation between the metal ion centers.

Of further interest is a correlation of exchange constant with metal ion. By fixing R at a constant value or constant bond length for the series of similar compounds, the ratio of J/k values for Mn^{2+}, Co^{2+} (as spin $\frac{3}{2}$) and Ni^{2+} was found as $1:3.6:7.7$. While these numbers are only valid for a particular $(180°)$ configuration in fluoride lattices, they provide a useful rule-of-thumb for other situations.

6.9 Field Dependent Phenomena

All of the magnetic phenomena discussed to this point with the exception of paramagnetic saturation, are assumed to occur at zero external field. Any applied magnetic field has been assumed to be a measuring field, but not one which changes the system. We now discuss such field-dependent magnetic phenomena [40].

6.9.1 Spin Flop

Recall that the direction in a crystal which is parallel to the direction of spin alignment is called the preferred or easy axis. Cooling a paramagnet in zero applied field leads to a phase transition to an antiferromagnetic state at some temperature, T_c. This yields one point on the $H=0$ axis of an H_a vs. T phase diagram (Fig. 6.14) corresponding to the magnetic ordering or Néel temperature, $T_c(H_a=0) \equiv T_c(0)$.

When a field H_a is applied parallel to the preferred axis of spin alignment in an antiferromagnet, it tends to compete with the internal exchange interactions, causing $T_c(H_a)$ to drop to a lower value as H_a increases. Thus a phase boundary between antiferromagnetic (AF) and paramagnetic (P) states begins to be delineated on the H_a–T diagram, as is illustrated in Fig. 6.14.

However, another phenomenon also can occur. When the system is in the antiferromagnetic state, i.e., at $T < T_c(0)$, and the field is again applied parallel to the preferred axis, a different kind of phase transition can occur. This is called spin flop, for when the field reaches a critical value, the moments flop perpendicular to the field (Fig. 6.14). This is then the thermodynamically favored state, and the transition is first order. That is, there is a discontinuity in the magnetization (net alignment of spins) on crossing the AF–SF phase boundary. If the susceptibility of the system is measured at constant temperature as the field is increased, a peak is usually observed on crossing this boundary (Fig. 6.15). As the field reaches high enough value, there is finally a transition from the spin-flop state to the paramagnetic state.

By contrast, the AF–P and the SF–P transitions are continuous or second-order transitions, as is implied by the moving of the moments in the SF phase suggested in

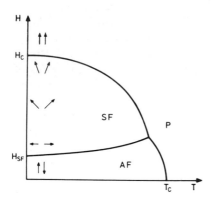

Fig. 6.14. Phase diagram (schematic) for a typical antiferromagnet with small anisotropy with the external field applied parallel to the preferred axis of spin alignment. The point marked T_c is the Néel temperature, $T_c(H_a=0)$. The bicritical or triple point is found where the three phase boundaries meet

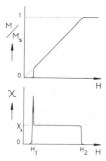

Fig. 6.15. The behavior as a function of field of the isothermal magnetization and differential susceptibility of a weakly anisotropic antiferromagnet, according to the MF theory for a temperature near $T=0\,\mathrm{K}$. The fields H_1 and H_2 correspond to the spin-flop transition field H_{SF} and the transition from the flopped to the paramagnetic phase (H_c) respectively. From Ref. [2]

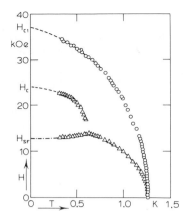

Fig. 6.16. Magnetic phase boundaries of [Ni(en)$_3$](NO$_3$)$_2$. The triangles refer to measurements with the applied field parallel to the easy axis while the circles refer to a perpendicular orientation. From Ref. [41]

Fig. 6.14. The magnetization changes continuously as the boundary is crossed, and the susceptibility at constant temperature only exhibits a change in slope at the phase transition field, as illustrated in Fig. 6.15. The phase diagram [41] of the well-known chemical [Ni(en)$_3$](NO$_3$)$_2$ is illustrated in Fig. 6.16 as an example. This type of behavior is typical of systems with small anisotropy, which is generally caused by zero-field splitting and magnetic dipole-dipole effects. Long-range antiferromagnetic order occurs in this compound at $T_c(0) = 1.25$ K at zero field. The bicritical point is that point where the three phase boundaries meet. Below the bicritical point (0.62 K and 1.39 T in this case), increasing an applied field which is parallel to the easy axis causes the sublattice magnetizations to flop perpendicular to the external field as the AF–SF boundary is crossed. This is a first-order transition and is characterized by a readily observed sharp peak in the ac susceptibility as the boundary is traversed. The other boundaries, antiferromagnetic to paramagnetic and spin flop to paramagnetic, are second-order transitions, and the field-dependent susceptibility exhibits a change in slope from which one can determine T_c. Similar phase diagrams have also recently been observed for a number of compounds such as Cs$_2$[FeCl$_5$(H$_2$O)] [42] and K$_2$[FeCl$_5$(H$_2$O)] [43].

There are a number of reasons for studying such phase diagrams, not the least of which is that the shape of the phase boundaries as they meet at the bicritical point is of intense current theoretical interest. Furthermore, several of the magnetic parameters can be evaluated independently and with greater accuracy than from fitting the zero-field susceptibilities, for example. An anisotropy field, H$_A$, can be defined [2] as the sum of all those (internal) factors which contribute to the lack of the ideal isotropic interactions in a magnetic system. The most important quantities are single-ion or crystal-field anisotropy and dipole-dipole interactions; if zero-field splitting alone contributes, then $g\mu_B H_A = 2|D|(\mathscr{S} - \frac{1}{2})$, where $\mathscr{S}$ is the spin of the magnetic ion. One may also define the exchange field, H$_E$, which is given by $g\mu_B H_E = 2z|J|\mathscr{S}$, and the ratio $\alpha = H_A/H_E$ is a useful relative measure of the ideality of the isotropic exchange interaction. One can show by molecular field theory that H$_{SF}(0)$, the value of the antiferromagnetic to spin-flop transition field extrapolated to 0 K, is given by H$_{SF}(0) = [2H_E H_A - H_A^2]^{1/2}$ and that H$_c(0)$, the field of the spin-flop to paramagnetic transition extrapolated to 0 K, equals $2H_E - H_a$. Thus, observation to low temperatures of these boundaries allows a determination of these quantities. All of these parameters have been evaluated for $[Ni(en)_3](NO_3)_2$, and the anisotropy field of 0.68 T could entirely be assigned to the zero-field splitting of the nickel ion. Indeed, since $T_c(0)=1.25$ K and $|D/k|=1$ K in this system, this method provides an unambiguous determination of the magnitude of the zero-field splitting.

Unfortunately, $H_c(0)$ is often very high, so that it is difficult to measure directly; it is estimated to be about 15 T for $Cs_2[FeCl_5(H_2O)]$, for example. In such cases, one can also make use of the relationship that $\chi_\perp(0)$, the zero-field susceptibility perpendicular to the easy axis extrapolated to 0 K, takes the value $\chi_\perp(0)=2M_s/(2H_E+H_A)$, where $M_s = Ng\mu_B\mathscr{S}/2$ is the saturation magnetization of one antiferromagnetic sublattice. This relationship, in combination with that for $H_{SF}(0)$, can be used to determine H_A and H_E.

If the applied field is perpendicular to the easy axis of spin alignment, then in a typical two-sublattice antiferromagnet there is simply a smooth boundary separating the antiferromagnetic and paramagnetic states. That is, there is of course no spin-flop phase, as illustrated by the highest set of data points in Fig. 6.16. The boundary, extrapolated to 0 K, yields $H_c'(0)$, which is equal to $2H_E+H_A$.

6.9.2 Field Induced Ordering

We return to $Cu(NO_3)_2 \cdot 2.5 H_2O$ to discuss one of the most elegant recent experiments [44] in magnetism, a field-induced magnetic ordering. Recall, from the discussion in the last chapter, that this compound acts magnetically in zero applied field as an assemblage of weakly-interacting dimers. The material cannot undergo long range ordering under normal conditions because the predominant pair-wise interaction offers a path to remove all the entropy as the temperature goes to zero. But, consider the behavior in a field, under the Hamiltonian

$$H = -2JS_1 \cdot S_2 + g\mu_B H \cdot (S_1+S_2)$$

which may be considered isotropic as long as J is itself isotropic. As illustrated in Fig. 6.17, the effect of a field on this system is quite straightforward, and notice in particular the level crossing that occurs as the field reaches the value $H_{1c}=2|J|/g\mu_B$. At this point, the pairs may be thought of in terms of a system with effective spin $\mathscr{S}'=\frac{1}{2}$, in

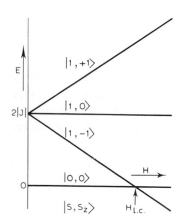

Fig. 6.17. Energy levels for an isolated pair of $\mathscr{S}=\frac{1}{2}$ spins. The quantum numbers indicating the total spin $\mathscr{S}$ and its component $\mathscr{S}_z$ in the direction of the external field are given in brackets

an effective field, $H_{eff} = 0$. If this happens, and it does [44], the salt can undergo a phase transition to long-range order by interpair interaction. The value of H_o actually differs from the above value, for it must be corrected for the interpair exchange interaction. The beauty of this system lies with the fact that H_o need only be about 35 kOe (3.5 T), which is easily accessible, in contrast with most other copper dimers known so far, where H_o must be some 2 orders of magnitude larger.

Van Tol [44] used both NMR and heat capacity measurements to observe the transition to long-range order. The specific heat of polycrystalline $Cu(NO_3)_2 \cdot 2\frac{1}{2} H_2O$ in an external field of 3.57 T is shown as a function of temperature in Fig. 6.18. A λ-like anomaly is observed at 175 mK, indicating the critical temperature; the broad maximum at higher temperatures is due to the short-range order that accumulates in this polymeric material, and is like those effects described in the next chapter. Only about 30% of the magnetic entropy change is in fact found associated with the long-range ordering.

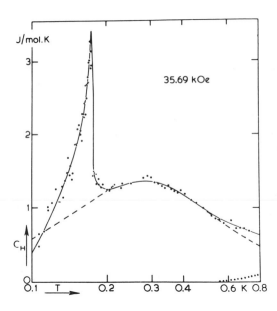

J/mol.K

35.69 kOe

C_H

0.1   T   0.2   0.3   0.4   0.6 K 0.8

Fig. 6.18. Specific heat vs. temperature of $Cu(NO_3)_2 \cdot 2\frac{1}{2} H_2O$ in an external magnetic field of 35.7 kOe (3.57 T). The dashed curve represents the short-range order contribution. The dotted curve is the contribution of the higher triplet levels to the specific heat. From Ref. [44]

Since $H_o = 35.7$ kOe (3.57 T) defines H_{eff} as zero, one last point of interest concerns the effect of a non-zero H_{eff}. As one would expect, in either the positive sense $(H_o > 35.7$ kOe (3.57 T)) or negative sense $(H_o < 35.7$ kOe (3.57 T)), the λ-peak shifts to lower temperature and the broad short-range order maximum shifts to higher temperature. That is, the influence of the inter-dimer interaction decreases as H_{eff} becomes non-zero. For $H_{eff} > \pm 6$ kOe (0.6 T), the λ-peak disappears and there is no more long-range order. These results may be thought of as tracing out the phase diagram of the copper compound in the H–T plane.

The anisotropy in $Cu(NO_3)_2 \cdot 2.5 H_2O$ is small enough so that the experiments described above may be carried out on randomly-oriented polycrystalline material. That there is some anisotropy is indicated by the fact that the transition temperature has a maximum value of 0.22 K for $H_a = 41$ kOe (4.1 T) when it is directed in the ac-plane. When H_a is directed along the b-axis, the maximum value is 0.175 K for $H_a = 37$ kOe (3.7 T). The phase diagram for the field applied parallel to the b-axis [45] is displayed in Fig. 6.19. The nature of the spin structure in the ordered state has been worked out by both NMR methods and neutron diffraction [45].

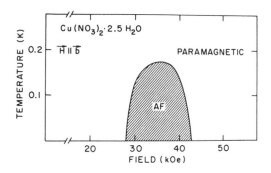

Fig. 6.19. Magnetic phase diagram of $Cu(NO_3)_2 \cdot 2.5 H_2O$, with the applied field parallel to the b-axis. From Ref. [45]

A recent reinterpretation [46] of these results will be discussed in the next chapter.

The interaction in copper nitrate which causes the formation of the singlet states is largely an isotropic one, and thus the level crossing experiments described above could be carried out on powdered samples. This is not true for the experiments which are about to be described, for as we have already seen, zero-field splitting in single-ion systems lead to large magnetic anisotropies.

In a typical paramagnet the Zeeman energy, $g\mu_B HS_z$, causes a small separation of the magnetic energy levels to occur, and it is the differing population among these levels caused by the Boltzmann distribution which gives rise to the normal paramagnetic susceptibility. However, consider the situation in an $\mathscr{S} = 1$ system as provided by either vanadium(III) or nickel(II) in octahedral coordination with energy levels as illustrated in Fig. 6.20. The experiment to be described requires an oriented single crystal. Indeed, we require a uniaxial crystal system so that the local crystal field or molecular (z) axes are all aligned parallel with respect to a crystal axis. This is a particularly stringent condition. It is also necessary that the energy separation D have the sign indicated, which is typical of V(III) compounds but happens randomly with compounds containing Ni(II). Then, in zero external field, typical paramagnetic susceptibilities which have been illustrated (Fig. 2.3) are obtained, with $\chi_\parallel$, (the susceptibility measured with the oscillating field parallel to the z-axis) having a broad maximum, and approaching zero at 0 K. Another way of saying this is that the magnetization for a small parallel field approaches zero at 0 K.

But, now, let a large magnetic field be applied parallel to the unique crystal axis as indicated in Fig. 6.20a [46]. One of the upper levels will descend in energy, and if the separation D is accessible to the available magnetic field, the two lower levels will cross and mix at some level crossing field, H_{lc}, and continue to diverge at higher fields as illustrated. In the absence of magnetic exchange interaction the separation is simply $g\mu_B H_{lc} = D$. In the presence of exchange interaction zJ, the field at which the levels cross becomes $g\mu_B H_{lc} = D + |zJ|$, where z is the magnetic coordination number and J

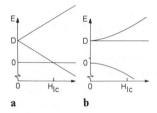

a   **b**

Fig. 6.20. The lowest energy levels of Ni(II) and V(III) in uniaxial crystalline fields as a function of external magnetic field parallel (a) and perpendicular (b) to the principal molecular magnetic axes

measures the strength of the interaction between neighboring ions. We are using a molecular field approximation here in which the individual interactions between the reference magnetic ion and the magnetic neighbors with which it is interacting are replaced by an effective internal magnetic or molecular field. The isothermal magnetization M (net alignment) with the applied field H_a parallel to all the z-molecular axes is $M = Ng\mu_B \langle S_z \rangle$, with

$$\langle S_z \rangle = (e^h - e^{-h})(e^d + e^h + e^{-h})^{-1}, \tag{6.24}$$

where $h = g_\parallel \mu_B H_e / kT$, $d = D/kT$, and k is the Boltzmann constant. The quantity H_e is an effective field $H_e = H_a + \lambda M$, where the molecular field approximation is used once again. Thus, λ is a molecular field constant, with value $\lambda = 2|zJ|/Ng_\parallel^2\mu_B^2$, and N is Avogadro's number. The magnetization calculated from Eq. (6.24) has a sigmoidal shape at low temperatures $(T \ll D/k)$, and a typical data set [48] of magnetization as a function of applied field for a system with large zero-field splitting is illustrated in Fig. 6.21. The level-crossing field is determined by the inflection point, and is about 50 kOe (5 T) for $[Ni(C_5H_5NO)_6](ClO_4)_2$.

Alternatively, the measured isothermal susceptibility, $\chi_T = \partial M/\partial H_a$, increases with field and attains a maximum value at the crossing field [49]. Other systems which have recently been found to behave in this fashion are $[C(NH_2)_3]V(SO_4)_2 \cdot 6H_2O$ [48, 50] and $Cs_3VCl_6 \cdot 4H_2O$ [51]. The analysis of the magnetization data has required the

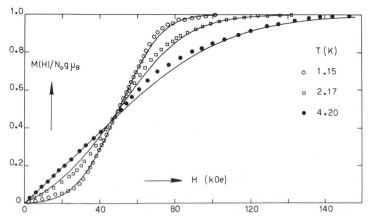

Fig. 6.21. Pulsed-field magnetization curves of $[Ni(C_5H_5NO)_6](ClO_4)_2$. Solid curves have been calculated for isothermal behavior with $D/k = 6.51$ K, $zJ/k = -1.41$ K, and $g_\parallel = 2.32$, using the mean field approximation in Eq. (1). From Ref. [48]

inclusion of the molecular field exchange interaction for all the nickel salts studied to date, but the exchange interaction has been found to be particularly weak in the vanadium salts.

If all the z molecular axes could not be oriented parallel simultaneously to the external field but some were, for example, because of the particular crystal structure, oriented perpendicular to the applied field, then the energy level scheme in Fig. 6.20b would apply. In this case, the energy levels diverge, there is no level crossing, and only paramagnetic saturation occurs, at high fields. This has been observed by applying the external field perpendicular to the unique axis of $Cs_3VCl_6 \cdot 4H_2O$.

Special attention attaches to those systems, $[Ni(C_5H_5NO)_6](ClO_4)_2$ and $[Ni(C_5H_5NO)_6](NO_3)_2$ being the best known examples, in which the exchange interaction is subcritical but moderately strong ($zJ/k \approx -1.5\,K$ for both of these isostructural salts). These rhombohedral materials do not undergo magnetic ordering

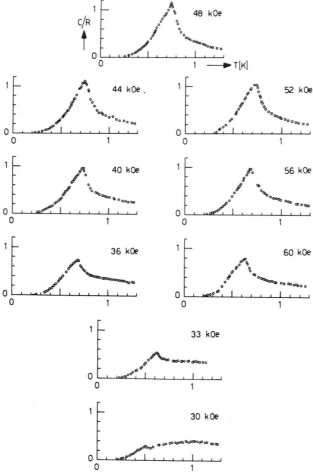

Fig. 6.22. Specific heat of $[Ni(C_5H_5NO)_6](ClO_4)_2$ as a function of temperature in various external fields applied parallel to the principal axis. From Ref. [55]

spontaneously in zero external field even as the temperature approaches 0 K, because D/k is about 6 K for both of them and that is much larger than the subcritical exchange interaction [49, 52]. Thus, the third law of thermodynamics is obeyed by single-ion processes, in which the total population lies in the $m=0$ level at $T=0$ K. But notice that a twofold spin degeneracy (effective spin $\mathscr{S}'=\frac{1}{2}$) has been induced (Fig. 6.20a) at H_{lc} when a field equal to H_{lc} in magnitude has been applied as described above. Since the exchange interactions are not negligible, maintaining the sample at H_{lc} and then lowering the temperature ought to lead to a field-induced magnetic ordering. Such is indeed the case, as has been shown by both susceptibility [48, 53] and specific heat measurements on $[Ni(C_5H_5NO)_6](ClO_4)_2$ [54, 55]. The latter measurements are illustrated in Fig. 6.22. Molecular field theory says that in the field induced ordered state the moments should lie in the plane perpendicular to the external field, that is the xy-plane perpendicular to the external field, which is the xy-plane of the rhombo-hedron. This has indeed been found to be the case [47], and thus these systems are unusual examples of the XY magnetic model.

Another feature of these results is that magnetic ordering is still observed even if the applied field moves away from the level crossing field. The levels diverge (Fig. 6.20a) on either side of H_{lc}, of course, but if exchange is strong enough, ordering can still occur. However, the further H_a is from H_{lc}, the lower the ordering temperature, T_c, will be. This is also illustrated in Fig. 6.22, where the loci of the specific heat maxima are tracing out a phase diagram in the H_a–T plane. The derived results are presented in Fig. 6.23.

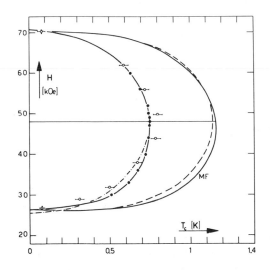

Fig. 6.23. Magnetic phase diagram of $[Ni(C_5H_5NO)_6]$ $(ClO_4)_2$. The points are experimental; the curve through them is calculated, taking into account the third component of the triplet state. The curves labeled MF ignore this level. The system is antiferromagnetic within the hemispherical region and paramagnetic without. From Ref. [55]

A similar phase diagram for tetragonal $[Ni(thiourea)_4Cl_2]$ has recently been measured [56]. The zero-field splitting is larger in this salt than in the pyridine N-oxide systems, and the exchange interactions are also stronger. Thus, the antiferromagnetic phases are shifted to both higher temperatures and stronger fields for this system. The phase boundaries were traced out in this case by the discontinuity in the susceptibility

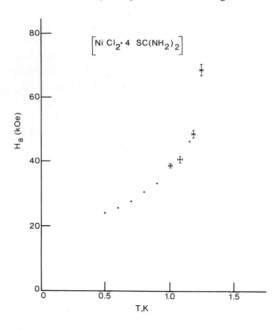

Fig. 6.24. Portion of the phase diagram of [Ni(thiourea)$_4$Cl$_2$]. From [56]

which was measured at constant temperature as the applied field was increased. The data are illustrated in Fig. 6.24.

When an external field is applied to a system with high anisotropy, the phenomenon of metamagnetism is found. This will be discussed in the next chapter.

6.10 Ferromagnets

Among the coordination compounds of the transition metals, many more have been found to order antiferromagnetically than ferromagnetically. Why this is so is not at all clear; perhaps the right compounds have not yet been examined. Furthermore, there seem to be no guides at all for the prediction of what sign an exchange constant will take in the extended lattice which is required for long-range ordering. An unusual instance that illustrates the dependence of the nature of the magnetic ordering upon changes in the superexchange path is provided by measurements [57] on (NEt$_4$)FeCl$_4$. This material is a typical Heisenberg antiferromagnet, ordering at 2.96 K, when it is cooled slowly. If, instead, the substance is cooled rapidly, the crystal undergoes several crystal or structural phase transitions, the nature of which are as yet undefined. Still, there must be small but significant changes in the superexchange paths, for now the material orders as a ferromagnet at 1.86 K! Ferromagnetism is better known among metals such as iron, cobalt and nickel, and many ferromagnetic alloys are known. A number of ferromagnetic compounds ("insulators") have nevertheless been discovered, and so it is important to discuss several of these here. A list of some ferromagnets is provided in Table 6.5.

The distinguishing features of ferromagnetism include the parallel alignment of magnetic moments within the ordered state, and the formation of domains. In the absence of any applied field (including the earth's), there is little reason to distinguish

Table 6.5. Some ferromagnetic insulators.

	T_c, K
$Rb_2CuCl_4 \cdot 2H_2O$	1.02
$K_2CuCl_4 \cdot 2H_2O$	0.88
$(NH_4)_2CuCl_4 \cdot 2H_2O$	0.70
$(NH_4)_2CuBr_4 \cdot 2H_2O$	1.83
$Rb_2CuBr_4 \cdot 2H_2O$	1.87
$FeCl[S_2CN(C_2H_5)_2]$	2.46
$[Cr(H_2O)(NH_3)_5][Cr(CN)_6]$	0.38
$[Cr(NH_3)_6][Cr(CN)_6]$	0.6
$NiSiF_6 \cdot 6H_2O$	0.135
$NiTiF_6 \cdot 6H_2O$	0.14
$NiSnF_6 \cdot 6H_2O$	0.164
$NiZrF_6 \cdot 6H_2O$	0.16
$CrCl_3$	16.8
$CrBr_3$	32.7
CrI_3	68
$GdCl_3$	2.20
$ErCl_3$	0.307
$ErCl_3 \cdot 6H_2O$	0.356
$DyCl_3 \cdot 6H_2O$	0.289
$Dy(OH)_3$	3.48
$Ho(OH)_3$	2.54
$Tb(OH)_3$	3.72
EuO	69

ferromagnetic ordering from antiferromagnetic ordering. In particular, as we saw at the beginning of this chapter, the nature of the specific heat λ-anomaly associated with 3D magnetic ordering does not depend on the sign ($+$, FM; $-$, AF) of the major interaction.

In an *ac* susceptibility measurement, no external field is required. A sample is placed within a coaxial set of coils and a low-frequency signal is applied. The change in mutual inductance (coupling) between the coils is proportional to the magnetic susceptibility. This method has the advantage that the *ac* (measuring) magnetic field imposed is generally only a few Oersteds and can be made very small if necessary.

A susceptibility measurement always requires a non-zero applied field and so the properties of ferromagnetic susceptibilities are different from those of antiferromagnetic susceptibilities. Indeed, as we shall see, the susceptibility changes with the size of the applied field, and we are faced with the problem that the nature of the susceptibility of a ferromagnet depends on how it is measured. This is in part due to the fact that an external field changes the size of the domains. Furthermore, these susceptibilities also depend on the shape and size of a sample and its purity; the latter term includes the presence of lattice defects.

The average field inside a substance, which is the only relevant magnetic field, is usually the same as the applied field when the substance is a paramagnet or simple antiferromagnet. This is not true for a ferromagnet because of what are called demagnetizing effects. With ferromagnets, the net internal magnetic moment is large and acts to repel the applied field. This is described by means of a sample-shape-dependent demagnetizing field which tends to counteract the external field, causing the internal field to differ from the external field.

The Curie-Weiss law may be written as $\chi = (dM/dH_i)_{H=0} = C/(T-T_c)^\gamma$, where H_i is the internal magnetic field and $\gamma = 1$ for the mean field theory and 1.3 to about 2 for real ferromagnets. One sees that χ should diverge as T_c is approached from higher temperatures. The experimentally measured susceptibility may be called χ_m, and is defined as dM/dH_a, where H_a is the applied (measuring) field. (The quantities χ and χ_m are the same whenever demagnetization effects are negligible, such as with a paramagnet or a normal antiferromagnet.) The net magnetization adjusts itself so as to shield out the applied field. Then the internal field differs from the applied field according to

$$H_i = H_a - NM,$$

where M is the magnetization of the sample and N is the demagnetization factor. The factor N depends on the shape of the sample, and is zero, for example, for infinitely long needles. It is $4\pi/3$ for a sphere. Then,

$$\chi_m = \frac{dM}{dH_a} = \frac{dM}{dH_i}\frac{dH_i}{dH_a} = \chi\left(1 - N\frac{dM}{dH_a}\right).$$

Rearranging,

$$\frac{dM}{dH_a} = \frac{\chi}{1 + N\chi},$$

or

$$\chi_m = \frac{1}{1/\chi + N} \tag{6.25}$$

which goes to the constant value of $1/N$ as χ goes to infinity. This is expressed by saying that χ_m is limited by the demagnetization factor of the sample.

We shall consider here only systems which do not exhibit hysteresis, and which are called soft ferromagnets. This means that they can be magnetized reversibly because the domain walls can move freely. Systems with hysteresis behave irreversibly, and the susceptibility is ill-defined.

Consider first a multidomain system, such as is found with a single crystal. Each individual domain is magnetized spontaneously, but since the sample breaks up into a large number of domains the net magnetization is zero when the applied field is zero. The magnetization as a function of applied field is illustrated schematically in Fig. 6.25. At temperature of T_c and above the magnetization increases smoothly as shown. For $T < T_c$, the magnetization increases according to $1/N$ but then there is a break in the curve at the point corresponding to the saturation magnetization. This point increases with decreasing temperature.

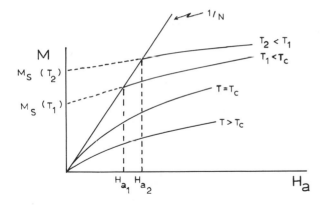

Fig. 6.25. The magnetization vs. applied field, for several temperatures. Both T_1 and T_2 are below T_c, and correspond to the saturation magnetization as indicated at the particular applied fields

A static susceptibility measured at low field, such as an *ac* susceptibility at low frequency and at essentially zero-applied field is measuring dM/dH_a, but since the magnetization is merely increasing with constant slope, the susceptibility remains constant at all temperatures below T_c. The net magnetization of the sample adjusts itself so as to shield out the applied field. Then $dM/dH_a = 1/N$, independent of temperature. The susceptibility [58] of $FeCl(S_2CNEt_2)_2$ in the [101] direction behaves in that fashion, as shown in Fig. 6.26. This is behavior characteristic of soft ferromagnets. Many ferromagnetic materials exhibit similar, though slightly rounded behavior, depending on the freedom of the domain wall movement.

Now consider a single domain particle. This is found with small samples, such as polycrystalline powders; the same effects are also observed when the hysteresis loop opens. A peak is observed in $dM(T)/dH_a$ vs. T. A data set [59] typical for this situation is illustrated for NdH_2 in Fig. 6.27. This kind of behavior occurs when the domain walls are not able to follow the applied *ac* field at all (see Chap. 11). There is also absorption of the signal, that is χ'' increases near T_c, often achieving a maximum at a temperature near T_c.

Since an applied field causes domain walls to move, domains tend to grow larger and align with the external field. Thus an applied field tends to destroy ferromagnetic ordering, and there is no phenomenon akin to spin-flop with an antiferromagnet.

Among the other ferromagnets which have been described, there are several series of interesting compounds. The first is the series $A_2CuX_4 \cdot 2H_2O$, where A is an alkali metal ion and X is chloride or bromide. These materials, which contain the *trans*-$[CuX_4(H_2O)_2]$ units are tetragonal, but their magnetic structure approximates the b.c.c. lattice [2]. The specific heat data, which were presented in Fig. 6.10, are considered to be those of a b.c.c. Heisenberg ferromagnet with mainly nearest-neighbor interactions. There is evidence in the critical behavior, however, that there are substantial further neighbor interactions [2].

A recently discovered series of ferromagnets is listed in Table 6.6, along with their isomorphous congeners which do not order [61]. All the substances listed are isostructural, belong to the space group $R\bar{3}$, and have one molecule in the unit cell. Those materials which order have three features in common: they contain the hexafluoro counterions, they have a negative ZFS, and they order as ferromagnets.

A careful examination of this table reveals several interesting but perplexing things. First, that the sign of the ZFS appears to correlate with the *a*-axis length, changing from

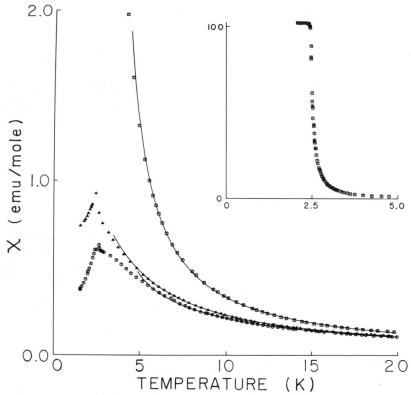

Fig. 6.26. Susceptibility of FeCl(S$_2$CNEt$_2$)$_2$. Circles, triangles, and squares are data along the [010], ($\bar{1}$01), and [101] axes, respectively. The inset shows the susceptibility in the [101] direction at low temperatures. From Ref. [58]

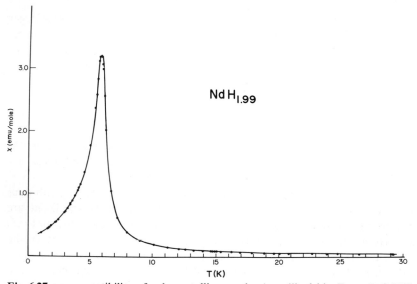

Fig. 6.27. *ac* susceptibility of polycrystalline neodymium dihydride. From Ref. [59]

Table 6.6. Some isostructural nickel salts.

	a (nm)	D/k (K)	T_c (K)	Comment
$NiSiF_6 \cdot 6 H_2O$	0.626	-0.168 ± 0.008	0.135	Ferromagnet
$NiTiF_6 \cdot 6 H_2O$	0.642	-1.66 ± 0.02	0.14	Ferromagnet
$NiSnF_6 \cdot 6 H_2O$	0.652	-2.52 ± 0.02	0.164	Ferromagnet
$NiZrF_6 \cdot 6 H_2O$	0.655	-3.14 ± 0.02	0.16	Ferromagnet
$NiSnCl_6 \cdot 6 H_2O$	0.709	0.59 ± 0.02	–	
$NiPtCl_6 \cdot 6 H_2O$	0.702	0.68	–	
$NiPdCl_6 \cdot 6 H_2O$	0.704	0.61	–	
$NiPtBr_6 \cdot 6 H_2O$	0.723	0.37	–	
$NiPtI_6 \cdot 6 H_2O$	0.782	0.22	–	

negative to positive as the unit cell increases in size. All the ZFS are relatively small, however, so that this correlation may actually result only from a small difference between large terms in the Hamiltonian. Whether or not this is a universal phenomenon remains to be discovered as more data become available on other series of isostructural systems. The correlation just seems too simple! For example, it has been pointed out [61] that |D| varies linearly with the M^{4+} ion radius (as well as the unit cell edge length) in these compounds, but the chemical nature of the M^{4+} ion has also been changed. Furthermore, one could as well correlate the sign and magnitude of D with the chemical nature of the ligands, fluoride vs. the other halides. Most perplexing of all is the fact that fluorides usually provide much the weakest superexchange path when compared with the other halides. Yet in this series of molecules, it is only the fluorides (to date!) which undergo magnetic ordering!

The two series of compounds $(RNH_3)_2MCl_4$ where M is Cu [2, 62] or Cr [62] order ferromagnetically. The chromium compounds order at much higher temperatures than do the copper compounds. Most of the interest in these materials centers on the fact that their major interactions are two-dimensional in nature (Chap. 7). A number of rare earth compounds described in Chap. 9 also order ferromagnetically.

Another example of three-dimensional ordering is provided by $CuL \cdot HCl$, where $H_2L = 2,3$-pyrazinedicarboxylic acid [63]. The crystal structure of this material consists of polymeric chains which lie in a criss-cross fashion. This structure led to the suggestion that the material was a magnetic linear chain, with a small ferromagnetic interaction. Susceptibility measurements down to 6 K yielded a positive Weiss constant. Further measurements down to 1.1 K could not, however, be fit by the calculations for a ferromagnetic Heisenberg linear chain, $\mathscr{S} = \frac{1}{2}$. An excellent fit of the data was obtained with the three-dimensional, (simple cubic), $\mathscr{S} = \frac{1}{2}$, ferromagnetic Heisenberg model. This unexpected result suggested that measurements should be carried out to lower temperatures, and when this was done, two sharp peaks, characteristic of long-range ferromagnetic ordering, were found, at 0.770 and 0.470 K. The behavior is typical of that of a powdered sample of a ferromagnetic material in which demagnetization effects become important.

The existence of two peaks suggests that perhaps a spin-reorientation occurs at the temperature of the lower one. The data and the analysis require a three-dimensional magnetic interaction while only a one-dimensional superexchange path is apparent in the crystal structure. These results are a good reminder that magnetic susceptibility measurements should be made in a temperature region where magnetic exchange makes a significant contribution to the measured quantity if inferences regarding the character of the magnetic exchange are to be made.

6.11 Ferrimagnetism

We have suggested until now that ordered substances have spins which are ordered exactly parallel (ferromagnets) or antiparallel (antiferromagnets) at 0 K. This is not an accurate representation of the true situation, for we are generally ignoring in this book the low energy excitations of the spin systems which are called spin-waves or magnons [4]. Other situations also occur, such as when the antiparallel lattices are not equivalent, and we briefly mention one of them in this section, the case of ferrimagnetism. Most ferrimagnets studied to date are oxides of the transition metals, such as the spinels and the garnets. But there is no apparent reason why transition metal complexes of the kind discussed in this book could not be found to display ferrimagnetism.

Ferrimagnetism is found when the crystal structure of a compound is more complicated than has been implied thus far, so that the magnitudes of the magnetic moments associated with the two AF sublattices are not exactly the same. Then, when the spontaneous anti-parallel alignment occurs at some transition temperature, the material retains a small but permanent magnetic moment, rather than a zero one. The simplest example is magnetite, Fe_3O_4, a spinel. The two chemical or structural sublattices are

　1) iron(III) ions in tetrahedral coordination to oxygen, and

　2) iron(II) and iron(III) ions in equal proportion to octahedral oxygen coordination.

The result is that the inequivalent magnetic sublattices cannot balance each other out, and a weak moment persists below T_c.

6.12 Canting and Weak Ferromagnetism

Certain substances that are primarily antiferromagnetic exhibit a weak ferromagnetism that is due to another physical phenomenon, a canting of the spins [64, 65]. This was first realized by Dzyaloshinsky [66] in a phenomenological study of α-Fe_2O_3, and put on a firm theoretical basis by Moriya [67]. Other examples of canted compounds include NiF_2 [67], $CsCoCl_3 \cdot 2H_2O$ [68], and $[(CH_3)_3NH]CoCl_3 \cdot 2H_2O$ [69]. The compound $CoBr_2 \cdot 6D_2O$ is apparently a canted magnet [70], while the hydrated analogue is not. As we shall see, there are two principle mechanisms, quite different in character, which cause canting, but there is a symmetry restriction that applies equally to both mechanisms. In particular, ions with magnetic moments in a unit cell cannot be related by a center of symmetry if canting is to occur. Other symmetry requirements have been discussed elsewhere [65].

Weak ferromagnetism is due to an antiferromagnetic alignment of spins on the two sublattices that are equivalent in number and kind but not exactly antiparallel. The

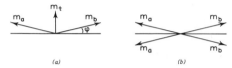

Fig. 6.28. Canting of sublattices: (a) two-sublattice canting which produces weak ferromagnetism; (b) "hidden" canting of a four-sublattice antiferromagnet. Four sublattices may also exhibit "overt" canting, with a net total magnetic moment. From Ref. [64]

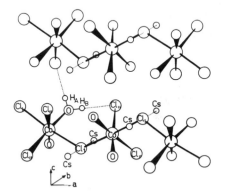

Fig. 6.29. Structure of $CsCoCl_3 \cdot 2H_2O$, according to Thorup and Soling. Only one set of hydrogen atoms and hydrogen bonds is shown. From Ref. [68]

sublattices may exist spontaneously canted, with no external magnetic field, only if the total symmetry is the same in the canted as in the uncanted state. The magnetic and chemical unit cells must be identical. The canting angle, usually a matter of only a few degrees, is of the order of the ratio of the anisotropic to the isotropic interactions. If only two sublattices are involved, the canting gives rise to a small net moment, m_t, and weak ferromagnetism occurs. This is sketched in Fig. 6.28a. If many sublattices are involved, the material may or may not be ferromagnetic, in which case the canting is called, respectively, overt or hidden. Hidden canting is illustrated in Fig. 6.28b, and just this configuration has indeed been found to occur at zero field in $LiCuCl_3 \cdot 2H_2O$ [71]. It has also been suggested that hidden canting occurs in $CuCl_2 \cdot 2H_2O$.

Another example is provided by $CsCoCl_3 \cdot 2H_2O$, whose molecular structure is illustrated in Fig. 6.29. Chains of chloride-bridged cobalt atoms run along parallel to the a-axis, and the compound exhibits a high degree of short-range order [68]. The proposed magnetic structure is illustrated in Fig. 6.30. This spin configuration was obtained from an analysis of the nuclear resonance of the hydrogen and cesium atoms in the ordered state (i.e., at $T < T_c$). Notice that the moments are more-or-less AF aligned along the chains (a-axis), but that they make an angle ϕ with the c-axis of some 15°, rather than the 0° that occurs with a normal antiferromagnet. The spins are ferromagnetically coupled along the c-axis, but moments in adjacent ac-planes are coupled antiferromagnetically. The spins lie in the ac-plane, and this results in a permanent, though small moment in the a-direction.

As the illustration of the structure of $CsCoCl_3 \cdot 2H_2O$ clearly shows, the octahedra along the chain in this compound are successively tilted with respect to each other. It has been pointed out that this is an important source of canting [72], and fulfills the symmetry requirement of a lack of a center of symmetry between the metal ions and

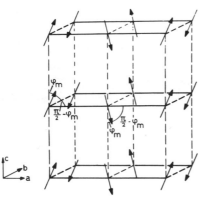

Fig. 6.30. Proposed magnetic-moment array of $CsCoCl_3 \cdot 2\,H_2O$ [68]. All spins lie in the *ac*-plane. The model suggests ferromagnetic coupling along the *c*-axis, antiferromagnetic coupling along the *b*-axis, and essentially antiferromagnetic coupling along the *a*-axis

thus the moments. This, along with the large anisotropy in the interactions, is the source of the canting in $CsCoCl_3 \cdot 2\,H_2O$, $[(CH_3)_3NH]CoCl_3 \cdot 2\,H_2O$, and α-$CoSO_4$.

Now the Hamiltonian we have used repeatedly to describe exchange,

$$H = -2J\Sigma S_i \cdot S_j \tag{6.26}$$

may be said to describe symmetric exchange, causes a normal AF ordering of spins, and does not give rise to a canting of the spins. The Hamiltonian

$$H' = D_{ij} \cdot [S_i \times S_j], \tag{6.27}$$

where D_{ij} is a constant vector, may then be said to describe what is called antisymmetric exchange. This latter Hamiltonian is usually referred to as the Dzyaloshinsky-Moriya (D-M) interaction. This coupling acts to cant the spins because the coupling energy is minimized when the two spins are perpendicular to each other. Moriya [65] has provided the symmetry results for the direction of D, depending on the symmetry relating two particular atoms in a crystal. The more anisotropic the system, the more important canting will be, for D is proportional to (g–2)/g. In $CsCoCl_3 \cdot 2\,H_2O$, with $g_a = 3.8$, $g_b = 5.8$ and $g_c = 6.5$, the required anisotropy is obviously present, and (g–2)/g can be as much as 0.7.

The geometry of β-MnS allows the D-M interaction to arise, and Keffer [73] has presented an illuminating discussion of the symmetry aspects of the problem. Cubic β-MnS has the zinc blend structure. The lattice is composed of an fcc array of manganese atoms interpenetrating an fcc array of sulfur atoms in such a way that every atom of one kind is surrounded tetrahedrally by atoms of the other kind. The local symmetry at one sulfur atom is displayed in Fig. 6.31; if the midpoint of a line

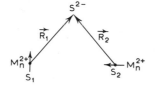

Fig. 6.31. Near-neighbor superexchange in β-MnS. The absence of an anion below the line of cation centers allows a Moriya interaction, with D taking the direction of $R_1 \times R_2$. From Ref. [73]

connecting the manganese atoms were an inversion center, that is, if another sulfur atom were present below the line of cation centers, then the D-M mechanism would not be allowed. As it is, the absence of such an anion, even though the total crystal symmetry is high, allows such a D-M coupling. The direction of D_{12} is normal to the plane of the paper, in the direction of the vector $R_1 \times R_2$, and Keffer finds that the energy associated with the D-M interaction will be minimum when S_1 and S_2 are orthogonal, as in the figure. Furthermore, Keffer goes on to analyze the spin structure of the lattice and concludes that the probable spin arrangement, illustrated in the Figs. 6.32 and 6.33, is not only consistent with the available neutron diffraction results, but is in fact determined primarily by the D-M interaction. Spins lie in an antiferromagnetic array in planes normal to the x-axis. The spin direction of these arrays turns by 90° from plane to plane, the x-axis being a sort of screw axis. Next-neighbor superexchange along the x-axis is not likely to be large, for it must be routed through two intervening sulfur ions. Thus, the D-M interaction alone is capable of causing the observed antiparallel orientation of those manganese ions which are next neighbors along x.

The symmetry aspects of the D-M interaction are also illustrated nicely by the spin structure of α-Fe_2O_3 and Cr_2O_3, as was first suggested by Dzyaloshinsky. The crystal and magnetic structures of these isomorphous crystals will be found in Fig. 6.34, and the spin arrangements of the two substances will be seen to differ slightly. Spins of the ions 1, 2, 3, and 4 differ in orientation only in sign and their sum in each unit cell is equal to zero. In α-Fe_2O_3, $S_1 = -S_2 = -S_3 = S_4$, while in Cr_2O_3, $S_1 = -S_2 = S_3 = -S_4$.

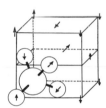

Fig. 6.32. Cubic cell of β-MnS. Only the sulfur in the lowest left front corner is shown. The proposed arrangement of manganese spins is indicated by the arrows. From Ref. [73]

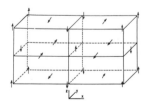

Fig. 6.33. Proposed spin arrangement in cubic β-MnS. From Ref. [73]

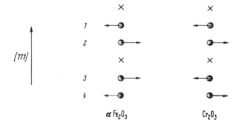

Fig. 6.34. Arrangement of spins along c-axis in unit cells of α-Fe_2O_3 and of Cr_2O_3 (schematic). Points marked $\times$ are inversion centers of the crystal-chemical lattices

Because of the three-fold crystallographic rotation axis, the vector D for any pairs along the trigonal axis will lie parallel to that trigonal axis. There are crystallographic inversion centers at the midpoints on the lines connecting ions 1 and 4, and 2 and 3, so there cannot be anti-symmetrical coupling between these ions; that is, $D_{23} = D_{14} = 0$. The couplings between 1 and 2 and 1 and 3 must be equal but of the opposite sign to those between 3 and 4, and 2 and 4, respectively, because of the glide plane present in the space group to which those crystals belong: $D_{12} = -D_{34} = D_{43}$; $D_{13} = D_{42}$. Thus, the expression for the couplings among the spins in a unit cell is

$$D_{12} \cdot [(S_1 \times S_2) + (S_4 \times S_3)] + D_{13} \cdot [(S_1 \times S_3) + (S_4 \times S_2)]. \qquad (6.28)$$

But, the spin arrangements in these crystals are

$$\alpha\text{-Fe}_2\text{O}_3 : S_1 \| S_4 \quad \text{and} \quad S_2 \| S_3,$$
$$\text{Cr}_2\text{O}_3 : S_1 \| S_3 \quad \text{and} \quad S_2 \| S_4.$$

For the iron compound, the cross products do not vanish, there is an antisymmetrical coupling, and it is therefore a weak ferromagnet. For the chromium compound, each of the cross products in the second term of Eq. (6.28) is zero, and those in the first term cancel each other. Thus, Cr_2O_3 cannot be and is found not to be a weak ferromagnet. The canting in $\alpha\text{-Fe}_2\text{O}_3$ is a consequence of the fact that the spins are arranged normal to the c-axis of the crystal, for if the spins had been arranged parallel to this axis, canting away from [111] alters the total symmetry and could therefore not occur.

The situation with NiF_2 is entirely different, and in fact Moriya shows [65] that the D-M interaction, Eq. (6.27), is zero by symmetry for this substance. But, on the other hand, the nickel(II) ion suffers a tetragonal distortion which results in a zero-field splitting. For the case where there are, say, two magnetic ions in a unit cell and the AF ordering consists of two sublattices, one can write the single-ion anisotropy energies at the two positions as $E_1(S)$ and $E_2(S)$. If it is found that $E_1(S) = E_2(S)$, as would happen if the spin-quantization or crystal field axes of the two ions were parallel, Moriya finds that a canted spin arrangement cannot be obtained because the preferred directions of the two spins are then the same and the AF ordering should be formed in this direction. When $E_1(S) \neq E_2(S)$, however, and the easiest directions for the spins at the two positions are different, canting of the sublattice magnetizations may take place. The crystal and magnetic structures of tetragonal MnF_2 (as well as FeF_2 and CoF_2) are illustrated in Fig. 6.35, where it will be seen that the c-axis is the preferred spin direction

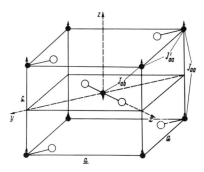

Fig. 6.35. Antiferromagnetism in MnF_2. Shown is the unit cell of this rutile-structure crystal. Open circles, F^- ions; closed circles, Mn^{2+} ions, indicating spin directions below T_c

and there is no canting. Any canting away from c would change the symmetry, in violation of the Dzyaloshinsky symmetry conditions. In NiF_2, however, the sublattices lie normal to the c axis and cant within the xy- or ab-plane. In this geometry there is no change of symmetry on canting.

Experimental results are in accordance [74]. Because of the weak ferromagnetic moment, and in agreement with Moriya's theory, the paramagnetic susceptibility in the ab ($\perp$) plane rises very rapidly, becoming very large as T_c (73.3 K) is approached. This is of course contrary to the behavior of a normal antiferromagnet. The behavior follows not so much from the usual parameter D of the spin-Hamiltonian (which nevertheless must be positive) but from a competition between the exchange parameter J and the rhombic zero-field splitting term, E. This is the term that causes the spins on different lattice sites to be perpendicular. The total energy is minimized through a compromise in which the spins are tilted away from the antiparallel towards a position in which they are perpendicular. The canting angle is not large, however, being only of the order of a degree.

This mechanism may also be one of the sources of the canting observed in $[(CH_3)_3NH]MnBr_3 \cdot 2H_2O$ [75], where, as usual, the g-value anisotropy of manganese(II) is so small as to make the existence of an important D-M interaction unlikely. Large zero-field splittings are suggested by the analysis of the specific heat, however, as well as by measurements [76] of the paramagnetic anisotropy. A zero-field splitting parameter D as large as -0.53 K was derived, along with a non-zero rhombic (E) term. As illustrated by the structure [77] of the chlorine analogue in Fig. 6.36, the adjacent octahedra are probably tilted, one with respect to the next, and so the principal axes of the distortion meet the same symmetry requirements illustrated above by NiF_2. On the other hand, it has been suggested [77] that biquadratic exchange [a term in $(S_i \cdot S_j)^2$] is important for this compound.

It is interesting to note that, although $CsCoCl_3 \cdot 2H_2O$ exhibits canting, the isomorphous $CsMnCl_3 \cdot 2H_2O$ does not [78]. By analogy with the cobalt compound, the requisite symmetry elements that do not allow canting must be absent but apparently, both g-value and zero-field anisotropies are either too small to cause observable weak ferromagnetism, or else are zero. There is also no evidence to suggest that hidden canting occurs. In the discussions of β-MnS, the sources of anisotropy considered were the D-M interaction and magnetic dipole anisotropy. Since that crystal is cubic, zero-field splittings are expected to be small, and g-value anisotropy is

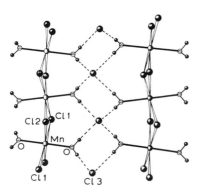

Fig. 6.36. The hydrogen-bonded plane of chains in $[(CH_3)_3NH]MnCl_3 \cdot 2H_2O$. From Ref. [77]

small. In those cases involving manganese where the zero-field splitting may be large, such as mentioned above, and the g-value anisotropy still small, this may be a more important factor in causing canting. It is difficult to sort out the different contributing effects, as will be seen in the following discussion.

Among the other canted 3D antiferromagnets there are NH_4CoF_3 [79], NH_4MnF_3 [80], and $Co(OAc)_2 \cdot 4H_2O$ [81]. The compound NH_4CoF_3 is of interest because the magnetic ordering transition and a structural transition from cubic symmetry to a tetragonal symmetry with small orthorhombic distortion occur at one and the same temperature, 124.5 K. In fact it is the reduction in symmetry caused by the crystal phase transition which allows the symmetry-restricted weak ferromagnetism to occur.

The manganese analog, NH_4MnF_3, in this case does exhibit canting [80]. This material is cubic at room temperature but undergoes an abrupt tetragonal elongation at 182.1 K, and the distortion increases in magnitude with decreasing temperature down to 4.2 K. The magnetic transition temperature is 75.1 K, and it is suggested that single-ion anisotropy is the origin of the canting.

Canting is symmetry-allowed in monoclinic $Co(OAc)_2 \cdot 4H_2O$ and the compound orders at 85 mK as an antiferromagnet [81]. There is, however, a permanent moment parallel to the b-axis below T_c, and the compound behaves (for measuring field along b) as an ordered ferromagnet which divides into domains and is without remanence. This is, then, an antiferromagnet with overt canting.

6.13 Characteristic Behavior of the 3d Ions

The representative paramagnetic properties of the iron series ions were summarized in Sect. 4.4. The characteristic ordering behavior will be summarized here briefly.

The T_c of $CsTi(SO_4)_2 \cdot 12H_2O$ has not been reported, but should be quite low. No T_c of any other discrete coordination compound of Ti(III) has been reported.

There appear to be no examples of magnetically-ordered vanadium(III) coordination compounds. This is because of the large zero-field splittings, and because exchange interactions seem to be particularly weak in the compounds studied to date [51]. Relatively few compounds of chromium(III) have also been found to order, the exchange again generally being found quite weak with this ion. A compound as concentrated magnetically as $[Cr(H_2O)(NH_3)_5][Cr(CN)_6]$ [82] orders (ferromagnetically) at only 38 mK, while several similar salts appear not to order at all [83]. The only chromium compound to which has been applied the model calculations described here appears to be $Cs_2CrCl_5 \cdot 4H_2O$ [84], but that material, which orders antiferromagnetically at 185 mK, is complicated because the zero-field splitting effects are comparable in magnitude to the exchange interaction.

Compounds of manganese(II) have been shown to exhibit, in one way or another, virtually all the magnetic phenomena described in this book. Many compounds have been prepared and studied, and several of them (MnF_2, $MnCl_2 \cdot 4H_2O$, $[(CH_3)_4N]MnCl_3$) have served as testing ground for almost every new theoretical development. Manganese offers the best example of an isotropic or Heisenberg ion. The compound $RbMnF_3$ is one of the most ideal Heisenberg systems [2]. Its anisotropy has been measured, and it is the very small value, $H_A/H_E \simeq 5 \times 10^{-6}$. This material, along with cubic $KNiF_3$, is the best approximation of the nearest-neighbor-only Heisenberg model currently available. It is interesting that, because H_A is so small, the critical field is only about 3 kOe (0.3 T). Thus, for fields exceeding this value measurements in any direction will yield the perpendicular susceptibility since also with the field parallel to the easy axis the moments will have swung to the perpendicular orientation [2]. This is the behavior of an ideally isotropic Heisenberg antiferromagnet, which does not differentiate between parallel and perpendicular, since the $\chi_\parallel$ is only defined for $H_A \neq 0$. With a transition temperature of 83.0 K, the only complaint one may have with $RbMnF_3$ is that specific heat measurements will involve a large lattice contribution near T_c.

One would expect compounds of iron(III), which is isoelectronic to manganese(II) to follow the Heisenberg model of magnetic ordering. Relatively few such compounds are, nevertheless, known. This is because, in part, of problems of synthesis and crystal growth of the desired materials. No good examples were available in 1974 for inclusion in the review of de Jongh and Miedema [2]. Since that time, there has been a number of studies [85–87] on the series of compounds $A_2[FeX_5(H_2O)]$, where A is an alkali metal ion and X may be chloride or bromide. These compounds are reviewed extensively in Chap. 10, but we point out here that $Cs_2[FeCl_5(H_2O)]$ in particular [86] provides a good example of the $\mathscr{S} = \frac{5}{2}$ Heisenberg model. With a transition temperature of 6.54 K, it has proved to be especially useful for studies of the specific heat.

There are only two compounds of octahedral iron(II) of interest to us here. The first is monoclinic $FeCl_2 \cdot 4H_2O$, which has been carefully studied by Friedberg and his coworkers [88, 89]. The analysis of the highly-anisotropic susceptibilities yielded the parameters $D/k = 1.83$ K and $E/k = -1.32$ K, which means that the D state, five-fold degenerate in spin, was completely resolved. The spin-quintet is split over an energy interval of about 10 K, into lower and upper groups of two and three levels, respectively. The fact that T_c is 1.097 K indicates that the exchange energy in this salt is roughly an order of magnitude smaller than the ZFS of the ferrous ion ground state and thus, of the single ion anisotropy. (A reorientation of the axes allows an alternative assignment of the single-ion parameters as $D/k = -2.895$ K and $E/k = -0.255$ K, which would imply that the spin-Hamiltonian has more axial symmetry than suggested by the complicated susceptibility behavior.) The parameters listed above also allow a fit of the broad specific-heat anomaly. The material has a complicated spin structure and, while the near-degeneracy of the two lowest levels implies that the system may follow the Ising model, it does not provide a good example of this magnetic model system.

The second compound, $[Fe(C_5H_5NO)_6](ClO_4)_2$ was discussed earlier as an excellent example of the 3D Ising model.

It has already been pointed out that octahedral cobalt(II) provides a number of good examples of the Ising model. This is because of the strong spin-orbit coupling that leaves a well-isolated doublet as the ground state. It has also been pointed out already that $[Co(C_5H_5NO)_6](NO_3)_2$ provides the best example of the 3D XY spin $\mathscr{S} = \frac{1}{2}$ magnetic model. It is interesting, however, that tetrahedral cobalt(II) probably provides more compounds which are better examples of the Ising model. A set of data on a representative set of cobalt compounds is listed in Table 6.7

The best-known tetrahedral cobalt(II) species, from our point of view, is the $CoCl_4^{2-}$ ion. This ion is always distorted in its crystalline compounds, and the nature of the distortion changes the ZFS. Thus the ion, as it occurs in Cs_3CoCl_5 has $2D/k$

Table 6.7. Magnetic parameters of some cobalt(II) compounds.*

	g_1	g_2	g_3	$2D/k$, K	T_c, K	Ref.
Octahedral						
$[Co(C_5H_5NO)_6](ClO_4)_2$	4.77	4.77	2.26	–	0.428	a
$[Co(C_5H_5NO)_6](NO_3)_2$	4.83	4.83	2.27	–	0.458	b
$[Co(\gamma\text{-}CH_3C_5H_4NO)_6](ClO_4)_2$	3.23	4.33	5.48	–	0.49	c
$[(CH_3)_3NH]CoCl_3 \cdot 2H_2O$	2.95	3.90	6.54	–	4.135	d
$Co(urea)_2Cl_2 \cdot 2H_2O$	2.9	4.0	7.3	–	2.585	e
$[Co(C_5H_5NO)_6](CoCl_4)$	–	–	–	–	0.95	f
Tetrahedral						
$(NEt_4)_2[CoCl_4]$	2.36	2.28	2.52	9.4	–	g
Cs_2CoCl_4	2.65	2.71	2.51	13.5	0.222	h
$CoCl_2 \cdot 2P(C_6H_5)_3$	–	–	7.33	Large, negative	0.21	i
$CoBr_2 \cdot 2P(C_6H_5)_3$	–	–	7.22	Large, negative	0.25	i

* $2D/k$ is the zero-field splitting parameter for the tetrahedral salts, while T_c is the long range magnetic ordering temperature.

References to Table 6.7

a Algra H.A., de Jongh L.J., Huiskamp W.J., and Carlin R.L., Physica B **83**, 71 (1976)
b Bartolome J., Algra H.A., de Jongh L.J., and Carlin R.L., Physica B **94**, 60 (1978)
c Carlin R.L., van der Bilt A., Joung K.O., Northby J., Greidanus F.J.A.M., Huiskamp W.J., and de Jongh L.J., Physica B **111**, 147 (1981)
d Losee D.B., McElearney J.N., Shankle G.E., Carlin, R.L., Cresswell P.J., and Robinson W.T., Phys. Rev. B **8**, 2185 (1973)
e Carlin R.L., Joung K.O., van der Bilt A., den Adel H., O'Connor C.J., and Sinn E., J. Chem. Phys. **75**, 431 (1981)
f Lambrecht A., Burriel R., Carlin R.L., Mennenga G., Bartolome J., and de Jongh L.J., J. Appl. Phys. **53**, 1891 (1982)
g McElearney J.N., Shankle G.E., Schwartz R.W., and Carlin R.L., J. Chem. Phys. **56**, 3755 (1972)
h Duxbury P.M., Oitmaa J., Barber M.N., van der Bilt A., Joung K.O., and Carlin R.L., Phys. Rev. B **24**, 5149 (1981)
i Carlin R.L., Chirico R.D., Sinn E., Mennenga G., and de Jongh L.J., Inorg. Chem. **21**, 2218 (1982)

$= -12.4$ K, while $2D/k = +13.5$ K is found for the $CoCl_4^{2-}$ ion in Cs_2CoCl_4 [90, 91]. The crystal structure of Cs_3CoCl_5 is the best-known of the compounds discussed here, having been studied by X-ray methods at room temperature and by neutron diffraction at 4.2 K [18]. The Co–Cl bond length is 0.2263 nm and the Cl–Co–Cl angles of the distorted tetrahedral cobalt environment are 107.22° and 110.61°. The remarkable aspect of the structural results is that the tetragonal distortion of the $CoCl_4^{2-}$ ion is independent of temperature. A crystal structure analysis of Cs_2CoCl_4 that is as accurate has not been published, but we may use the careful analysis [92] of Cs_2ZnCl_4, for the two materials are isostructural and the lattice constants differ at most by 0.2%.

The Zn–Cl bond lengths are 0.2249, 0.2259, and 0.2252 (twice) nm, while the Cl–Zn–Cl bond angles are 115.34, 109.62, 106.52, and 109.03°. Thus the $CoCl_4^{2-}$ ion is more distorted in Cs_2CoCl_4 than it is in Cs_3CoCl_5. It has been suggested that the non-bonded Cl---Cl intermolecular contact may be the origin of the distortion of the $CoCl_4^{2-}$ tetrahedron, and so this may also be related to the change in sign of D. A calculation has been carried out [93] within the angular overlap model which seems to explain the change in the sign of D. The tetrahedral ion in Cs_3CoCl_5 is more elongated than the ideal tetrahedron, while it is more compressed in Cs_2CoCl_4. A change in the sign of D is calculated to occur when the Cl–Co–Cl angle passes through 109.47°, which is consistent with the results mentioned above. The magnitude of D is less easily calculated with accuracy, since the structure of the tetrahedron must be idealized in order to do the calculation at all. But the calculation is useful because it uses one set of parameters to describe both systems.

Furthermore, the same ion, albeit distorted differently, is found to have $2D/k = +9.4$ K in $[NEt_4]_2CoCl_4$ [94]; the crystal undergoes a phase transition as it cools, and the low-temperature structure is unknown.

The change in the nature of the ground state alters the nature of the magnetic ordering that the different substances undergo [95].

The compounds of tetrahedral cobalt that are known to undergo magnetic ordering have two characteristics in common. First, the magnetic exchange is weak, and thus the compounds are found to order at temperatures below 1 K. Secondly, the ZFS are all relatively large, being 10 to 15 K. That means that the state involved in the ordering is but one doublet of the $^4A_2$ ground state, and this doublet is relatively well-isolated from the higher one. If the ZFS were much smaller than the transition temperature, then the cobalt would order as a spin $\mathscr{S} = \frac{3}{2}$ ion and would follow the Heisenberg magnetic model. The g-values would be nearly isotropic, with values 2.2 to 2.4. No such example has yet been found.

Since only the lower doublet is involved in magnetic ordering, the metal ion acts as an effective spin $\mathscr{S} = \frac{1}{2}$ ion, with the g-values depending on which component is the lower one. When the ZFS is negative, as in Cs_3CoCl_5 for example, then the $|\pm\frac{3}{2}\rangle$ state needs to be examined. Since $\Delta m_s = 3$, the effective $g_\parallel$ is $3 g_\parallel$ or about 7; there are no matrix elements for the $S_\pm$ operators that connect the two states, so $g_\perp$ is zero to first order. Since the exchange constant J is proportional to the moments of the ions, J is proportional to g^2, and so the large anisotropy in the g-values is reflected in highly-anisotropic exchange. That is, these are the conditions for a system to behave as an Ising system, and that is precisely what has been found.

The situation changes when the $|\pm\frac{1}{2}\rangle$ state is the ground state. Then the effective $g_\parallel$ is approximately the same as the "real" $g_\parallel$, or about 2.4. The effective $g_\perp$ is twice the real $g_\perp$, or about 5.4. So the exchange constants $J_\parallel$ and $J_\perp$ are in the ratio of the square of the effective g-values, or $(g_\parallel/g_\perp)^2 \approx \frac{1}{4}$. This ratio leads to XY-model behavior; of course, one must know the g-values from experiment to evaluate this ratio for any real system, and the smaller the ratio, the more closely is the XY model followed [96]. There is known but one example of this situation, and that is found with the compound Cs_2CoCl_4. The analysis of the magnetic data for this compound is complicated by the fact that the material exhibits substantial short-range, and it is better discussed in the next chapter.

Some nickel salts have ordering temperatures that are high with respect to the ZFS. One example would be anhydrous nickel chloride, $NiCl_2$, which has an ordering

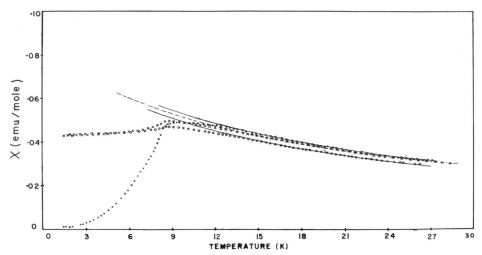

Fig. 6.37. Crystal axes susceptibilities of $NiBr_2 \cdot 6 H_2O$, illustrating the small but measureable anisotropy. The points are the experimental data for measurements along three orthogonal axes, and the curves represent the fitted susceptibilities, as described in Ref. [98]

temperature of 52 K. The ZFS may be assumed to be much smaller than that. Many other compounds have a ZFS which is comparable in magnitude to the ordering temperature. Several examples would be $NiCl_2 \cdot 6 H_2O$ ($T_c = 5.34$ K, $D/k = -1.5$ K; Ref. [97]), $NiBr_2 \cdot 6 H_2O$ ($T_c = 8.30$ K, $D/k = -1.5$ K; Ref. [98]), and $[Ni(en)_3](NO_3)_2$ ($T_c = 1.25$ K, $D/k = -1$ K; Ref. [41]). In each of these cases, the ZFS introduces some anisotropy into the system, which causes the paramagnetic susceptibilities to be anisotropic and influences both the long-range ordering temperature and the magnetic phase diagram in H–T space; a typical data set is illustrated in Fig. 6.37. But these ZFS cannot be said to *control* the kind of magnetic ordering these substances undergo. By and large, this is true whenever the ZFS of nickel is negative in sign.

The ZFS has probably been determined for more compounds of nickel than for those of any other metal ion. Furthermore, many of the ideas concerning the effects of ZFS on magnetic ordering originally arose from studies on nickel compounds. A recent exposition on $[Ni(H_2O)_6]SnCl_6$ has been provided [99].

An example of a compound in which ZFS actually prevents magnetic ordering from occurring is offered by $Ni(NO_3) \cdot 6 H_2O$ [100, 101]. The crystal field has such an important rhombic component that all the degeneracy of the ground state is resolved; the parameters are $D/k = 6.43$ K and $E/k = 1.63$ K. This places the three levels at, successively, 0, 4.8, and 8.1 K; since the exchange interactions are much weaker than this splitting (~ 0.6 K), the substance will not spontaneously undergo a spin ordering even at 0 K. This is because the magnetic levels have already been depopulated at low temperatures. The exchange interactions are present but subcritical; they affect the susceptibility measurements, because of the presence of the measuring field, but not the heat capacity measurements. Spontaneous ordering in the $[Ni(H_2O)_6]MF_6$ series of crystal has already been discussed above, as well as field-induced ordering in such compounds as $[Ni(C_5H_5NO)_6](NO_3)_2$.

A typical example of a system with large ZFS which still undergoes spontaneous magnetic ordering is provided by nickel chloride tetrahydrate [102]. This monoclinic crystal orders antiferromagnetically at 2.99 K, and exhibits an anisotropy in the susceptibilities in the paramagnetic region that is larger than that found with the hexahydrates of either nickel chloride [97] or bromide [98]; the latter compounds have ZFS of about -1.5 K. An unresolved Schottky anomaly was also discovered in the specific heat data for $NiCl_2 \cdot 4 H_2O$; these facts are consistent with the presence of a ZFS larger than T_c and of negative sign. All the magnetic data could in fact be interpreted in terms of a D/k of -11.5 K, that is, the $|\pm 1\rangle$ components are the ground state. (However, recent measurements [103] on $NiCl_2 \cdot 4 H_2O$ of the phase boundaries in a transverse field have suggested that D/k is only -2 K.)

Copper compounds usually provide good examples of Heisenberg magnets; interestingly, many copper compounds exhibit ferromagnetic interactions. The compound $CuCl_2 \cdot 2 H_2O$ is one of the most-studied antiferromagnets, and is a good example of the spin $\mathscr{S} = \frac{1}{2}$, 3D antiferromagnet [2]. In many ways, copper is better known as a source of dimers (Chap. 5) and of linear chain and planar magnets (Chap. 7). This is probably because of the Jahn-Teller distortions which copper compounds suffer so frequently.

6.14 References

1. Morrish A.H., (1965) Physical Principles of Magnetism, John Wiley and Sons, New York
2. de Jongh L.J. and Miedema A.R., Adv. Phys. **23**, 1 (1974)
3. Rosenberg H.M., (1963) Low Temperature Solid State Physics, Oxford University Press, Oxford
4. Martin D.H., (1967) Magnetism in Solids, MIT Press, Cambridge, Mass., (USA)
5. Danielian A., Proc. Phys. Soc. (London) **80**, 981 (1962)
6. Connolly T.F. and Copenhaver E.D., (1972) Bibliography of Magnetic Materials and Tabulation of Magnetic Transition Temperatures, IFI/Plenum, New York, (USA)
7. Rives J.E., Transition Metal Chem. **7**, 1 (1972)
8. O'Connor C.J., Prog. Inorg. Chem. **29**, 203 (1982)
9. Robinson W.K. and Friedberg S.A., Phys. Rev. **117**, 402 (1960)
10. Cracknell A.P. and Tooke A.O., Contemp. Phys. **20**, 55 (1979)
11. Domb C. and Miedema A.R., Progr. Low Temp. Phys. **4**, 296 (1964)
12. Fisher M.E., Proc. Roy. Soc. (London) **A 254**, 66 (1960); Fisher M.E., Phil. Mag. **7**, 1731 (1962)
13. Skalyo, Jr. J., Cohen A.F., Friedberg S.A., and Griffiths R.B., Phys. Rev. **164**, 705 (1967)
14. Nordblad P., Lundgren L., Figueroa E., Gäfvert U., and Beckman O., Phys. Scripta **20**, 105 (1979)
15. Miedema A.R., Wielinga R.F., and Huiskamp W.J., Physica **31**, 1585 (1965)
16. Wielinga R.F., Blöte H.W.J., Roest J.W., and Huiskamp W.J., Physica **34**, 223 (1967)
17. Blöte H.W.J. and Huiskamp W.J., Phys. Lett. **A 29**, 304 (1969); Blöte H.W.J., Thesis, Leiden, 1972
18. Figgis B.N., Gerloch M., and Mason R., Acta Cryst. **17**, 506 (1964); Figgis B.N., Mason R., Smith A.R.P., and Williams G.A., Acta Cryst. **B 36**, 509 (1980); Williams G.A., Figgis B.N., and Moore F.H., Acta Cryst. **B 36**, 2893 (1980); Reynolds P.A., Figgis B.N., and White A.H., Acty Cryst. **B 37**, 508 (1981); Figgis B.N. and Reynolds P.A., Aust. J. Chem. **34**, 2495 (1981)
19. Algra H.A., Bartolome J., de Jongh L.J., Carlin R.L., and Reedijk J., Physica B **93**, 114 (1978)
20. van der Bilt A., Joung K.O., Carlin R.L., and de Jongh L.J., Phys. Rev. **B 22**, 1259 (1980)

21. van Amstel W.D., de Jongh L.J., and Matsuura M., Solid State Comm. **14**, 491 (1974)
22. Carlin R.L., O'Connor C.J., and Bhatia S.N., J. Am. Chem. Soc. **98**, 685 (1976)
23. Algra H.A., de Jongh L.J., Huiskamp W.J., and Carlin R.L., Physica **83 B**, 71 (1976)
24. Joung K.O., O'Connor C.J., and Carlin R.L., J. Am. Chem. Soc. **99**, 7387 (1977)
25. Bartolome J., Algra H.A., de Jongh L.J., and Carlin R.L., Physica **B 94**, 60 (1978)
26. van der Bilt A., Joung K.O., Carlin R.L., and de Jongh L.J., Phys. Rev. B **24**, 445 (1981)
27. Gonzalez D., Bartolomé J., Navarro R., Greidanus F.J.A.M., and de Jongh L.J., J. Phys. (Paris) **39**, C6-762 (1978)
28. Carlin R.L., Carnegie, Jr. D.W., Bartolome J., Gonzalez D., and Floria L.M., Phys. Rev. B., in press
29. Sinn E., O'Connor C.J., Joung K.O., and Carlin R.L., Physica B **111**, 141 (1981)
30. Carlin R.L., van der Bilt A., Joung K.O., Northby J., Greidanus F.J.A.M., Huiskamp W.J., and de Jongh L.J., Physica B **111**, 147 (1981)
31. Hudson R.P., (1972) Principles and Application of Magnetic Cooling, North-Holland, Amsterdam
32. Hudson R.P. and Pfeiffer E.R., J. Low Temp. Phys. **16**, 309 (1974)
33. Zalkin A., Forrester J.D., and Templeton D.H., J. Chem. Phys. **39**, 2881 (1963)
34. Mess K.W., Lubbers J., Niesen L., and Huiskamp W.J., Physica **41**, 260 (1969)
35. Lagendijk E., Thesis, Leiden, 1972; Lagendijk E. and Huiskamp W.J., Physica **65**, 118 (1973)
36. Keffer F., Phys. Rev. **87**, 608 (1952); Walker L.R., Dietz R.E., Andres K., and Darack S., Solid State Comm. **11**, 593 (1972); See also Smith T. and Friedberg S.A., Phys. Rev. **177**, 1012 (1969) for a case where the dipolar term appears not to be the source of the anisotropy
37. Watanabe T., J. Phys. Soc. Japan **17**, 1856 (1962)
38. McElearney J.N., Losee D.B., Merchant S., and Carlin R.L., Phys. Rev. B **7**, 3314 (1973)
39. de Jongh L.J. and Block R., Physica **79 B**, 568 (1975)
40. Carlin R.L. and van Duyneveldt A.J., Acc. Chem. Res. **13**, 231 (1980)
41. O'Connor C.J., Bhatia S.N., Carlin R.L., van der Bilt A., and van Duyneveldt A.J., Physica B + C (Amsterdam) **95 B**, 23 (1978)
42. Paduan-Filho A., Palacio F., and Carlin R.L., J. Phys. Lett. (Orsay, Fr.), **39**, L-279 (1978); Carlin R.L., Bhatia S.N., and O'Connor C.J., J. Am. Chem. Soc. **99**, 7728 (1977); O'Connor C.J., Deaver, Jr. B.S., and Sinn E., J. Chem. Phys. **70**, 5161 (1979)
43. Palacio F., Paduan-Filho A., and Carlin R.L., Phys. Rev. B **21**, 296 (1980)
44. van Tol M.W., Diederix K.M., and Poulis N.J., Physica **64**, 363 (1973)
45. Diederix K.M., Groen J.P., and Poulis N.J., Physica B **86–88**, 1151 (1977); Diederix K.M., Groen J.P., and Poulis N.J., Phys. Lett. A **60**, 247 (1977); Diederix K.M., Groen J.P., Henkens L.S.J.M., Klaassen T.O., and Poulis N.J., Physica B **93**, 99 (1978); Diederix K.M., Groen J.P., Henkens L.S.J.M., Klaassen T.O., and Poulis N.J., Physica B **94**, 9 (1978); Eckert J., Cox D.E., Shirane G., Friedberg S.A., and Kobayashi H., Phys. Rev. B **20**, 4596 (1979)
46. Bonner J.C., Friedberg S.A., Kobayashi H., Meier D.L., and Blöte H.W.J., Phys. Rev. B **27**, 248 (1983)
47. Diederix K.M., Groen J.P., Klaassen T.O., Poulis N.J., and Carlin R.L., Physica B + C (Amsterdam) **97 B**, 113 (1979)
48. Smit J.J., de Jongh L.J., de Klerk D., Carlin R.L., and O'Connor C.J., Physica B + C (Amsterdam) **86–88 B**, 1147 (1977)
49. Carlin R.L., Joung K.O., Paduan-Filho A., O'Connor C.J., and Sinn E., J. Phys. C **12**, 293 (1979)
50. Mahalingam L.M. and Friedberg S.A., Physica B **86–88**, 1149 (1977); Ashkin J. and Vanderven N.S., Physica B **95**, 1 (1978)
51. Smit J.J., van Wijk H.J., de Jongh L.J., and Carlin R.L., Chem. Phys. Lett. **62**, 158 (1979)
52. Carlin R.L., O'Connor C.J., and Bhatia S.N., J. Am. Chem. Soc. **98**, 3523 (1976)
53. Diederix K.M., Algra H.A., Groen J.P., Klaassen T.O., Poulis N.J., and Carlin R.L., Phys. Lett. A **60**, 247 (1977)
54. Algra H.A., Bartolome J., Diederix K.M., de Jongh L.J., and Carlin R.L., Physica B + C (Amsterdam) **85 B**, 323 (1977)

55. Algra H.A., Bartolome J., de Jongh L.J., O'Connor C.J., and Carlin R.L., Physica B + C (Amsterdam) **93 B**, 35 (1978)
56. Paduan-Filho A., Chirico R.D., Joung K.O., and Carlin R.L., J. Chem. Phys. **74**, 4103 (1981)
57. Puertolas J.A., Navarro R., Palacio F., Gonzalez D., Carlin R.L., and van Duyneveldt A.J., J. Mag. Mag. Mat. **31–34**, 1067 (1983)
58. DeFotis G.C., Palacio F., Carlin R.L., Phys. Rev. B **20**, 2945 (1979)
59. Carlin R.L., Krause L.J., Lambrecht A., and Claus H., J. Appl. Phys. **53**, 2634 (1982)
60. Carnegie, Jr. D.W., Tranchita C.J., and Claus H., J. Appl. Phys. **50**, 7318 (1979)
61. Karnezos M., Meier D.L., and Friedberg S.A., Phys. Rev. B **17**, 4375 (1978); Lecomte G.V., Karnezos M., and Friedberg S.A., J. Appl. Phys. **52**, 1935 (1981); Warner J.D., Karnezos M., Friedberg S.A., Jafarey S., and Wang, Y.-L., Phys. Rev. B **24**, 2817 (1981); Suter R.M., Karnezos M., and Friedberg S.A., J. Appl. Phys. **55**, 2444 (1984)
62. Steijger J.J.M., Frikkee E., de Jongh L.J., and Huiskamp W.J., Physica B **123**, 271, 284 (1984); Day P., Acc. Chem. Res. **12**, 236 (1979); Stead M.J. and Day P., J. Chem. Soc. Dalton **1982**, 1081
63. O'Connor C.J., Burriel R., and Carlin R.L., Inorg. Chem., in press
64. Keffer F., (1966) Handbuch der Physik, Vol. VXIII, part 2, p.1. Springer, Berlin, Heidelberg, New York
65. Moriya T., (1963) in: Magnetism, (Eds. Rado G.T. and Suhl H.), Vol. 1, Chapt. 3. Academic Press, New York
66. Dzyaloshinsky I., J. Phys. Chem. Solids **4**, 241 (1958)
67. Moriya T., Phys. Rev. **117**, 635 (1960); **120**, 91 (1960)
68. Herweijer A., de Jonge W.J.M., Botterman A.C., Bongaarts A.L.M., and Cowen J.A., Phys. Rev. B **5**, 4618 (1972); Kopinga K., van Vlimmeren Q.A.G., Bongaarts A.L.M., and de Jonge W.J.M., Physica B **86–88**, 671 (1977)
69. Losee D.B., McElearney J.N., Shankle G.E., Carlin R.L., Cresswell P.J., and Robinson W.T., Phys. Rev. B **8**, 2185 (1973)
70. Hijmans J.P.A., de Jonge W.J.M., van den Leeden P., and Steenland M., Physica **69**, 76 (1973)
71. Metselaar J.W. and de Klerk D., Physica **69**, 499 (1973)
72. Silvera I.F., Thornley J.H.M., and Tinkham M., Phys. Rev. **136**, A 695 (1964)
73. Keffer F., Phys. Rev. **126**, 896 (1962)
74. Cooke A.H., Gehring K.A., and Lazenby R., Proc. Phys. Soc. **85**, 967 (1965)
75. Merchant S., McElearney J.N., Shankle G.E., and Carlin R.L., Physica **78**, 308 (1974)
76. Yamamoto I. and Nagata K., J. Phys. Soc. (Japan) **43**, 1581 (1977)
77. Caputo R.E., Willett R.D., and Muir J., Acta Cryst. B **32**, 2639 (1976)
78. Kobayashi H., Tsujikawa I., and Friedberg S.A., J. Low Temp. Phys. **10**, 621 (1973)
79. Bartolomé J., Navarro R., Gonzalez D., and de Jongh L.J., Physica B **92**, 45 (1977)
80. Bartolomé J., Burriel R., Palacio F., Gonzalez D., Navarro R., Rojo J.A., and de Jongh L.J., Physica B **115**, 190 (1983)
81. Karnezos M., Lecomte G., and Friedberg S.A., J. Appl. Phys. **50**, 1856 (1979); Simizu S., Vander Ven N.S., and Friedberg S.A., J. Phys. Chem. Solids **43**, 373 (1982)
82. Carlin R.L., Burriel R., Fina J., and Casabo J., Inorg. Chem. **21**, 2905 (1982). See also Burriel R., Casabo J., Pons J., Carnegie, Jr. D.W., and Carlin R.L., Physica B **132**, 185 (1985)
83. Carlin R.L., Burriel R., Pons J., and Casabo J., Inorg. Chem. **22**, 2832 (1983)
84. Carlin R.L. and Burriel R., Phys. Rev. B **27**, 3012 (1983)
85. Carlin R.L., Bhatia S.N., and O'Connor C.J., J. Am. Chem. Soc. **99**, 7728 (1977)
86. Puertolas J.A., Navarro R., Palacio F., Bartolome J., Gonzalez D., and Carlin R.L., Phys. Rev. B **26**, 395 (1982)
87. Carlin R.L. and Palacio F., Coord. Chem. Revs. **65**, 141 (1985)
88. Schriempf J.T. and Friedberg S.A., Phys. Rev. **136**, A 518 (1964)
89. Raquet C.A. and Friedberg S.A., Phys. Rev. B **6**, 4301 (1972)
90. McElearney J.N., Merchant S., Shankle G.E., and Carlin R.L., J. Chem. Phys. **66**, 450 (1977)
91. Algra H.A., de Jongh L.J., Blöte H.W.J., Huiskamp W.J., and Carlin R.L., Physica B **82**, 239 (1976)

92. McGinnety J.A., Inorg. Chem. **13**, 1057 (1974)
93. Horrocks, Jr. W. De W. and Burlone D.A., J. Am. Chem. Soc. **98**, 6512 (1976)
94. McElearney J.N., Shankle G.E., Schwartz R.W., and Carlin R.L., J. Chem. Phys. **56**, 3755 (1972)
95. Carlin R.L., J. Appl. Phys. **52**, 1993 (1981); Carlin R.L., Science **227**, 1291 (1985)
96. Smit J.J. and de Jongh L.J., Physica B **97**, 224 (1979)
97. Hamburger A.I. and Friedberg S.A., Physica **69**, 67 (1973)
98. Bhatia S.N., Carlin R.L., and Paduan-Filho A., Physica B **92**, 330 (1977)
99. Meier D.L., Karnezos M., and Friedberg S.A., Phys. Rev. B **28**, 2668 (1983)
100. Berger L. and Friedberg S.A., Phys. Rev. **136**, A 158 (1964)
101. Herweijer A. and Friedberg S.A., Phys. Rev. B **4**, 4009 (1971)
102. McElearney J.N., Losee D.B., Merchant S., and Carlin R.L., Phys. Rev. B **7**, 3314 (1973)
103. Figueiredo W., Salinas S.R., Becerra C.C., Oliveira, Jr. N.F., and Paduan-Filho A., J. Phys. C.: Solid State Phys. **15**, L 115 (1982)

7. Lower Dimensional Magnetism

7.1 Introduction

Much of what has gone heretofore could come under the classification of short-range order. Certainly the intracluster magnetic exchange in Chap. 5 falls within this heading, and a significant part of the discussion of Chap. 6 also dealt with this subject. By "short-range order," we shall mean the accumulation of entropy in a magnetic system above the long-range ordering temperature. Though this occurs with all magnetic systems, the physical meaning implied by the chapter title is that magnetic ions are assumed here to interact only with their nearest neighbors in a particular spatial sense. For operational purposes, the term will be restricted to magnetic interactions in one and two (lattice) dimensions. In that regard, it is interesting to note that the discussion is thereby restricted to the *paramagnetic* region. Let it be clear right at the beginning that this magnetic behavior follows directly from the structure of the various compounds.

The study of magnetic systems which display large amounts of order in but one or two dimensions has been one of the most active areas recently in solid state physics and chemistry [1–20]. The reason for this seems to be that, though Ising investigated the theory of ordering in a one-dimensional ferromagnet as long ago as 1925 [21], it was not until recently that it was realized that compounds existed that really displayed this short-ranged order. The first substance recognized as behaving as a linear chain or one-dimensional magnet is apparently $Cu(NH_3)_4SO_4 \cdot H_2O$ (CTS). Broad maxima at low temperatures were discovered in measurements of both the susceptibilities and the specific heat [22]. Griffith [23] provided the first quantitative fit of the data, using the calculations for a linear chain of Bonner and Fisher [24]. Fortunately, the theoreticians had been busy already for some time [4] in investigating the properties of one-dimensional systems, so that the field has grown explosively once it was recognized that experimental realizations could be found.

7.2 One-Dimensional or Linear Chain Systems

There are very good theories available which describe the thermodynamic properties of one-dimensional (1D) magnetic systems, at least for $\mathcal{S} = \frac{1}{2}$; what may be more surprising is that there are extensive experimental data as well of metal ions linked into uniform chains.

The first point of interest, discovered long ago by Ising [21], is that an infinitely long 1D system undergoes long-range order only at the temperature of absolute zero.

Though Ising used a Hamiltonian that is essentially

$$H = -2J \sum_i S_{z,i} S_{z,i+1},\tag{7.1}$$

where J is the exchange or interaction constant, it has been shown subsequently that Heisenberg systems, with the Hamiltonian

$$H = -2J \sum_i S_i \cdot S_{i+1}\tag{7.2}$$

likewise do not order at finite temperatures. Both of these Hamiltonians are restricted in their practical application to nearest-neighbor interactions. This has repeatedly been shown to be satisfactory for lower-dimensional systems. In reality, of course, interchain actions become more important as the temperature is lowered, and all known 1D systems ultimately interact and undergo long-range order. But, with $Cu(NH_3)_4SO_4 \cdot H_2O$ as a typical example, $T_c = 0.37\,K$, while the characteristic broad maximum in the specific heat reaches its maximum value at about $3\,K$. The low ordering temperature makes available a wide temperature interval where the short-range order effects can be observed.

The specific heat and susceptibilities of an Ising $\mathcal{S} = \frac{1}{2}$ chain have been calculated exactly in zero-field [25, 26]. The calculation results in the following expression for the molar specific heat,

$$c = R(J/2kT)^2 \operatorname{sech}^2 \left(\frac{J}{2kT}\right).\tag{7.3}$$

Note that this function is the same as that in Eq. (3.14) and plotted in Fig. 3.2, if $g\mu_B H$ is replaced by J. The curve is broad and featureless. Equation (7.3) is an even function of J, and so the measured specific heat of an Ising system does not allow one to distinguish ferromagnetic $(J > 0)$ from antiferromagnetic $(J < 0)$ behavior. The compound $CoCl_2 \cdot 2py$, where py is pyridine (C_5H_5N), offers a good example of an Ising spin $\mathcal{S} = \frac{1}{2}$ linear chain. The structure [27] of this compound is sketched in Fig. 7.1, where it will be seen to consist of a chain of cobalt atoms bridged by two chlorine atoms. In *trans*-position are found the two pyridine molecules, forming a distorted octahedral configuration. This basic structure is quite common and will occupy a large part of our discussion of 1 D systems. The specific heat of the compound [1, 28] is illustrated in Fig. 7.2; the compound seems to undergo ferromagnetic interaction, and the long-

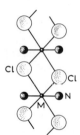

CL

CL

N

M

Fig. 7.1. The $-MCl_2-$ chain in the *ac*-plane. The N atoms are from the pyridine rings. From Ref. [28]

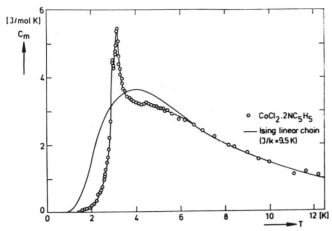

Fig. 7.2. Magnetic specific heat versus temperature of $CoCl_2 \cdot 2C_5H_5N$. The curve is the theoretical prediction for the $\mathscr{S} = \frac{1}{2}$ Ising chain calculated with $J/k = 9.5\,K$. From Ref. [28]

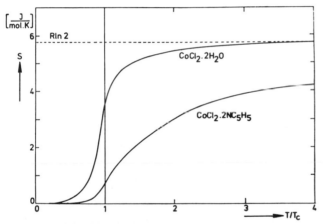

Fig. 7.3. Comparison of the entropy versus temperature curves of two Co salts. The large enhancement of the 1D character by substituting the C_5H_5N molecules for the H_2O molecules can be clearly seen from the large reduction in the amount of entropy gained below T_c. From Ref. [1]

range ordering at 3.15 K complicates the situation, but clearly the breadth of the peak suggests that a large amount of (Ising) short-range order is present.

The one-dimensional effect is illustrated nicely upon comparing in Fig. 7.3, on a reduced temperature scale, the magnetic entropies obtained from the specific heats of $CoCl_2 \cdot 2py$ and $CoCl_2 \cdot 2H_2O$. As with all ordering phenomena, the entropy $\Delta S_M = R \ln(2\mathscr{S} + 1)$ must be acquired, and lower dimensional systems acquire much of this entropy above T_c. The hydrate has a structure [29] similar to that of the pyridine adduct, but with water molecules in place of the pyridine molecules. Replacement of water by pyridine would be expected to enhance the magnetic 1D character, for the larger pyridine molecules should cause an increased interchain separation. Any hydrogen bonding interactions between chains would also be diminished. Not only does T_c drop from 17.2 K for the hydrate to 3.15 K for the pyridinate, but it will be seen that 60% of the entropy ($R\ln 2$) is obtained by $CoCl_2 \cdot 2\,H_2O$ below T_c, while only 15% of the entropy occurs below T_c for $CoCl_2 \cdot 2py$. The exchange interaction within the chains is essentially the same in the two compounds, and so these effects must be ascribed to a more ideal 1D (that is, less 3D) character in the pyridine adduct.

This entropy argument is an important one, to which we shall return repeatedly. An ideal Ising chain which ordered only at 0 K would necessarily acquire all its entropy of ordering above T_c. This is what we mean by short-range order. Since all real systems do order at some finite temperature, due to the weak but not zero interchain interactions, the fraction of the entropy acquired above T_c, $(S_\infty - S_c)/R$, is a relative measure of how ideal a system is.

7.2.1 Ising Systems

The zero-field susceptibilities have been derived by Fisher [30] for the $\mathscr{S} = \frac{1}{2}$ Ising chain. They are

$$\chi_{\|} = \frac{Ng_{\|}^2\mu_B^2}{2J}\,(J/2kT)\exp(J/kT),\tag{7.4}$$

$$\chi_{\perp} = \frac{Ng_{\perp}^2\mu_B^2}{4J}\,[\tanh(J/2kT)+(J/2kT)\,\mathrm{sech}^2(J/2kT)].\tag{7.5}$$

It should be noticed that $\chi_{\|}$ is an odd function of J, but $\chi_{\perp}$ is an even function. This means that the sign of the exchange constant in an Ising system can only be determined from the parallel susceptibility (within the limits of zero-field specific heat and susceptibility measurements).

More importantly, the meaning of the symbols "parallel" and "perpendicular" should be emphasized. In the present context, they refer to the external (measuring) magnetic field direction with respect to the direction of spin-quantization or alignment within the chains, rather than to the chemical or structural arrangement of the chains. In most of the examples studied to date, the spins assume an arrangement perpendicular to the chain direction; this is thought to be due to dipole-dipole interaction. The specific heat, Eq. (7.3) and susceptibilities of Eqs. (7.4) and (7.5) are illustrated for a representative set of parameters in Fig. 7.4. A characteristic feature of

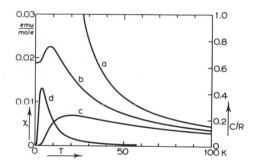

Fig. 7.4. Calculations of the specific heat and susceptibilities of Ising linear chains according to Eqs. (7.3)–(7.5). In all cases, $|J/k| = 10\,K$; $\mathscr{S} = \frac{1}{2}$, $g_{\|} = g_{\perp} = 2$. a: $\chi_{\|}(J > 0)$; b: $\chi_{\perp}$; c: $\chi_{\|}$ $(J < 0)$; d: c_p

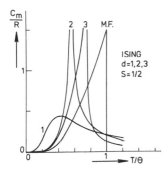

Fig. 7.5. Theoretical magnetic specific heats c_m of the $\mathscr{S} = \frac{1}{2}$ Ising model for a 1, 2, and 3D lattice. The chain curve has been obtained by Ising (1925), who first performed calculations on the model that bears his name. The 2D curve is also an exact result, derived by Onsager (1944) for the quadratic lattice. The 3D curve has been calculated by Blöte and Huiskamp (1969) and Blöte (1972) for the simple cubic lattice from the high and low-temperature series expansions of c_m given by Baker et al. (1963) and Sykes et al. (1972). For comparison the molecular field prediction (MF) has been included. From Ref. [1]

the curves is a broad maximum at temperatures comparable to the strength of the exchange interaction; the lone exception is the susceptibility for a measuring field parallel to a ferromagnetic spin alignment.

Figure 7.5, which is taken from the review of de Jongh and Miedema [1], illustrates the effect of lattice dimensionality on the specific heat for the $\mathscr{S} = \frac{1}{2}$ Ising model. As was noted earlier, the dimensionality usually exerts a greater influence on the thermodynamic functions than does the variation of lattice structure within a given dimensionality. It will be seen that a decrease in dimensionality increases the importance of short-range order effects. The molecular field calculation, which is based on an infinite number of neighbors, takes no account at all of short-range order. In the 1D case, all of the entropy must be acquired by short-range order, for T_c in the ideal case is 0 K. In a simple quadratic (2D) lattice, 44% of the entropy is acquired below T_c, while 81% of the entropy is acquired below T_c in the case of the 3D, simple cubic lattice.

Several other Ising linear chains have been studied since the important review of de Jongh and Miedema [1]. One of these is $CsCoCl_3 \cdot 2H_2O$, in which the intrachain coupling is antiferromagnetic. Related systems are $RbCoCl_3 \cdot 2H_2O$, $CsFeCl_3 \cdot 2H_2O$, and $RbFeCl_3 \cdot 2H_2O$. An extensively-studied system with ferromagnetically aligned chains is $[(CH_3)_3NH]CoCl_3 \cdot 2H_2O$. All these systems have complex arrangements of spins in the ordered state, and several of them were described in Sect. 6.12. Further discussion will be delayed until Sect. 7.9.

7.2.2 Heisenberg Systems

There are no exact or closed-form solutions for the Heisenberg model, even for an $\mathscr{S} = \frac{1}{2}$ one-dimensional system. Nevertheless, machine calculations are available which characterize Heisenberg behavior to a high degree of accuracy, particularly in one-dimension [24, 31]. The calculated specific heat of the Heisenberg model in 1, 2, and 3 dimensions is illustrated in Fig. 7.6. It will be noticed that not only does the 1D system not have a non-zero T_c, but even the 2D Heisenberg system does not undergo long-range order at a non-zero temperature. Comparison of Figs. 7.5 and 7.6 shows that changing the type of interaction from the anisotropic Ising to the isotropic Heisenberg form has the effect of enhancing the short-range order contributions. Furthermore, one-dimensional short-range order effects are extended over a much larger region in temperature for the more or less isotropic systems than for the Ising systems.

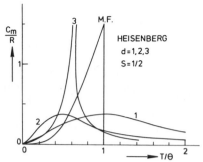

Fig. 7.6. Specific heats of the $\mathscr{S} = \frac{1}{2}$ Heisenberg model in 1, 2, and 3 dimensions. The 1D curve is the result for the antiferromagnetic chain obtained by Bonner and Fisher (1964), from approximate solutions. The 2D curve applies to the ferromagnetic quadratic lattice and has been constructed by Bloembergen (1971) from the predictions of spin-wave theory ($T/\theta < 0.1$), from the high-temperature series expansion ($T/\theta > 1$), and from the experimental data on approximants of this model ($0.1 < T/\theta < 1$). The 3D curve follows from series expansions for the bcc ferromagnet given by Baker et al. (1967). Also included is the molecular field prediction. From Ref. [1]

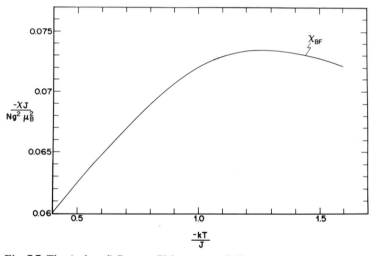

Fig. 7.7. The (reduced) Bonner-Fisher susceptibility

The susceptibility results for a spin $\mathscr{S} = \frac{1}{2}$ Heisenberg linear chain are due to Bonner and Fisher [24]. Since the Heisenberg model is an isotropic one, the susceptibility is calculated to be isotropic; the major exception to this rule is found with some copper compounds where a small g-value anisotropy may enter. A broad maximum in the susceptibility is predicted, as illustrated in Fig. 7.7. For an infinite linear chain, Bonner and Fisher calculate that the susceptibility maximum will have value

$$\frac{\chi_{max}}{(Ng^2\mu_B^2/|J|)} \simeq 0.07346$$

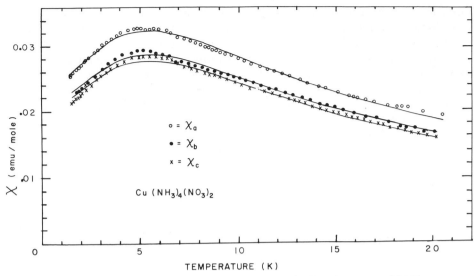

Fig. 7.8. Experimental susceptibilities of $[Cu(NH_3)_4](NO_3)_2$ (CTN). From Ref. [32]

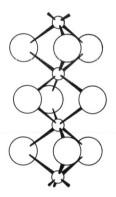

Fig. 7.9. A sketch of the linear chains found in $[N(CH_3)_4]MnCl_3$. The octahedral environment about the manganese atom is slightly distorted, corresponding to a lengthening along the chain

at the temperature

$$kT_{max}/|J| \simeq 1.282.$$

This theory has been applied to a very large number of data, as illustrated in Fig. 7.8 for $[Cu(NH_3)_4](NO_3)_2$ [32]. The chemical aspects have been reviewed [17].

The best example of an antiferromagnetic Heisenberg chain compound is $[N(CH_3)_4]MnCl_3$, TMMC, the structure [33] of which is illustrated in Fig. 7.9. The crystal consists of chains of $\mathscr{S} = \frac{5}{2}$ manganese atoms bridged by three chloride ions; the closest distance between manganese ions in different chains is 0.915 nm. The first report [34] of 1 D behavior in TMMC has been followed by a flood of papers, reviewed in [1] and [2], for the compound is practically an ideal Heisenberg system. The intrachain exchange constant J/k is about -6.7 K, which is strong enough to cause a broad peak in the magnetic heat capacity at high temperatures. Unfortunately, this relatively large intrachain exchange causes the magnetic contribution to overlap severely with the lattice contribution. Nevertheless, a broad peak with maximum at about 40 K has been separated out [35] by empirical procedures, which depend in part on comparing the measured specific heat of TMMC with its isomorphic, diamagnetic cadmium analog. Long-range order sets in at 0.83 K, as determined both by heat capacity [35, 36] and susceptibility [34, 37] measurements. One measure of the ideality of a linear chain system is the ratio of J' to J, where J'/k represents interchain exchange. This ratio is only 10^{-4} for TMMC. Another measure of the ideality is the critical entropy, a mere 1% of the total for TMMC. One of the problems faced in analyzing the data is that the theoretical calculations have been hindered by the relatively high value of the spin. One procedure that has been used is based on Fisher's [38] calculation for a classical or infinite spin linear chain, scaled to a real spin of $\frac{5}{2}$ [34]. This procedure, which was introduced in the first report on linear chain magnetism in $CsMnCl_3 \cdot 2H_2O$ [39], makes use of the equation

$$\chi = \frac{Ng^2\mu_B^2\mathscr{S}(\mathscr{S}+1)}{3kT} \cdot \frac{1-u}{1+u}, \tag{7.6}$$

where $u = (T/T_o) - \coth(T_o/T)$ with $T_o = 2J\mathscr{S}(\mathscr{S}+1)/k$. Weng [40] has carried out some calculations that take more explicit cognizance of the spin value, but these remain unpublished. Perhaps the most useful calculations are the numerical ones of Blöte [31, 41] and de Neef [42], which not only are applicable to the $\mathscr{S}=\frac{5}{2}$ chain but also include zero-field splitting effects.

Since the spin quantum number takes discrete values of $\frac{1}{2}, 1, \frac{3}{2}$, -----, spin is clearly a quantum phenomenon. As the spin gets larger in value, however, the discrete spectrum of states gets closer together and begins to approach a continuous or classical spin. It turns out that systems with large values of the spin, $\frac{5}{2}$ and above, may often be treated with a good deal of accuracy with this approximation. A spin of infinity corresponds to a true classical spin, and is found to be a useful subterfuge often used by theorists.

As with *all* short-range ordered antiferromagnetic systems, the susceptibility of TMMC shows [34] a broad maximum, in this case at about 55 K. A system with only Heisenberg exchange should exhibit isotropic susceptibilities, but in weak fields TMMC begins to show anisotropic susceptibilities parallel and perpendicular to the chain (c) axis of the crystal below 60 K. A small dipolar anisotropy was found to be the cause of the magnetic anisotropy [37] and to favor alignment of the spins perpendicular to the chain axis.

The compound TMMC is one of a large series of ABX_3 systems, such as $CsCuCl_3$, $CsCoCl_3$ [43] and $[(CH_3)_4N]NiCl_3$, (TMNC), which display varying degrees of short-range order. Interestingly, TMNC, which is isostructural with TMMC, displays ferromagnetic intrachain interactions, which causes a noncancellation of long-range dipolar fields [44]. This result, along with the zero-field splitting of about 3 K, causes TMNC to be a less perfect example of a linear chain magnet than is TMMC. Many of the other physical properties of these anhydrous systems have been reviewed [45].

A system somewhat related to TMMC is DMMC, $[(CH_3)_2NH_2]MnCl_3$ [46]. The magnetic chains are similar, and J/k drops slightly, to -5.8 K. Long range order occurs at 3.60 K. The ratio J'/J is 1.2×10^{-3} and the critical entropy is only 3% of the total. It follows the Heisenberg magnetic model more ideally than $CsMnCl_3 \cdot 2H_2O$ (see below) but not quite as well as does TMMC. The dipolar interaction again causes the spin orientation to be perpendicular to the chemical chain direction.

A major research effort has gone into the study of a variety of hydrated metal halides, for they are easily prepared and yield many quasi-one-dimensional systems of varying degrees of ideality. The compound $CsMnCl_3 \cdot 2H_2O$ was mentioned above; it is far from being an ideal system, the critical entropy being 13% of the total. Nevertheless, it has been studied extensively as an example of the Heisenberg spin $\mathscr{S} = \frac{5}{2}$ linear chain antiferromagnet. Its structure, like the cobalt analogue, has a μ-chloro (single halide) bridge between *cis*-octahedra, as was illustrated in Fig. 6.28. Two substances which are isostructural with $CsMnCl_3 \cdot 2H_2O$ are α-$RbMnCl_3 \cdot 2H_2O$ and $CsMnBr_3 \cdot 2H_2O$ [47]. These salts behave similarly to the Cs/Cl salt, as expected, with exchange constants of -3 K and -2.6 K, respectively. If the lack of ideality is ascribed only to interchain interactions, then the order of ideality is

$$\alpha\text{-}RbMnCl_3 \cdot 2H_2O > CsMnCl_3 \cdot 2H_2O > CsMnBr_3 \cdot 2H_2O \,.$$

Other related materials include $(CH_3NH_3)MnCl_3 \cdot 2H_2O$ [48], $[(CH_3)_2NH_2]MnCl_3 \cdot 2H_2O$ [49], $[(CH_3)_3NH]MnCl_3 \cdot 2H_2O$ [50], and $[C_5H_5NH]MnCl_3 \cdot H_2O$ [51], where one can see the chemist's hand at work here in the successive substitution on the organic cation. For better or worse, these substances do not all have the same crystal structure. The $(CH_3NH_3)^+$ salt is isotypic to the Cs^+ salt but not isomorphous, the cesium salt being orthorhombic and the meth-ylammonium salt being monoclinic. The susceptibility of the methylammonium compound [48] is displayed in Fig. 7.10; the susceptibility of the cesium compound [39] is quite similar. Note the broad maximum around 30 K, the high-degree of isotropy above 15 K, and the fact that long-range order occurs at $T_c = 4.12$ K. This is caused by interchain interactions. The exchange constants for the two salts are the

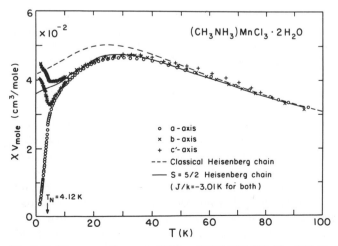

Fig. 7.10. The magnetic susceptibility of $(CH_3NH_3)MnCl_3 \cdot 2H_2O$ along three orthogonal axes. From Ref. [48]

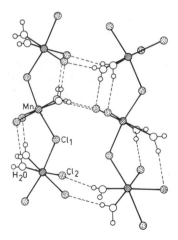

Fig. 7.11. Schematic representation of a hydrogen-bonded plane of chains in $[(CH_3)_2NH_2]MnCl_3 \cdot 2H_2O$. The chains, which run in the c-direction, are hydrogen bonded in the b-direction. In the a-direction, the chains are separated by dimethylammonium cations. From Ref. [52]

same, -3 K, within experimental error. The fact that $T_c = 4.89$ K for $CsMnCl_3 \cdot 2H_2O$ therefore means that interchain interactions are stronger in the cesium salt.

The compound $[(CH_3)_2NH_2]MnCl_3 \cdot 2H_2O$ also contains a μ-chloro bridge [52]. There is strong hydrogen-bonding, as evidenced by the crystal structure which is illustrated in Fig. 7.11. The susceptibility is similar to that of the two salts described above, and may be fit in the same fashion by $J/k = -2.65$ K. Some 19% of the magnetic entropy is acquired below $T_c = 6.36$ K, and the ratio of interchain to intrachain (J'/J) interactions was estimated as 2×10^{-2}.

Thus the three compounds, with $A = Cs^+$, $CH_3NH_3^+$, and $[(CH_3)_2NH_2]^+$, all contain cis-$[MnCl_4(OH_2)_2]$ octahedra linked into crooked chains, and the exchange constants remain almost the same despite small differences in local distortions.

There is a series of salts of manganese which contain μ-dihalide bridges. They are $[(CH_3)_3NH]MnCl_3 \cdot 2H_2O$, $[(CH_3)_3NH]MnBr_3 \cdot 2H_2O$ [50], (pyridinium)$MnCl_3$ $\cdot H_2O$, (quinolinium)$MnCl_3 \cdot H_2O$ [51], $MnCl_2 \cdot 2H_2O$ [53], and $MnCl_2 \cdot 2py$ [54]. The change in structure from the $AMnX_3 \cdot 2H_2O$ substances changes the magnetic properties substantially, and each of these materials exhibits different properties. Both $[(CH_3)_3NH]^+$ salts, the structure of which was illustrated in Fig. 6.36, have $trans$-$[MnX_4(H_2O)_2]$ coordination spheres. The chains are linked together, weakly, into planes by a halide ion which is hydrogen-bonded to the water molecules in two adjoining planes. The $[(CH_3)_3NH]^+$ ions lie between the planes and act to separate them both structurally and magnetically; these two compounds are perhaps better described as anisotropic two-dimensional materials rather than as linear chains. A broad maximum is observed in the specific heat of each, as expected, but the exchange constants drop to the relatively small values of -0.3 to -0.4 K. They are also canted antiferromagnets, as was discussed in Sect. 6.12.

The pyridinium and quinolinium salts have a similar structure for the chains, but are not as extensively hydrogen-bonded. They are not as good examples of the one-dimensional antiferromagnet as some of the other materials described above. Thus, for the $[C_5H_5NH]^+$ salt, it has been found that $T_c = 2.38$ K, $J/k \simeq -0.7$ K, $J'/J = 7 \times 10^{-2}$ and the critical entropy is as large as 39% of the total. Less is known of the quinolinium salt, but its exchange constant is about -0.36 K.

It is interesting to observe that the μ-halide chain systems have exchange constants about an order of magnitude larger than the μ-dihalide systems.

The compounds $MnCl_2 \cdot 2H_2O$ and MnX_2py_2 (X=Cl, Br) are isomorphous with their respective cobalt analogs which were described at the beginning of this chapter. The enhancement of linear chain character on replacing water by pyridine is evidenced here as well; indeed, interchain interactions are comparable in magnitude to intrachain interactions in $MnCl_2 \cdot 2H_2O$ [53] and so the material appears to behave as a three-dimensional antiferromagnet! The chained compounds [54] MnX_2L_2 (X=Cl, Br; L=pyridine, pyrazole) all exhibit broad maxima in the specific heat but, as with $CoCl_2 \cdot 2py$, the long-range ordering anomaly overlaps seriously with the short-range order feature.

With all the effort that has gone into manganese linear chain systems, it is interesting to note how little is known about systems derived from the isoelectronic iron(III) ion. Since the ion is more acidic, fewer compounds are known with bases such as pyridine as a ligand, and the syntheses of iron(III) compounds are generally less thoroughly explored.

One such compound which is of current interest [55] is $K_2[FeF_5(H_2O)]$. Discrete $[FeF_5(H_2O)]^{2-}$ octahedra are hydrogen-bonded in zig-zag chains, and this is the source of the chainlike superexchange path. Since this is not a strong superchange path in the first place, and since the fluoride ion provides generally a weaker superexchange path than chloride or bromide, it is no surprise that the exchange constant is quite weak, being only -0.40 K. Long-range order occurs at 0.80 K, and it is estimated that $|J'/J| \simeq 1.4 \times 10^{-2}$. This compound is discussed further in Chapter 10.

7.2.3 XY Systems

The XY model [56] is obtained from the Hamiltonian

$$H = -2J \sum_i (S_{x,i}S_{x,i+1} + S_{y,i}S_{y,i+1}) \tag{7.7}$$

which is again anisotropic; the anisotropy is increased (planar Heisenberg model) by adding a term of the form $D[S_z^2 - \mathscr{S}(\mathscr{S}+1)/3]$ to the above Hamiltonian, for then the spins are constrained to lie in the xy-plane. The reader should be careful to distinguish this spin anisotropy from dimensionality anisotropy. There are to date only two examples of one-dimensional compounds that follow this model, $(N_2H_5)_2Co(SO_4)_2$ [57] and Cs_2CoCl_4 [58–61]. Considerable anisotropy in the g-values, with $g_\perp \gg g_\parallel$, is the prerequisite for the applicability of the XY model [11, 57–62] and the hydrazinium compound, for example, meets this restriction, with $g_\perp = 4.9$, $g_\parallel = 2.20$. The compound clearly has a linear chain structure (Fig. 7.12) with metal atoms bridged by two sulfate groups. The magnetic specific heat of $(N_2H_5)_2Co(SO_4)_2$ is illustrated in Fig. 7.13, along with a very good fit to the calculated behavior of the XY model linear chain for $\mathscr{S} = \frac{1}{2}$; the fitting parameter is $J/k = -7.05$ K. The deviations at lower temperatures are due to the onset of a weak coupling between the chains which causes long-range spin-ordering at 1.60 K.

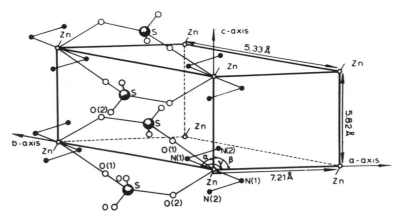

Fig. 7.12. Structure of triclinic $Zn(N_2H_5)_2(SO_4)_2$ according to Prout and Powell. The chemical linear chain axis (*b*-axis) is approximately 10 percent shorter than the *a*-axis. The compounds $M(N_2H_5)_2(SO_4)_2$ with $M=Ni$, Co, Fe, and Mn are isomorphous with $Zn(N_2H_5)_2(SO_4)_2$. From Ref. [57]

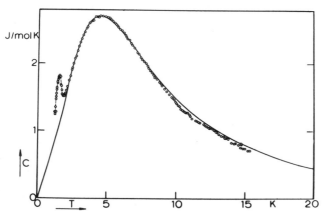

Fig. 7.13. The magnetic specific heat, c_m, of $Co(N_2H_5)_2(SO_4)_2$ as a function of temperature. The drawn line presents the results of the transverse coupled linear chain model $(J_{||}=0; J_\perp=J)$ as calculated by Katsura for $\mathscr{S}=\frac{1}{2}$; $|J/k|=7.05\,\mathrm{K}$. From Ref. [57]

The material Cs_2CoCl_4 has been more thoroughly studied because it is available in single crystal form. Furthermore, the crystal structure consists of discrete tetrahedra (Fig. 7.14) rather than the more obvious linear superexchange path caused by bridging ligands.

Specific heat measurements on this compound [58] in Fig. 7.15 have been interpreted with the following results:

a) the zero-field splitting is large – of the order of 14 K;

b) a broad maximum at about 0.9 K is ascribed to linear chain antiferromagnetism, and can be fit by the $\mathscr{S}=\frac{1}{2}$, XY magnetic model Hamiltonian with J_{xy}/k of about $-1.4\,\mathrm{K}$;

c) long range order occurs at 0.222 K.

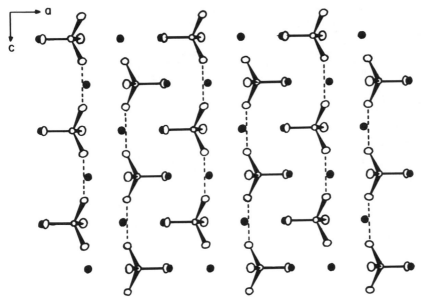

Fig. 7.14. Crystal structure of Cs_2CoCl_4

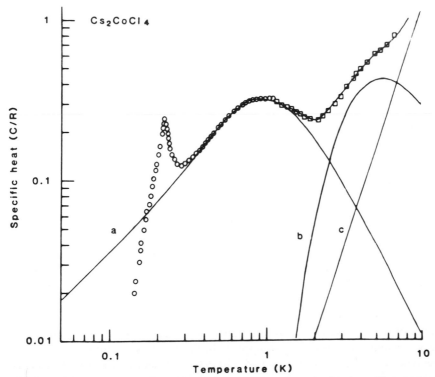

Fig. 7.15. Specific heat data on Cs_2CoCl_4. Curve a represents the fitted specific heat for the XY linear chain, while curve b represents the Schottky contribution. Curve c is the fitted lattice specific heat. From Ref. [58]

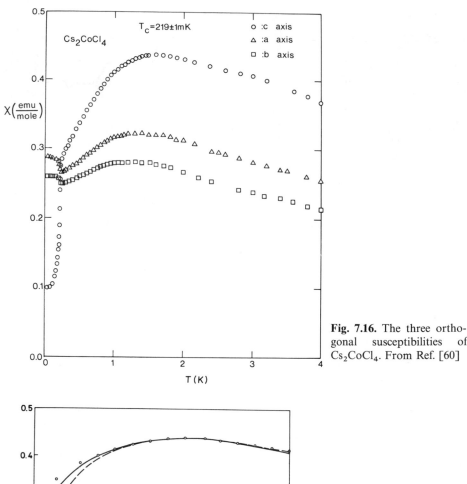

Fig. 7.16. The three orthogonal susceptibilities of
Cs_2CoCl_4. From Ref. [60]

Fig. 7.17. Fit to the c-axis susceptibility of Cs_2CoCl_4. From
Ref. [60]

Susceptibility measurements down to 1.5 K were consistent with these results, and furthermore required that the $|\pm\frac{1}{2}\rangle$ component be the ground state. Susceptibility data [58] below 4 K are presented in Fig. 7.16. The crystal structure, with two equivalent molecules per unit cell, of Cs_2CoCl_4 is such [58–60] that χ_z, the susceptibility parallel to the z-axis of the linear chain cannot be realized physically. The c-axis susceptibility displayed is the susceptibility in the XY-plane, χ_{xy}, and exhibits a broad maximum, along with the long-range magnetic order at 0.22 K. These data show that the c-axis is the easy axis. Duxbury, Oitmaa, and Barber [60] have recently calculated χ_{xy} for the $\mathcal{S}=\frac{1}{2}$, XY antiferromagnetic chain, and their results agree well with experiment. An exchange constant of -1.6 K within the pure XY model fits the χ_c data well; the introduction of small Ising-like anisotropy, as anticipated [58] from the analysis of the specific heat results, would be expected to provide an even better fit to the data, with an exchange constant closer to the previously reported one. The fit of the easy-axis data to the calculated behavior is illustrated in Fig. 7.17.

The a-axis and b-axis data are consistent with this analysis, and also behave as the anticipated perpendicular susceptibilities on the ordered state. Since these suscepti- bilities are mixtures of the linear chain $\chi_{\parallel}$ and $\chi_{\perp}$ above T_c, there is no way to fit these data. These results have been confirmed by neutron scattering studies [61].

O'Connor [9] presents a survey of a number of other experimental data on magnetic linear chains.

7.2.4 Some Other Aspects

An ever-present question concerns the procedure one should use to determine whether a system is following the Heisenberg, Ising, or XY models (to list only the three limiting cases). The answer lies not only with the amount of anisotropy exhibited by the available data, but also with the fits of the data to the different quantities calculated within each model. Quantities calculated by Bonner and Fisher have been listed above; a broader selection of the theoretical values of empirical parameters for linear chain systems is presented in Table 7.1.

Several other figures taken from de Jongh and Miedema [1] illustrate some of the trends anticipated with 1D magnets. Figure 7.18 illustrates the heat capacities of a number of chains with $\mathcal{S}=\frac{1}{2}$, while the perpendicular susceptibilities of the Ising chain and XY models are compared in Fig. 7.19. Note that the specific heat for the ferromagnetic Heisenberg linear chain is rather flat and differs substantially from that for the antiferromagnetic chain. The dependence upon spin-value of the susceptibility and specific heat of the Ising chain is illustrated in Fig. 7.20, while this dependence for

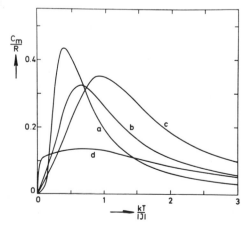

Fig. 7.18. Theoretical heat capacities of a number of magnetic chains with $\mathcal{S}=\frac{1}{2}$. Curves a and b correspond to the Ising and the XY model, respectively (ferro- and antiferromagnetic). Curves c and d are for the antiferromagnetic and ferromagnetic Heisenberg chains, respectively. From Ref. [1]

Table 7.1. Numerical results for the 1D Heisenberg, XY and Ising models. Listed are the spin value $\mathscr{S}$; the temperatures at which the maxima in the specific heat and the susceptibility occur, divided by the exchange constant $|J|/k$; the heights of the specific heat and susceptibility maxima in reduced units; the ratio of $T(\chi_{max})$ and $T(c_{max})$; the quantity $(1/g)^2 \chi_{max} T(\chi_{max})$. The plus and minus signs in brackets denote whether the interaction is ferro or antiferromagnetic, respectively. From Ref. [1].

Interaction	$\mathscr{S}$	Specific heat		Antiferromagnetic susceptibility			
		$\dfrac{kT(c_{max})}{|J|}$	$\dfrac{c_{max}}{R}$	$\dfrac{kT(\chi_{max})}{|J|}$	$\dfrac{\chi_{max}|J|}{Ng^2\mu_B^2}$	$\dfrac{T(\chi_{max})}{T(c_{max})}$	$\dfrac{\chi_{max}T(\chi_{max})}{g^2}$
Heisenberg	$\tfrac{1}{2}$	0.962(−) / 0.70(+)	0.350(−) / 0.134(+)	1.282(−) / —	0.07346(−) / —	1.33(−) / —	0.0353(−) / —
Heisenberg	1	1.8(−) / 1.6(+)	0.52(−) / 0.28(+)	2.70(−) / —	0.088(−) / —	1.5(−) / —	0.089(−) / —
Heisenberg	$\tfrac{3}{2}$			$\simeq$ 4.75(−)	$\simeq$ 0.091(−)		$\simeq$ 0.16(−)
Heisenberg	2	$\simeq$ 4.25(−) / $\simeq$ 3.5(+)	$\simeq$ 0.67(−) / $\simeq$ 0.46(+)	$\simeq$ 7.1(−)	$\simeq$ 0.094(−)	$\simeq$ 1.7(−)	$\simeq$ 0.25(−)
Heisenberg	$\tfrac{5}{2}$			$\simeq$ 10.6(−)	$\simeq$ 0.095(−)		$\simeq$ 0.38(−)
Heisenberg	3	$\simeq$ 6.6(−) / $\simeq$ 5.7(+)	$\simeq$ 0.74(−) / $\simeq$ 0.56(+)	$\simeq$ 13.1(−)	$\simeq$ 0.096(−)	$\simeq$ 2.0(−)	$\simeq$ 0.47(−)
XY	$\tfrac{1}{2}$	$\simeq$ 0.64	$\simeq$ 0.326	$\simeq$ 0.64 $(\chi_\perp)$	$\simeq$ 0.174 $(\chi_\perp)$	$\simeq$ 1.0	$\simeq$ 0.0417
Ising	$\tfrac{1}{2}$	0.416	0.445	1 $(\chi_\parallel)$ / 0.4168 $(\chi_\perp)$	0.09197 $(\chi_\parallel)$ / 0.2999 $(\chi_\perp)$	2.40 / 1.0	0.0345 / 0.0469
Ising	1	$\simeq$ 1.22	$\simeq$ 0.94	$\simeq$ 2.55 $(\chi_\parallel)$	$\simeq$ 0.098 $(\chi_\parallel)$	$\simeq$ 2.09	$\simeq$ 0.094
Ising	$\tfrac{3}{2}$	$\simeq$ 2.32	$\simeq$ 1.26	$\simeq$ 4.70 $(\chi_\parallel)$	$\simeq$ 0.100 $(\chi_\parallel)$	$\simeq$ 2.03	$\simeq$ 0.18
Ising	2	$\simeq$ 3.72	$\simeq$ 1.47	$\simeq$ 7.46 $(\chi_\parallel)$	$\simeq$ 0.101 $(\chi_\parallel)$	$\simeq$ 2.01	$\simeq$ 0.28
Ising	$\tfrac{5}{2}$	$\simeq$ 5.41	$\simeq$ 1.61	$\simeq$ 10.8 $(\chi_\parallel)$	$\simeq$ 0.1015	$\simeq$ 2.00	$\simeq$ 0.41
Ising	3	$\simeq$ 7.41	$\simeq$ 1.70	$\simeq$ 14.8 $(\chi_\parallel)$	$\simeq$ 0.102 $(\chi_\parallel)$	$\simeq$ 2.00	$\simeq$ 0.57

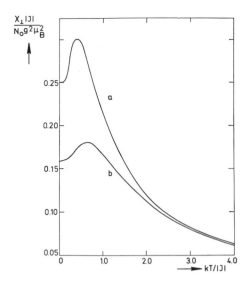

Fig. 7.19. Theoretical curves for the perpendicular susceptibility of the $\mathscr{S}=\frac{1}{2}$ Ising (a) and XY (b) chain model. The values for $\chi_\perp$ at $T=0$ are $Ng^2\mu_B^2/2z|J|$ and $Ng^2\mu_B^2/\pi z|J|$, respectively. From Ref. [1]

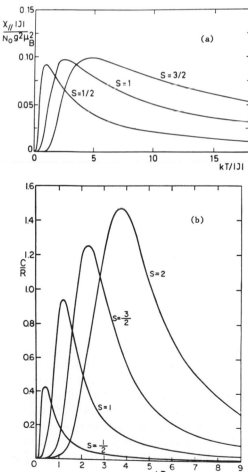

Fig. 7.20. Dependence on the spin value of the parallel susceptibility (a) and the specific heat (b) of the Ising chain. (After Suzuki et al. 1967 and Obokata and Oguchi, 1968.) From Ref. [1]

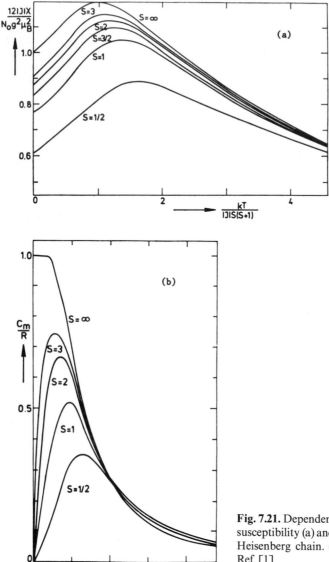

Fig. 7.21. Dependence on the spin value of the susceptibility (a) and the specific heat (b) of the Heisenberg chain. (After Weng, 1969.) From Ref. [1]

the Heisenberg model is shown in Fig. 7.21. The experimenter is seen to be faced with a large number of curves of similar behavior when he seeks to fit experimental data to one of the available models. Recall also (Chap. 5) that isolated dimers and clusters offer specific heat and susceptibility curves not unlike those illustrated. The different curves appear so similar that, for example, it has even been proposed that CTS, the first supposed linear chain magnet, actually behaves more like a two-dimensional magnet. Part of the problem lies with the fact that the physical properties calculated for Heisenberg one- and two-dimensional systems do not differ that much [63].

Furthermore, there are not chains or planes of magnetic ions which are self-evident in the structure of CTS.

Clearly, the measurement of one thermodynamic quantity is unlikely to characterize a system satisfactorily. Powder measurements of susceptibility are uniformly unreliable as indicators of magnetic behavior. Crystallographic structures are usually necessary in order to aid in the choice of models to apply, though even here the inferences may not be unambiguous. Thus, $MnCl_2 \cdot 2\,H_2O$ is a chemical or structural chain but does not exhibit chain magnetism [50], while $Cu(NH_3)_4(NO_3)_2$ [32] and Cs_2CoCl_4 [58] each consists of isolated monomeric polyhedra but acts like a magnetic chain! The isostructural series $CuSO_4 \cdot 5\,H_2O$, $CuSeO_4 \cdot 5\,H_2O$, and $CuBeF_4 \cdot 5\,H_2O$, which has been studied extensively by Poulis and co-workers [64] provides an interesting example in which two independent magnetic systems are found to co-exist. The copper ions at the $(0,0,0)$ positions of the triclinic unit cell have strong exchange interactions in one dimension, leading to short-range ordering in antiferromagnetic linear chains at about 1 K. The other copper ions at $(\frac{1}{2}, \frac{1}{2}, 0)$ have much smaller mutual interactions and behave like an almost ideal paramagnet with Curie-Weiss θ of about 0.05 K. Another example of the sort of complications that can occur is provided by recent studies [65] on $[(CH_3)_3NH]_3Mn_2Cl_7$. The structure of this material consists of linear chains of face-centered $MnCl_6$ octahedra, such as are found in TMMC, and discrete $MnCl_4^{2-}$ tetrahedra; the trimethylammonium cations serve to separate the two magnetic systems. The magnetic susceptibility could best be explained by assuming that the $MnCl_4^{2-}$ ions are formed into linear chains with the intrachain exchange achieved through Cl–Cl contacts. Since the TMMC-like portion of the substance also behaves as a magnetic linear chain, the material in fact consists of two independent linear chain species. Independent studies of the magnetization to high fields [66] confirm this picture. The exchange constant within the TMMC-like chains is about -8 K, while the tetrahedra interact (through a Mn–Cl---Cl–Mn path, similar to that in Cs_2CoCl_4) with an exchange constant of about -0.2 K. The analogous bromide behaves somewhat similarly [67], and exhibits considerable anisotropy between $\chi_{\parallel}$ and $\chi_{\perp}$. This has been shown to be due to dipolar intrachain interactions.

Copper(II), always a spin $\mathscr{S} = \frac{1}{2}$ ion, is also well-known as a Heisenberg ion and forms many linear chain antiferromagnets [17] and some one-dimensional ferromagnets [18]. Cs_2CuCl_4, which is isomorphous to Cs_2CoCl_4, is a Bonner-Fisher antiferromagnet.

It was pointed out in Chap. 4 that copper usually suffers a Jahn-Teller distortion. There are several cases where this affects the magnetic interaction in a compound. This is because the distortions can become cooperative [68]. Thus, the compound $[Cu(C_5H_5NO)_6](ClO_4)_2$ is strictly octahedral at room temperature, but its specific heat, in the vicinity of 1 K, is that of a linear chain antiferromagnet [69]. Similarly, $K_2Pb[Cu(NO_2)_6]$ is face-centered cubic at room temperature, which implies three-dimensional magnetism, but it undergoes several structural phase transitions as it is cooled and it too acts as an antiferromagnetic linear chain [70]. This has been explained [68, 71] in terms of Jahn-Teller distortions which are propagated cooperatively throughout the crystal lattice.

Certain generalizations are useful: copper(II) is likely to be a Heisenberg ion while manganese(II) always is; cobalt(II) is likely to be an Ising or XY ion, depending on the geometry of the ion and the sign and magnitude of the zero-field splitting. These facts were described in earlier chapters. At the least, qualitative comparisons of the shape of theoretical and experimental data are inadequate in order to characterize the magnetic behavior of a substance. A quantitative analysis of the data and quantitative fit to a model are imperative.

The dynamic properties of one-dimensional systems have been reviewed [2, 72]. The subjects discussed include spin waves, neutron scattering and resonance results, especially as applied to TMMC, $CsNiF_3$ and $CuCl_2 \cdot 2C_5H_5N$.

7.3 Long-Range Order

As has been implied above repeatedly, all real systems with large amounts of short-range order eventually undergo long-range ordering at some (low) temperature. This is because ultimately, no matter how small the deviation from ideality may be, eventually it becomes important enough to cause a three dimensional phase transition. Naturally, any interaction between chains, no matter how weak, will still cause such a transition. Furthermore, any anisotropy will eventually cause long-range ordering. The latter can be due to a small zero-field splitting in manganese(II), for example, which at some temperature becomes comparable to kT, but even effects such as anisotropic thermal expansion are sufficient to cause the transition to long-range order.

Nevertheless, the nature of the ordered state when it is derived from a set of formerly independent antiferromagnetic linear chains differs from the usual 3D example. This is particularly so in the case of $\mathscr{S} = \frac{1}{2}$ isotropic Heisenberg chains [64], as a result of the extensive spin-reduction $\Delta\mathscr{S}$ that occurs in these systems. Spin-reduction is a consequence of the low-lying excitations of an antiferromagnetic system which are called spin waves [1]. Though we shall not explore these phenomena any further, the fully antiparallel alignment of the magnetic moments is an eigenstate of the antiferromagnetic exchange Hamiltonian only in the Ising limit. An effect of zero-point motions or spin waves is that at $T = 0$ the magnetic moment per site is no longer $g\mu_B\mathscr{S}$, but $g\mu_B(\mathscr{S} - \Delta\mathscr{S})$. The spin-reduction $\Delta\mathscr{S}(T, H)$ depends on the anisotropy in the exchange interaction and reduces to zero in the Ising limit. The spin-reduction depends inversely on the number of interaction neighbors z and on the spin value $\mathscr{S}$. So, although $\Delta\mathscr{S}$ is small for 3D systems, it increases for 2D systems, and is maximal for linear-chain ($z = 2$), $\mathscr{S} = \frac{1}{2}$ Heisenberg antiferromagnets.

Finally, it was pointed out in Chap. 6 that for antiferromagnets the ordering temperature, T_c, lies below the temperature at which the susceptibility has its maximum value. In fact, for 3D ordering, the maximum is expected to lie some 5% above T_c. One of the most noticeable characteristics of lower-dimensional ordering is that the broad maximum in the susceptibility occurs at a temperature substantially higher than T_c. This is well-illustrated by the c-axis data of Cs_2CoCl_4, shown in Fig. 7.16.

As with other ordered systems with small anisotropy, ordered Heisenberg linear chain systems also exhibit a spin-flop phase. There is an important difference, however, in the observed behavior, and that is that the critical temperature of Heisenberg linear chain systems tends to increase drastically when an external field is applied. An early example is provided by the results [73] on $CsMnCl_3 \cdot 2H_2O$, the phase diagram of which is illustrated in Fig. 7.22. Or, consider the phase diagram of $[(CH_3)_2NH_2]MnCl_3 \cdot 2H_2O$, a linear chain system [49] with $T_c(0) = 6.4$ K and

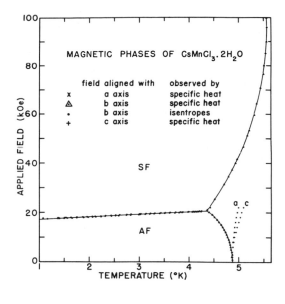

Fig. 7.22. Magnetic phase diagram of $CsMnCl_3 \cdot 2H_2O$. From Ref. [73]

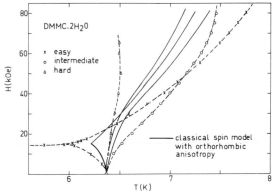

Fig. 7.23. Phase diagram of $[(CH_3)_2NH_2]MnCl_3 \cdot 2H_2O$. From Ref. [49]

$J/k = -3.7$ K. The phase boundaries above the bicritical point [at 6.0 K and 14.7 kOe (1.47 T)] bulge out to temperatures nearly 2 K above $T_c(0)$. This is illustrated in Fig. 7.23. This behavior is common in linear chain systems [74] and is due to the characteristic susceptibility behavior of the individual isolated chains. This is true even though the phenomenon being measured is the three-dimensional transition temperature, that is, the long-range ordering temperature describing the interaction between the chains. Anisotropy, which is probably due to dipole-dipole interactions favoring spin orientation perpendicular to the chain axis, causes the phase boundaries to differ for each orientation of the sample with respect to the field.

Oguchi [75] has presented a relationship between the ratio $|J'/K|$, the transition temperature T_c and the value of the spin, where J'/k is the interchain coupling constant. This relationship has been derived for a tetragonal lattice structure with isotropic interaction, but it has been used frequently for other systems as a guide to the order of magnitude of $|J'/J|$. Most of the values of this quantity referred to earlier were obtained by Oguchi's method.

7.4 Alternating Linear Chains

The model discussed thus far for linear chain systems is called a uniform one, for the intrachain exchange constant has been assumed not to vary with position. That is, the Hamiltonian is taken as

$$H = -2J \sum_i [S_{i-1} \cdot S_i + \alpha S_i \cdot S_{i+1}] \tag{7.8}$$

with $\alpha = 1$. An alternating chain is defined by letting α be less than one. For a Heisenberg alternating linear chain, $-2J$ is then the exchange interaction between a spin and one of its nearest neighbors, and $-2\alpha J$ is the exchange constant between the same spin and the other nearest neighbor in the chain. When $\alpha = 0$, the model reduces to the dimer model with pairwise interactions. The alternating linear chain antiferromagnet with spin $\mathcal{S} = \frac{1}{2}$ has been studied in detail lately, both theoretically and experimentally [10, 17, 19, 20, 76–84].

The important susceptibility results [77] for an alternating Heisenberg linear chain ($\mathcal{S} = \frac{1}{2}$) are illustrated in Fig. 7.24. Broad maxima are observed for all values of the alternation parameter α. The susceptibility curves vanish exponentially as temperature $T \to 0$ for all $\alpha < 1$; the uniform (Bonner-Fisher chain, $\alpha = 1$) alone reaches a finite value at $T = 0$. The case with $\alpha = 0$ is of course simply the isolated dimer (Bleaney-Bowers) equation. Likewise, broad maxima in the specific heat are also predicted [82].

The interesting question is, how can one prepare a true alternating system, with $0 < \alpha < 1$? The answer is not clear, for several such systems appear to be uniform structurally at room temperature, but undergo a crystal phase transition as they are cooled. This yields alternating linear chains with [Cu(γ-picoline)$_2$Cl$_2$], for example [79, 81], the susceptibility of which is illustrated in Fig. 7.25. The data exhibit a sharper

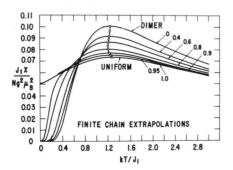

Fig. 7.24. The calculated susceptibility of alternating linear chains, as a function of the alternation parameter, α. From Ref. [77]

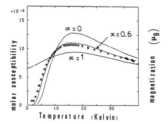

Fig. 7.25. Powder-susceptibility of $CuCl_2(\gamma$-picoline)$_2$. From Ref. [81]

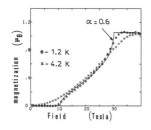

Fig. 7.26. High-field magnetization of $CuCl_2(\gamma\text{-picoline})_2$ powder. From Ref. [82]

maximum than is predicted by the uniform chain results ($\alpha = 1$), and may be fit by the alternating-chain model with $J/k = -13.4\,K$, $g = 2.14$, and $\alpha = 0.6$. The magnetization results (Fig. 7.26) agree; the calculated curve is a $T = 0$ approximation, and the lower the temperature, the more the data agree with experiment. Two critical fields are observed; the magnetization is zero below H_{c1}, and increases to saturation at H_{c2}. Similar behavior has been observed with [Cu(N-methylimidazole)$_2$Br$_2$] [78]. The organic molecule is of lower symmetry in the latter case, and the lower value of $\alpha = 0.4$ is observed.

The most thoroughly studied alternating linear chain antiferromagnet is $Cu(NO_3)_2 \cdot 2.5\,H_2O$ [76, 82]. The reader will recall our earlier discussion (Sect. 5.2) of this molecule, in which the first studies of this system suggested that it was primarily a dimer, i.e., a chain with $\alpha = 0$. These data have been reinterpreted, and new data obtained.

There is, to begin with, a systematic discrepancy of about 5% in the data illustrated in Fig. 5.5 and the calculated specific heat for a dimer [82]. This and other evidence suggested that there is a weak antiferromagnetic interdimer interaction in $Cu(NO_3)_2 \cdot 2.5\,H_2O$. Diederix and co-workers [76] studied the proton magnetic resonance spectra of the material and concluded that it contained alternating linear chains. Bonner et al. [82] in turn reinterpreted the earlier specific heat data. The zero-field specific heat maximum depends on α only, i.e., is independent of J (and g) while the position in temperature of the maximum depends only on the exchange constant J. The (corrected) c_p data are illustrated in Fig. 7.27, where the excellent fit of theory and experiment is obtained over the whole temperature range of ~ 0.5–$4.2\,K$. The position of the maximum fixes J/k at $-2.58\,K$, in agreement with the value of $-2.6\,K$ of Diederix. The discrepancy at the maximum of less than 1% could be removed by correcting for interchain interactions. The parameter $\alpha = 0.27$ is found from all the data analyses (recall that the dimer model described in Chap. 5 took $\alpha = 0$). Likewise, the susceptibility, field-dependent specific heat and magnetization could be described by the same set of parameters.

Another example of an alternating linear chain is provided [84] by the adduct between the hexafluoroacetylacetonate of Cu and the nitroxide, 4-hydroxy-(2,2,6,6)-tetramethylpiperidinyl-N-oxy. The compound is called $Cu(hfa)_2 \cdot TEMPOL$. There is a strong (19 K) ferromagnetic interaction between the copper ion and the nitroxide, evident in the data taken above 4.2 K, and the susceptibility rises rapidly as the temperature is lowered below 1 K. Then a broad maximum, characteristic of an antiferromagnetic linear chain, is found at about 80 mK. These data have been analysed in terms of a spin $\mathscr{S} = 1$ linear chain, with a weak ($2J = -78\,mK$) antiferromagnetic interaction between the $\mathscr{S} = 1$ units. That is, the alternating

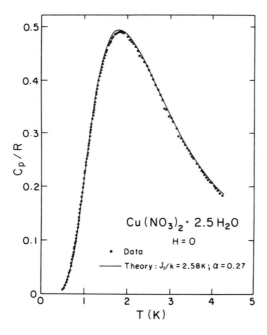

Fig. 7.27. Heat capacity at zero field of $Cu(NO_3)_2 \cdot 2.5\,H_2O$ vs. temperature. Data of Friedberg and Raquet, as revised in Ref. [83]

character is obtained by a repeating strong ferromagnetic and weak antiferromagnetic interaction. Similar results were also obtained from magnetization data on (4-benzylpiperidinium)$CuCl_3$ [81].

7.5 Spin-Peierls Systems

The spin-Peierls model refers to an alternating linear chain system of another sort. Here a system of uniform, quasi-one-dimensional $\mathscr{S} = \frac{1}{2}$, linear Heisenberg antiferromagnetic chains, coupled to the three-dimensional vibrations of the lattice, distorts to become a system of alternating linear Heisenberg antiferromagnetic chains, well-described by the Hamiltonian, Eq. (7.8), with temperature dependent alternation. In other words, this is a progressive spin-lattice dimerization [19, 20, 77, 85–88], which occurs below a characteristic temperature, T_{sp}. There is no long-range magnetic order in such a system, even at $T=0$.

Experimental examples of the spin-Peierls transition were first observed in the donor-acceptor complexes TTF–M$[S_2C_2(CF_3)_2]_2$, where M = Cu or Au, with $T_{sp} = 12$ and 2.1 K, respectively. TTF is tetrathiafulvene and the ligand is a tetrathiolene. A number of other systems have since been found to exhibit a spin-Peierls transition [88], and experimental data include magnetic susceptibility, specific heat, EPR, NMR and neutron diffraction measurements.

The occurrence of the spin-Peierls transition is most dramatically illustrated in the temperature dependence of the susceptibility (Fig. 7.28). Below T_{sp} (identified by the knee in the data at 12 K), the ground state is a nonmagnetic singlet (with a widening separation from the magnetic excited state), so that the powder susceptibility decreases

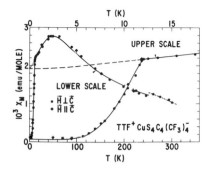

Fig. 7.28. Magnetic susceptibility of TTF $CuS_4C_4(CF_3)_4$ along two directions. Solid lines are calculated from a spin-Peierls theory which contains AF chains with uniform exchange above 12 K and temperature-dependent alternating exchange below. From Ref. [85]

exponentially to zero. This contrasts to the finite value found for the powder susceptibility at $T=0$ for uniform linear antiferromagnets, and provides strong evidence for the dimerization.

It is of interest to point out that because of the electron transfer between the cation and anion, that the $Cu[S_2C_2(CF_3)_2]_2^-$ ion is spin-paired in this compound, and the only spin carrier is the $(TTF)^+$ ion!

The transition temperature T_{sp} decreases as an external field is applied [86, 87]. An H-T phase diagram has been elucidated in part for $TTF–Au[S_2C_2(CF_3)_2]_2$.

7.6 Two-Dimensional or Planar Systems

The specific heats of one-dimensional and two-dimensional systems in the Ising and Heisenberg limits have already been compared in Figs. 7.5 and 7.6, respectively. Several of the thermodynamic quantities which have been calculated are listed in Table 7.2.

Table 7.2. Estimated values for the temperatures at which the maxima occur in the antiferromagnetic susceptibility of the quadratic Heisenberg lattices. The values attained by the susceptibility at these temperatures are also given. Note that the product $\chi_{max}T(\chi_{max})/C$ is only very slowly varying with $\mathscr{S}$. For comparison the result for the parallel susceptibility of the quadratic Ising lattice ($\mathscr{S}=\frac{1}{2}$) has been included. After de Jongh [1].

	Quadratic Heisenberg lattice						Ising $(\chi_{\parallel})$
$\mathscr{S}$	$\frac{1}{2}$	1	$\frac{3}{2}$	2	$\frac{5}{2}$	∞	$\frac{1}{2}$
$\dfrac{kT(\chi_{max})}{\|J\|\mathscr{S}(\mathscr{S}+1)}$	2.53	2.20	2.10	2.07	2.05	2.01	2.325
$\dfrac{\chi_{max}\|J\|}{Ng^2\mu_B^2}$	0.0469	0.0521	0.0539	0.0547	0.0551	0.0561	0.5370
$\dfrac{\chi_{max}T(\chi_{max})}{C}$	0.356	0.344	0.340	0.340	0.339	0.338	0.375

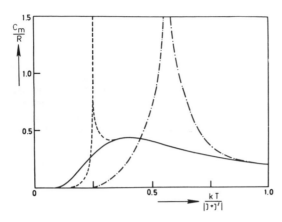

Fig. **7.29.** Specific heat of the 2D quadratic Ising lattice (Onsager) with different exchange interactions J and J' along the two crystallographic axes;   J'/J = 1,   dot-dash   curve; J'/J = 0.01, dashed curve; J' = 0, J ≠ 0, solid curve. From Ref. [1]

The important calculation for two-dimensional systems within the Ising limit is that of Onsager [89] for a $\mathscr{S} = \frac{1}{2}$ system. Letting J and J' be the exchange parameters in the two orthogonal directions in the plane, he showed that the quadratic lattice, with J = J', does have a phase transition and exhibits a λ-type anomaly. In the other limit, as soon as any anisotropy is introduced to the one-dimensional system, in which it was pointed out above that no phase transition can occur, a phase transition is now found to occur. That is, with a rectangular lattice even with, say, J/J' = 100, a sharp spike indicative of ordering is found on the heat capacity curve. The specific heats of the linear, rectangular, and quadratic lattice as calculated by Onsager are illustrated in Fig. 7.29.

The compound Cs_3CoBr_5 offers a good example of a quadratic lattice Ising system [1]. This is particularly interesting because the compound is isomorphic to Cs_3CoCl_5, the structure of which was illustrated in Fig. 6.11, and which acts as a three-dimensional magnet. Apparently in the bromide there is an accidental cancellation of interactions in the third dimension. The specific heat near T_c (0.282 K) is plotted in Fig. 7.30, and compared with Onsager's exact solution for this lattice [90]. The planar square Ising

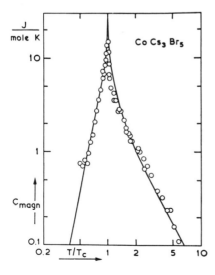

Fig. **7.30.** Magnetic specific heat of $CoCs_3Br_5$ plotted versus the temperature relative to T_c. The full curve represents Onsager's exact solution (1944) of the heat capacity of the quadratic, $\mathscr{S} = \frac{1}{2}$, Ising lattice. From Ref. [90]

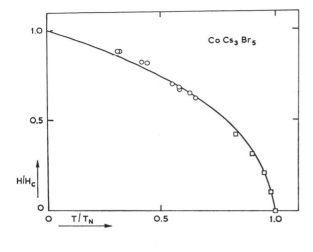

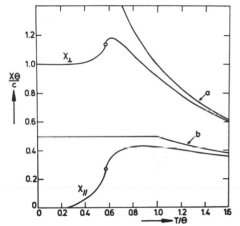

Fig. 7.31. Phase diagram of Cs_3CoBr_5 on a reduced temperature scale, taking $H_c = 2100$ Oe (0.210 T) at $T = 0$ K. The drawn line is computed for a planar square Ising antiferromagnet. From Ref. [90]

Fig. 7.32. A comparison of the perpendicular ($\chi_\perp$) and the parallel ($\chi_\parallel$) susceptibility of the quadratic Ising lattice (Fisher 1963; Sykes and Fisher 1962) with $\mathcal{S} = \frac{1}{2}$. Curve a is the susceptibility of a paramagnetic substance. Curve b is the molecular field prediction for the antiferromagnetic susceptibility in the paramagnetic region and for the perpendicular part below the transition temperature. From Ref. [1]

antiferromagnet has a particularly simple phase diagram, as is illustrated in Fig. 7.31 for Cs_3CoBr_5 [90], because the spin-flop phase does not appear in a system with the large anisotropy characterized by the Ising model.

There are no exact solutions for the susceptibilities of the planar Ising lattice, although computer calculations coupled with experimental examples have provided a good approximation for the kind of behavior to be expected. We reproduce in Fig. 7.32 a drawing that compares the susceptibilities of a quadratic Ising lattice with both molecular field and paramagnetic susceptibilities. Notice that $\chi_\perp$, reflecting the great anisotropy of the Ising model, does not differ from the paramagnetic susceptibility over a large temperature interval.

Calculations of the susceptibility of the rectangular antiferromagnetic Ising lattice have recently been published [91]. The procedure used is an approximate one, but estimated to be in error by less than a per cent from the true value, and leads to the curves illustrated in Fig. 7.33. The linear chain ($\gamma = 0$) is seen to be a sharper curve than those for the planar systems.

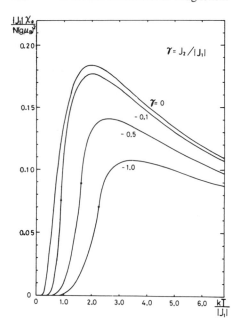

Fig. 7.33. Temperature dependence of the zero-field magnetic susceptibility of the rectangular antiferromagnet with negative values of γ, where γ denotes the ratio of the exchange-interaction constants, i.e., $\gamma = J_2/|J_1|$. The case $\gamma = 0$ corresponds to the exact solution for the antiferromagnetic linear chain, and the result for the case of the square antiferromagnet or $\gamma = -1.0$ (i.e., $|J_1| = |J_2|$) is practically indistinguishable on the curve from the most reliable result by the series-expansion method. The spot on each curve indicates the critical point. From Ref. [91]

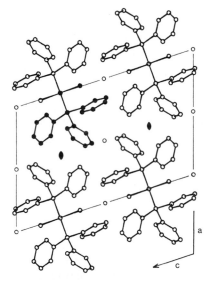

Fig. 7.34. Projection of the structure of $CoX_2 \cdot 2\,P(C_6H_5)_3$ onto the ac-plane. The asymmetric unit is indicated by filled circles. Molecules with the metal atom at $y = \frac{1}{4}$ are related to molecules with $y = \frac{3}{4}$ by inversion centers. From Ref. [92]

These calculations have recently been applied to the pseudo-tetrahedral molecules, $CoX_2 \cdot 2\,P(C_6H_5)_3$, $X = Cl$, Br [92]. A projection of the structure in Fig. 7.34 shows the pronounced two-dimensional character of the lattice.

The temperature dependences of the susceptibilities parallel to the c-axes of both compounds are presented in Fig. 7.35. Broad maxima are found at about 0.37 K for the chloride and 0.52 K for the bromide. At lower temperatures, the susceptibilities decrease toward zero, which may be interpreted in terms of long-range antiferromagnetic ordering, the c-axis being the easy or preferred axis of spin alignment in both

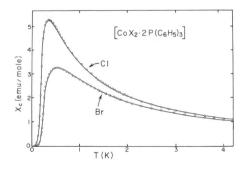

Fig. 7.35. Susceptibilities parallel to the c-axis of $CoCl_2 \cdot 2P(C_6H_5)_3$ and of $CoBr_2 \cdot 2P(C_6H_5)_3$. The data points are experimental, and the fitted curves are described in the text. From Ref. [92]

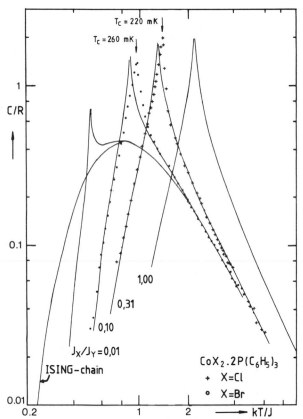

Fig. 7.36. Measured specific heats of $CoCl_2 \cdot 2P(C_6H_5)_3(+)$ and $CoBr_2 \cdot 2P(C_6H_5)_3(\cdot)$. The curves are explained in the text. The interaction constants, J, used in obtaining the relative temperature scales are defined by $|J| = |J_y|$. From Ref. [92]

compounds. The ordering temperature as derived from the maximum slope of the χ vs. T curve is 0.21 K for the chloride, while the bromide orders at 0.25 K.

The specific heats of both compounds are presented in Fig. 7.36 (on a reduced temperature scale). No lattice (phonon) contribution was observed below 1 K. The ordering temperatures are 0.22 and 0.26 K for the chloride and bromide, respectively, in good agreement with those found from the susceptibility measurements. Furthermore, the entropy found by taking the integrals $\int_0^\infty (c/T)\mathrm{d}T$ for both salts equals $R\ln 2$ within the experimental errors of about 2%. This indicates that the Co^{2+} ion is in an effective spin $\mathscr{S}'=\frac{1}{2}$ state in the temperature range ($T<1$ K) covered in the experiments. The solid curves in Figs. 7.35 and 7.36 are best fits to this rectangular Ising model of the susceptibility and specific heat data and agree quite well. For the chloride, the best-fit parameters to the theory of Tanaka and Uryû for the susceptibility are $J_x/k=0.046$ K, $J_y/k=0.150$ K, $J_x/J_y=0.31$, and $g_c=7.33$. For the bromide, these parameters are $J_x/k=0.026$ K, $J_y/k=0.265$ K, $J_x/J_y=0.10$, and $g_c=7.22$.

The analysis of the specific heat is presented in Fig. 7.36 together with the data. Solid curves represent the calculated specific heats for rectangular Ising systems having different J_x/J_y values; calculations were done according to the theory of Onsager. Curves for the linear chain and quadratic planar lattice are also illustrated.

These systems actually lie closer to being one-dimensional antiferromagnets rather than two-dimensional ones, and the bromide compound is more one-dimensional than is the chloride. This must be due to the larger size of bromide and the shorter Br–Br distance.

These systems provide quite interesting examples of the rectangular Ising model. Although the magnitude of the zero-field splitting cannot be determined from the present data, its negative sign is established without doubt. This leaves the $|\pm\frac{3}{2}\rangle$ doublet as the ground state, and the specific heat data indicate that the $|\pm\frac{1}{2}\rangle$ state is already sufficiently depopulated at temperatures below 1 K so as not to affect significantly the thermodynamic behavior.

Turning now to isotropic systems, the first point of interest is that it can be proved [1] that a two-dimensional Heisenberg system does not undergo long range order. Thus, as was illustrated above in Fig. 7.6, the specific heat of a planar Heisenberg system again consists only of a broad maximum. The shape of the curve differs from that for a one-dimensional system, but nevertheless there is no λ-type anomaly. Of course, just as with the experimental examples of one-dimensional systems, non-ideal behavior is observed as the temperature of a two-dimensional magnet is steadily decreased. Anisotropy can arise from a variety of sources, and interlayer exchange, though weak, can also eventually become important enough at some low temperature to cause ordering. In many cases, however, the λ-anomaly appears as only a small spike, superimposed on the two-dimensional heat capacity, and the system still acts largely as a two-dimensional lattice even to temperatures well below T_c. Unfortunately, most of the planar magnets investigated to date exhibit their specific heat maxima at high temperatures, and the separation of the lattice contribution has been difficult.

There has been a large amount of success in finding systems that are structurally, and therefore magnetically, two-dimensional. One system investigated extensively at Amsterdam [1,93] has the stoichiometry $(C_nH_{2n+1}NH_3)_2CuCl_4$, with n having the values 1,2,3,..., all the way to 10. The crystals consist of ferromagnetic layers of copper ions, separated by two layers of non-magnetic alkyl ammonium groups. The representative structure is illustrated in Fig. 7.37. By varying n, the distance between copper ions in neighboring layers may be increased from 0.997 nm in the methyl compound to 2.58 nm when n = 10, while the configuration within the copper layers is not appreciably changed. The Cu–Cu distance within the layers is about 0.525 nm.

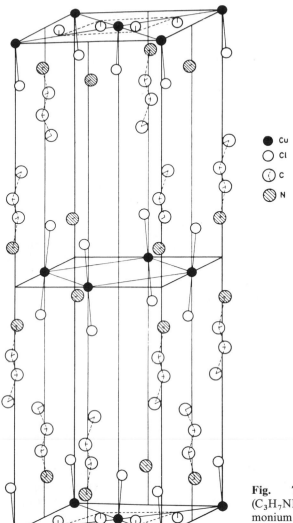

● Cu
○ Cl
◉ C
◍ N

Fig. 7.37. Crystal structure of $(C_3H_7NH_3)_2CuCl_4$. Part of the propylam-monium groups and H atoms have been omitted for clarity. From Ref. [1]

Thus, the interlayer separation can be varied almost at will, while the intralayer separations remain more or less the same. The manganese analogues are isostructural [1], studies of which allow this to be one of the best understood two-dimensional structures.

The high-temperature susceptibilities of the copper compounds yield positive Curie-Weiss constants, suggesting a ferromagnetic interaction in the plane. A series expansion calculation has provided the susceptibility behavior down to $T \simeq 1.5\,J/k$, and this calculation is compared in Fig. 7.38 with the powder susceptibility of $(C_2H_5NH_3)_2CuCl_4$. Here $C/\chi T$ is plotted vs. θ/T, where C is the Curie constant, so that the Curie-Weiss law appears as a straight line, $C/\chi T = 1 - \theta/T$ (Curve 3). The curve 1, which requires only one parameter to scale the experimental points, is the calculation

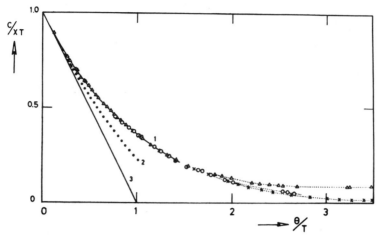

Fig. 7.38. The susceptibility of the ferromagnetic layer compound $(C_2H_5NH_3)_2CuCl_4$ in the high-temperature region ($T \gg T_c$; $\theta/T_c \simeq 3.6$). The full curve 1 drawn for $\theta/T < 1.4$ has been calculated from the high-temperature series expansion for the quadratic Heisenberg ferromagnet with $\mathscr{S} = \frac{1}{2}$. The exchange constant J/k was obtained by fitting the data to this prediction. The dotted curve 2 represents the series expansion result for the bcc Heisenberg ferromagnet. The straight line 3 is the MF prediction $C/\chi T = 1 - \theta/T$ for the quadratic ferromagnet. (After de Jongh et al. 1972). $\triangle : H = 10\,\text{kOe}$ (1 T); $\circ : H = 4\,\text{kOe}$ (0.4 T); $\times : H = 0$ (ac susceptibility measurements). From Ref. [1]

for a planar $\mathscr{S} = \frac{1}{2}$ Heisenberg system; the parameter mentioned is the exchange constant, positive in sign, $J/k = 18.6$ K. The dotted line, curve 2, represents the series expansion result for the (3D) b.c.c. Heisenberg ferromagnet for comparison. Figure 7.39 presents susceptibility data for 11 compounds of the series $(C_nH_{2n+1}NH_3)_2CuX_4$. Since the results for the various compounds coincide, although they differ in strength and type of interplane interaction (J') and anisotropy, the curve through the data may be considered to represent the susceptibility of the ideal quadratic Heisenberg ferromagnet, at least in the region $0.5 < J/kT < 1.1$, where $J/k = \theta/2$. Though the intraplane exchange is ferromagnetic for all the salts, J' is antiferromagnetic except for the cases of

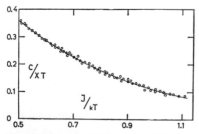

Fig. 7.39. Susceptibility data of eleven different members of the series $(C_nH_{2n+1}NH_3)_2CuX_4$ in the region $0.5 < J/kT < 1.1$ ($J/k = \theta/2$). Since the results for the various compounds coincide, although they differ in strength and type of the interlayer interaction as well as in anisotropy, the line through the data may be considered to represent the χ of the ideal quadratic Heisenberg ferromagnet. As such this result may be seen as an extension of the series expansion prediction, which is trustworthy up to $J/kT < 0.6$ only. From Ref. [1]

$n=1$ and 10. The susceptibility in the ordered state of ferromagnetic $(CH_3NH_3)_2CuCl_4$ has recently been studied [94]. For the ethyl derivative $|J'/J|$ is about 8.5×10^{-4}, which illustrates how weak the interplanar interaction is, with respect to the intraplanar one; furthermore, $T_c = 10.20$ K.

The planar systems $[(CH_2)_n(NH_2)_2]CuCl_4$, $n=2,3,5$ have been studied recently [95]. For $n=2$, $[NH_3(CH_2)_2NH_3]CuCl_4$, ferromagnetically-coupled layers are found, but since by chance the interlayer coupling is also strong, the material behaves as a three-dimensional ordered substance. Several model calculations have been reported on the superexchange interaction in these layered solids [96]. The results were found to depend on the angle between two neighboring copper ions in adjacent layers a and b, $Cu_a-Cl_a---Cl_b-Cu_b$.

Another series of two-dimensional systems that has been widely studied [1] is based on the K_2NiF_4 structure. As illustrated in Fig. 7.40, octahedral coordination of the nickel ion is obtained in the cubic perovskite, $KNiF_3$. The tetragonal K_2NiF_4 structure can be looked upon as being derived from the $KNiF_3$ lattice by the addition of an extra layer of KF between the NiF_2 sheets. By this simple fact a 3D antiferromagnetic lattice $(KNiF_3)$ is transformed into a magnetic layer structure (K_2NiF_4). It is of importance that the interaction within the layer is antiferromagnetic, since this causes a cancellation of the interaction between neighboring layers in the ordered state. Similar compounds have been examined with the nickel ion replaced by manganese, cobalt or iron, the potassium by rubidium or cesium, and in some cases fluoride by chloride. The correlation between crystallographic structure and magnetic exchange which the chemist would like to observe is in fact rather remarkably displayed by compounds such as these.

The antiferromagnetic susceptibility of planar magnets exhibits a behavior similar to that of isotropic AF chains, because the two systems have in common the absence of long range order and of anisotropy. Then, a broad maximum due to short range order

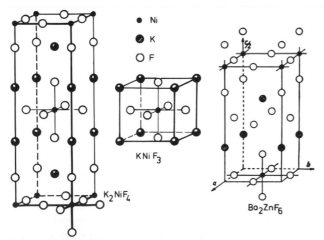

Fig. 7.40. Comparison of three related crystal structures, two of which are 2D in magnetic respect. In the middle the cubic perovskite structure of $KNiF_3$, on the left the tetragonal K_2NiF_4 unit cell. On the right the structure of Ba_2ZnF_6 (Von Schnering 1967). These crystal structures offer the possibility of comparing the 2D and 3D properties of compounds which are quite similar in other respects. From Ref. [1]

effects is expected at higher temperatures, whereas at $T=0$ the susceptibility should attain a finite value. There is no closed-form theory available.

The susceptibilities of six samples that approximate the quadratic antiferromagnetic Heisenberg layer, with different spin values are illustrated in Fig. 7.41; note the large deviation from the molecular field (MF) result for the susceptibility in the paramagnetic region and below the transition temperature, where $\chi_\perp$ has been plotted. Note also that the curves deviate more from the MF result, the lower the value of the spin, and therefore the more important are quantum effects. In Fig. 7.42, the susceptibilities of K_2MnF_4 are illustrated over a wide temperature range [97]. At the transition temperature, $T_c = 42.3$ K, the parallel and perpendicular susceptibilities will be seen to diverge; the exchange constant, which is obtained by fitting the data to a series expansion prediction, is $J/k = -4.20$ K. The anisotropy in the plane was estimated as about 3.9×10^{-3}, while the interplanar exchange is small, $|J'/J| \simeq 10^{-6}$.

The potential variety of magnetic phenomena is illustrated by recent measurements on the compounds $Rb_3Mn_2F_7$, $K_3Mn_2F_7$, and $Rb_3Mn_2Cl_7$ [98, 99]. The structure is one of essentially double layers, or thin films: sheets of $AMnX_3$ unit cells, of one unit cell thickness, are formed, separated from each other by non-magnetic AX layers. That is, two quadratic layers are placed upon each other at a distance equal to the lattice spacing within each layer, and are in turn well-separated from the adjoining layers. The consequence of this is that a broad maximum is found in the susceptibility while, furthermore, neutron diffraction shows the behavior expected of the two-dimensional Heisenberg antiferromagnet. Each manganese ion is antiparallel to each of its neighbors. Interestingly, both theory and experiment show that in a thin magnetic film the high-temperature properties are not appreciably different from bulk (i.e., 3D) behavior. Only as the temperature is lowered does the effect of the finite film thickness become apparent. Indeed, theory (high-temperature series) shows that the susceptibility for a Heisenberg film consisting of four layers is already very close to the bulk susceptibility in the temperature region studied.

The important role that hydrogen-bonding plays in determining lattice dimensionality should not be ignored, and illustrates the importance of a careful examination of the crystal structure. Though crystallographic chains are present in the compounds $[(CH_3)_3NH]MX_3 \cdot 2H_2O$ referred to earlier, hydrogen-bonding between the chains causes a measurable planar magnetic character for these molecules, which are described more fully below. Two other compounds investigated recently which obtain planar character largely through hydrogen-bonding that links metal polyhedra together are the copper complex of L-isoleucine [100] and $Rb_2NiCl_4 \cdot 2H_2O$ [101]. In the case of $Cu(L-isoleucine)_2 \cdot 2H_2O$, the material consists of discrete five-coordinate molecules of this stoichiometry, but they are hydrogen-bonded together in two crystalline directions, while well-isolated in the third. The material acts as a two-dimensional Heisenberg ferromagnet. Appreciable short range order was not anticipated in the investigation of $Rb_2NiCl_4 \cdot 2H_2O$, for earlier investigations of the manganese isomorphs had not revealed significant lower-dimensional behavior. Yet, in fact, this material, which consists of the discrete $[NiCl_4(H_2O)_2]$ octahedra which usually lead to normal 3D ordering, exhibits important low-dimensional character. A careful examination of the crystal structure shows that the octahedra are linked by hydrogen-bonds into sheets, which in turn lead to the observed magnetic behavior. It is not yet clear why the manganese compounds behave in a different fashion.

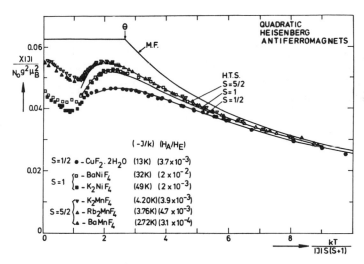

Fig. 7.41. The susceptibility of six examples of the quadratic Heisenberg antiferromagnet. The experimental data in the high-temperature region ($T > T_c : kT \simeq \mathscr{S}(\mathscr{S} + 1)|J|$) have been fitted to the theoretical (solid) curves by varying the exchange constants J/k. These curves have been calculated from the high-temperature series expansions for the susceptibility (H.T.S.). Note the large deviation from the molecular field result (MF) for the susceptibility in the paramagnetic region and below the transition temperature ($\chi_\perp$). Below T_c the measured perpendicular susceptibilities of two $\mathscr{S} = \frac{5}{2}$ and two $\mathscr{S} = 1$ compounds have been included. The differences between $\chi_\parallel$ for compounds with the same $\mathscr{S}$ reflect the difference in anisotropy. From Ref. [1]

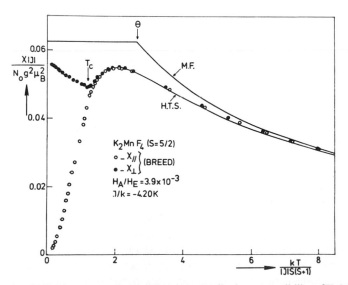

Fig. 7.42. The measured parallel and perpendicular susceptibility of K_2MnF_4, which is an example of the quadratic $\mathscr{S} = \frac{5}{2}$ Heisenberg antiferromagnet. The value of J/k has been determined by fitting the high-temperature susceptibility to the series expansion prediction (H.T.S.). The value of the anisotropy parameter H_A/H_E and the transition temperature T_c have been indicated. From Ref. [97]

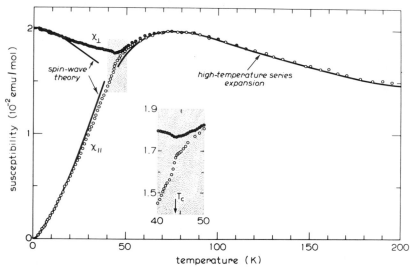

Fig. 7.43. Recent susceptibility measurements on K_2MnF_4. From Ref. [103]

Planar copper compounds tend to order ferromagnetically, but there are several two-dimensional antiferromagnets of copper. The first, $[Cu(C_5H_5NO)_6](BF_4)_2$, consists of discrete octahedra at room temperature but distorts cooperatively as it is cooled. Specific heat studies [69] were consistent with two-dimensional behavior. The specific heat of (benzylammonium)$_2$[Cu(oxalate)$_2$] likewise [102] exhibits a broad maximum characteristic of planar interactions. The exchange constant is only $-0.145\,K$, and long-range order occurs at $0.116\,K$.

The magnetic phase diagrams of relatively few two-dimensional Heisenberg antiferromagnets have been determined, one recent example being K_2MnF_4 [103]. The differential susceptibility was remeasured at zero-applied field, the data being illustrated in Fig. 7.43. The broad maximum characteristic of lower-dimensional ordering is apparent, along with the fairly isotropic behavior above 50 K. The ordering temperature was determined as 43.9 K, and the exchange constant, obtained from fitting the high temperature data, is 4.20 K. The low-temperature data were analyzed by what is called spin-wave theory, valid only for $T \ll T_c$, and resulted in a temperature independent, uniaxial anisotropy parameter $\alpha = H_A/H_E = 3.9 \times 10^{-3}$. (Recall that H_A is the anisotropy field and H_E is the exchange field.) Measurements of the susceptibility as the applied field was varied allowed the phase boundary (Fig. 7.44) to be elucidated. No evidence for the spin-flop-paramagnetic boundary was apparent in the susceptibility data, though both first order and second order boundaries were found. A more complete phase diagram was determined by neutron scattering [103].

The material $CuF_2 \cdot 2H_2O$ is a planar antiferromagnet which undergoes long-range order at 10.9 K. The anisotropy and exchange fields have been estimated as $H_A = 1.3\,kOe$ (0.13 T) and $H_E = 370\,kOe$ (37 T), and the spin-flop boundary has been determined [104].

The only apparent examples known to date of planar or 2D magnets that follow the XY model are $CoCl_2 \cdot 6H_2O$ and $CoBr_2 \cdot 6H_2O$ [62, 105, 106]. Again theory predicts

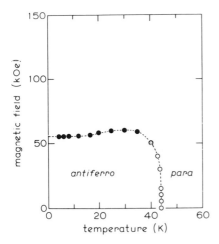

Fig. 7.44. Phase diagram of K_2MnF_4. From Ref. [103]

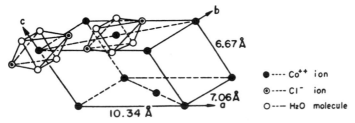

6.67 Å

●---- Co^{++} ion

◎---- Cl$^-$ ion

○--- H$_2$O molecule

7.06 Å

10.34 Å

Fig. 7.45. Structure of monoclinic $CoCl_2 \cdot 6 H_2O$ projected on [100]. There are two formula units per unit cell. After Ref. [108]

that long-range order will set in for this system only at $T=0$, but non-ideality in the form of in-plane anisotropy $(J_x, J_y,$ and $J_z)$ and interplanar interactions (J') causes three-dimensional ordering to occur.

The structure of both these crystals consists of discrete *trans*-$[Co(H_2O)_4X_2]$ octahedra linked by hydrogen-bonds in a face-centered arrangement as illustrated in Fig. 7.45; the two-dimensional character arises in the ab (XY) plane, for the molecules are closer together in this direction and provide more favorable superexchange paths than along the c-axis. The g-values are approximately $g_\perp = 5.1$ and $g_\parallel = 2.5$ for the chloride, so there is a strong preference for the moments to lie within a plane that nearly coincides with the $Co(H_2O)_4$ plane. This is a precondition, of course, for the applicability of the XY model. Furthermore, the large degree of short-range order required for the two dimensional model to be applicable may be inferred from the fact that some 40 to 50% of the magnetic entropy is gained above T_c.

Thus, the specific heat of $CoCl_2 \cdot 6 H_2O$ (Fig. 7.46) at high temperatures follows the theoretical curve for the XY model, but a λ-type peak is superimposed on the broad maximum. The parameters are [62] $T_c = 2.29$ K and $J_x/k = -2.05$ K, with only about a 4% anisotropy in the plane, while it is found that $J_z/J_x = 0.35$, which illustrates the substantial intraplane anisotropy. Furthermore $|J'/J|$, where J' is the interplane exchange, is about 10^{-2}.

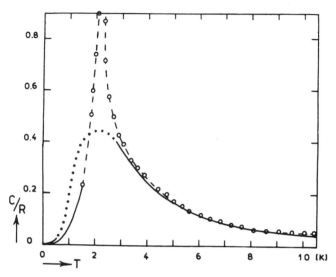

Fig. 7.46. Specific heat data on $CoCl_2 \cdot 6\,H_2O$ compared to the prediction of the quadratic, $\mathscr{S} = \frac{1}{2}$, XY model. From Ref. [62]

The magnetic structures of $CoCl_2 \cdot 6\,D_2O$ and $CoBr_2 \cdot 6\,D_2O$ have been investigated by neutron diffraction [107]. Part of the interest in the system derived from the fact that earlier nuclear resonance results had suggested the unusual result that both the magnetic and crystallographic structures of these molecules differed from their protonic analogs. In fact the neutron work showed that the deuterates undergo a crystallographic transition as they are cooled, going from monoclinic to a twinned triclinic geometry. Although the magnetic structure of the deuterohydrates differs slightly from that of the hydrates by a rotation of the moments away from the ac-plane over an angle of 45°, both compounds remain as examples of the 2D XY model.

7.7 CaCu(OAc)$_4 \cdot 6\,H_2O$

This compound has had a tortuous, recent history in magnetic studies. The problem with it is that it is an interesting compound from a structural point of view. This has led several investigators to postulate the presence of measurable magnetic interactions which other investigations have never been able to verify.

Thus, a portion of the tetragonal structure [109] is illustrated in Fig. 7.47. The structure consists of polymeric chains of alternating copper and calcium ions, which are bridged by bidentate acetate groups. These chains, aligned along the c-axis, are bound together by solvent cages of 12 water molecules. Early magnetic susceptibility measurements extended over the temperature region above 80 K, and the results of one study [110] were interpreted on the basis of Fisher's linear chain Ising model. This model with a value of $J/k = -1.4$ K for the exchange parameter, was reported to fit the data very well. It was also noted that the data yielded an isotropic value of -10 K for the Weiss constant.

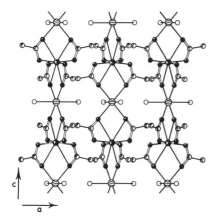

Fig. 7.47. Portion of a polymeric chain of CaCu(OAc)$_4 \cdot 6$H$_2$O. After Ref. [109]

It will be observed that in the high temperature limit, Eqs. (7.4) and (7.5) reduce to the Curie law for an $\mathscr{S} = \frac{1}{2}$ system. Since both of the parameters listed above indicate a significant amount of exchange, and since copper is not generally an Ising ion, the single crystal susceptibilities were measured [111] in the temperature region below 20 K. The purpose was to study the magnetic properties in a temperature region where the exchange interaction makes a more significant contribution to the measured quantity.

In Fig. 7.48 is shown the behavior calculated from Fisher's equation in the low temperature region with the exchange parameter $J/k = -1.4$ K. Only the parallel susceptibility is illustrated, but the following conclusion also applies to the perpendicular direction. Two sets of data measurements [111] through the temperature region 1–20 K are illustrated, and it will be seen that experiment does not agree in any fashion with the calculated behavior. That is, Curie-Weiss behavior with a θ of only -0.03 K is observed [111–113] and thus only a weak exchange interaction is in fact present.

Specific heat studies below 1 K provided no evidence for significant magnetic interactions [113], but single crystal susceptibilities at very low temperatures finally clarified the situation [114]. The susceptibilities parallel and perpendicular to the principal axis exhibit broad maxima at about 40 mK. The data were successfully fit, not to the linear chain model, but to a nearly-Heisenberg two-dimensional antiferromagnet! Only the slightest anisotropy was observed, with $J_{c\parallel}/k = -21.4$ mK and

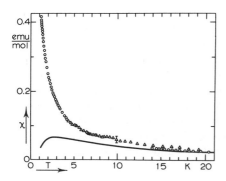

Fig. 7.48. Experimental parallel susceptibility of CaCu(OAc)$_4 \cdot 6$H$_2$O compared to the theoretical prediction for a linear chain with $J/k = -1.4$ K. From Ref. [111]

$J_{c\perp}/k = -22.5\,\text{mK}$. Long-range order has not been observed above 27 mK. The very weak exchange interactions are really quite consistent with the geometrical isolation of the copper ions from one another.

A final note concerns the specific heat of $CaCu(OAc)_4 \cdot 6\,H_2O$ in the helium region [113, 115] which exhibits a broad, Schottky-like maximum that evidently, from the discussion above, must be non-magnetic in origin. Reasoning by analogy to the reported [116] behavior of some hexammine nickel salts, which provide specific heat maxima which are due to hindered rotation of the NH_3 groups about the $M–NH_3$ axis, this maximum was analyzed in terms of hindered rotation of the methyl groups of the acetate ligand. Substitution of deuterium on the methyl group gave consistent results, and illustrates nicely the fact that all phenomena observed with magnetic compounds at low temperature need not be magnetic in origin.

7.8 Metamagnetism

Antiferromagnets with a large anisotropy do not show a spin-flop phase but, in the presence of competing interactions, they may undergo a first-order transition to a phase in which there exists a net magnetic moment. Such compounds are usually referred to as metamagnets [117] and are among the more complicated ones in analyzing the field-dependent ac susceptibility (Chap. 11). Let us emphasize this topic by discussing the relatively simple example, $[(CH_3)_3NH]CoCl_3 \cdot 2\,H_2O$ [118]. This metamagnet [119] consists of ferromagnetically coupled chains of cobalt ions along the b-axis ($J_0/k \simeq 15.4\,\text{K}$) which are coupled by a ferromagnetic interaction $J_1/k \simeq 0.17\,\text{K}$ in the c-direction. Then these ferromagnetically coupled planes order antiferromagnetically below $T_c = 4.18\,\text{K}$ due to a weak interaction between the bc-planes ($J_2/k \simeq -0.007\,\text{K}$). If an external magnetic field is applied parallel to the easy axis of antiferromagnetic alignment, then at a certain field value the antiferromagnetic arrangement is broken up. The system exhibits a first-order phase transition toward a ferromagnetic state and the transition field (extrapolated to $T = 0\,\text{K}$) is in fact a direct measure for J_2. For $[(CH_3)_3NH]CoCl_3 \cdot 2\,H_2O$ examining $\chi(H)$ for $H\|c$ leads [120] to a phase diagram as shown in Fig. 7.49. The figure shows three distinct regions, which meet at a temperature of 4.13 K. This diagram resembles strongly the spin-flop examples given earlier, but the phase lines have a completely different meaning. The difference in meaning is as follows.

With ferromagnets (recall, from Chap. 6), the net internal magnetic moment is large and acts to repel the external field. This is described by means of a sample-shape-dependent demagnetizing field which tends to counteract the external field, causing the internal field to differ from the external field.

In the weakly anisotropic systems that exhibit spin-flop, demagnetizing effects are often small enough to be ignored, and one can equate the applied field H_a with the more relevant internal field H_i ($= H_a - \lambda M$, λ being the demagnetizing factor). This is generally not possible with metamagnetic systems, however, as ΔM at the first-order transition is no longer negligible. For such a system it is not possible for all spins to change orientation at the value of H_a where H_i reaches its critical value, because as soon as a number of spins are reoriented, H_i is reduced and no further transitions are possible. As a result there will be a region of H_a, of magnitude $\lambda\Delta M$, where the phase

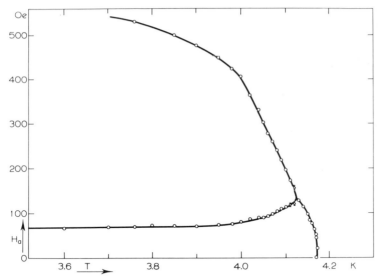

Fig. 7.49. Phase diagram of the metamagnet, $[(CH_3)NH]CoCl_3 \cdot 2H_2O$. The ordinate indicates the applied field. From Ref. [120]

transition takes place and where H_i remains constant at its critical value. Thus, the isothermal susceptibility, χ_T, over the "seemingly spin-flop" phase in Fig. 7.49 remains constant at the value $1/\lambda$, and this phase is in fact a mixed (paramagnetic and antiferromagnetic) phase in which the transition takes place gradually with increasing H_a. This discontinuity in the magnetization disappears at the tricritical temperature T_t, and the mixed phase of Fig. 7.49 joins the second-order antiferromagnetic-paramagnetic boundary at this temperature (4.13 K).

If plotted vs. the internal field, the phase diagram appears as a single smooth curve (Fig. 7.50), consisting of a line of first-order points at temperatures below T_t, going over to a line of second-order points above T_t. Tricritical points have been of enormous interest recently to both theorists and experimentalists. The latter group should be aware that in the above it is assumed that the ac susceptibility equals χ_T, the isothermal susceptibility, which is often not true. In the above-mentioned experiments large

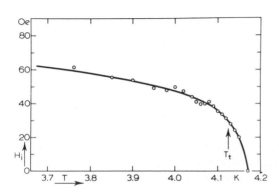

Fig. 7.50. Magnetic phase diagram of $[(CH_3)_3NH]CoCl_3 \cdot 2H_2O$. The ordinate is the internal field, in orientation parallel to the c-axis of the crystal. The tricritical point is indicated. From Ref. [120]

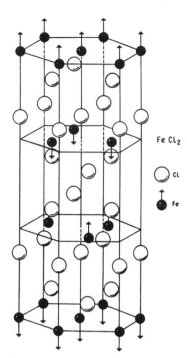

Fe Cl₂

○ Cl

● Fe

Fig. 7.51. The crystal and magnetic structures of FeCl₂

relaxation effects occurred in the mixed-phase region and the experimental determination of χ_T becomes very difficult (cf. Chap. 11).

Metamagnets, then, are found as systems with large magnetic anisotropy, such as in Ising systems and in systems with strong ferromagnetic interactions. Note that metamagnetism is defined only in terms of the behavior of a substance in the presence of an applied magnetic field, and that the phase diagram differs substantially from that found with weakly anisotropic systems.

The compound $FeCl_2$ is perhaps the best-known example of a metamagnet. The crystal and magnetic structures of $FeCl_2$ are illustrated in Fig. 7.51, where the compound will be seen to consist of layers of iron atoms, separated by chloride ions. The coupling between the ions within each layer is ferromagnetic, which is the source of the large anisotropy energy that causes the system to behave as a metamagnet, and the layers are weakly coupled antiferromagnetically. Because of the large anisotropy, the transition from the AF to paramagnetic phase occurs in the fashion described above with, for $T \ll T_c$, a discontinuous rise of the magnetization at the transition field to a value near to saturation. The isothermal magnetizations illustrated in Fig. 7.52 are qualitatively different in behavior from those in Fig. 6.15. The phase diagram that results is also given in Fig. 7.52, where it will also be noted that the boundary above about 20.4 K is dashed rather than solid. This is because the tricritical point of $FeCl_2$ occurs at about 20.4 K.

A tricritical point is found to occur in the phase diagram with metamagnets as well as in certain other systems, such as $^3He-^4He$. The metamagnet must have important ferromagnetic interactions, even though the overall magnetic structure will be antiferromagnetic. In the H-T plane, then, although a continuous curve is observed as

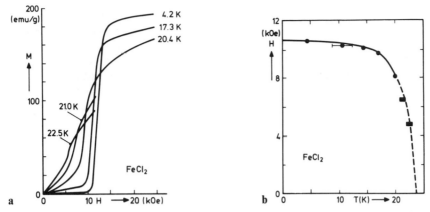

Fig. 7.52. a) Magnetization isotherms of the metamagnet $FeCl_2$ as measured by Jacobs and Lawrence (1967). The transition temperature is 23.5 K. b) The metamagnetic phase diagram of $FeCl_2$ showing the first-order transition line (solid curve) and the second-order line (dashed curve). The tricritical point is estimated to be at 20.4 K. From Ref. [1]

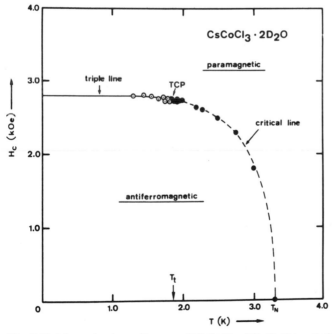

Fig. 7.53. Magnetic phase diagram of $CsCoCl_3 \cdot 2 D_2O$. From Ref. [123]

in Fig. 7.53b, it consists of a set of first-order points at low temperature which, at the tricritical point, goes over into a set of second-order points. Tricritical points have been observed in $FeCl_2$ by means of magnetic circular dichroism [121, 122] and in $CsCoCl_3 \cdot 2 D_2O$ by neutron measurements [123]. The phase diagram of $CsCoCl_3 \cdot 2 D_2O$ is illustrated in Fig. 7.53.

One last point worth mentioning about metamagnets is that it has been found that a few metamagnets undergo transitions at quite low fields. In particular, the compound $[(CH_3)_3NH]CoCl_3 \cdot 2H_2O$ undergoes a transition at 64 Oe (0.0064 T) while the transition field increases to 120 Oe (0.012 T) when bromide replaces chloride in this compound. More impressively, $Mn(OAc)_2 \cdot 4H_2O$ undergoes a transition at a mere 6 Oe (0.0006 T) [124]! The reason why the transition fields are so low in these compounds is not clear yet, but they all have a crystal lattice in which substantial hydrogen bonding occurs, and they all exhibit canting (cf. below). The dipolar interaction may also be important. Certainly these results suggest that caution be applied by doing experiments in zero-field before they are carried out in the presence of a large external field.

7.9 Canting and Weak Ferromagnetism

This subject was introduced in Chap. 6. Many lower-dimensional systems exhibit canting, and a number of linear chain systems have already been discussed. The system $RbCoCl_3 \cdot 2H_2O$ behaves [125] much like $CsCoCl_3 \cdot 2H_2O$, in that it is a canted antiferromagnetic chain, with a canting angle of about $17°$ in both compounds, yet the two are not isomorphous. The chain in $CsCoCl_3 \cdot 2H_2O$ consists of singly-bridged cis-$[CoCl_4(H_2O)_2]$ octahedra, while singly-bridged trans-$[CoCl_4(H_2O)_2]$ octahedra are found in the rubidium salt. Hydrogen bonding is extensive in both. There is a striking difference between the two compounds in their field dependent behavior which may arise from the structural differences. In $RbCoCl_3 \cdot 2H_2O$ a metamagnetic phase transition is found at 37 Oe (0.0037 T), but the transition occurs at 2.85 kOe (0.285 T) in the cesium salt. The metamagnetic transitions involve the same reversal of moments in alternate planes of the structures.

The compounds $RbFeCl_3 \cdot 2H_2O$ and $CsFeCl_3 \cdot 2H_2O$ are isomorphous to $CsMnCl_3 \cdot 2H_2O$ and $CsCoCl_3 \cdot 2H_2O$. They are again linear chain canted antiferromagnets, but the exchange is strong, causing the one-dimensional properties to occur at higher temperatures. Thus, $J/k = 39$ K, and $T_c = 11.96$ K for the Rb salt. They are especially good examples of the Ising model for the crystal field resolves most of the degeneracy and leaves a pseudo-doublet as the ground state with $g_\parallel \simeq 9.6$ [126, 127].

What is most interesting about these isomorphous salts is their magnetic structure below T_c and their field-dependent behavior for there is a difference between the two salts [126]. As illustrated in Fig. 7.54, the interactions are antiferromagnetic along the a-axis, which is the chain axis; the moments actually are canted and lie in the ac-plane at an angle ϕ_m which is some $15°$. The coupling along the c-direction is ferromagnetic in $RbFeCl_3 \cdot 2H_2O$ but antiferromagnetic in $CsFeCl_3 \cdot 2H_2O$. Both salts undergo two metamagnetic transitions in fields applied along the c-axis, first to a ferrimagnetic phase and then to a ferromagnetically-ordered phase. The phase diagram is illustrated in Fig. 7.55, and the likely spin arrays in Fig. 7.56. In the ferrimagnetic phase there are three different arrays with the same energy.

Canted spin structures seem to occur more commonly with linear chain systems than with two or three dimensional materials, but this may only be a result of the intense recent study of one-dimensional systems. A two-dimensional system that has recently been shown to exhibit canting is $(n\text{-}C_3H_7NH_3)_2MnCl_4$ [128]. Octahedral

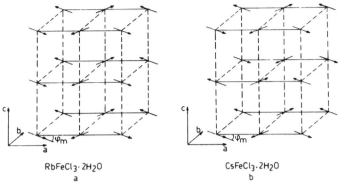

Fig. 7.54. Array of the magnetic moments in the ordered state for $RbFeCl_3 \cdot 2H_2O$ (a) and $CsFeCl_3 \cdot 2H_2O$ (b). All moments are located in the ac-plane at an angle ϕ_m from the a-axis. From Ref. [126]

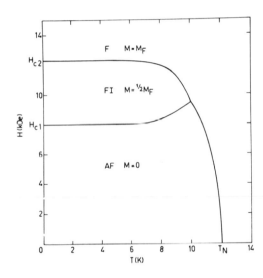

Fig. 7.55. Schematic phase diagram of $RbFeCl_3 \cdot 2H_2O$. From Ref. [126]

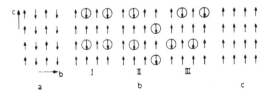

Fig. 7.56. Arrays of the ferromagnetic components of the chains. (a) Antiferromagnetic array of the chains. (b) Three possible ferrimagnetic arrays (the down moments are encircled). (c) Ferromagnetic array. From Ref. [126]

$MnCl_6$ units are linked together into a planar structure by Mn–Cl–Mn bonds, as illustrated in Fig. 7.57; an important feature of the structure is that the octahedra are canted alternatingly towards the c- and $-c$-directions, making an angle of 8° with the b-axis. The susceptibility (Fig. 7.58) exhibits a broad maximum at about 80 K, which is as expected for a two-dimensional antiferromagnet. A sharp peak in the zero-field susceptibility not only defines accurately the three-dimensional ordering temperature,

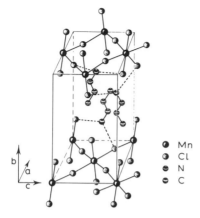

● Mn
◑ Cl
● N
● C

Fig. 7.57. Three dimensional view of the crystal structure of $(n\text{-}C_3H_7NH_3)_2MnCl_4$. Only half a unit cell has been drawn. The canted $MnCl_6$ octahedra form a two-dimensional network in the ac-plane. For clarity only three of the propylammonium groups are drawn. From Ref. [128]

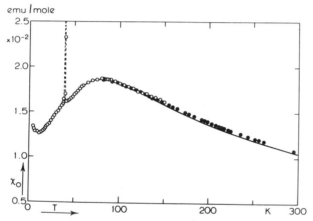

Fig. 7.58. The susceptibility vs. T for a powdered sample of $(n\text{-}C_3H_7NH_3)_2MnCl_4$. O: ac susceptibility measurements; ●: susceptibilities obtained with a Faraday balance. Drawn line: susceptibility calculated for the 2D quadratic Heisenberg antiferromagnet with $\mathscr{S} = \frac{5}{2}$ and $J/k = -4.45\,\text{K}$. From Ref. [128]

$T_c = 39.2\,\text{K}$, but also indicates that the material is a weak ferromagnet. The high-temperature data can be fit to Navarro's calculation for the $\mathscr{S} = \frac{5}{2}$ Heisenberg quadratic layer antiferromagnet, with $J/k = -4.45\,\text{K}$ (at 80 K). The ratio $\alpha_A = H_A/H_E = 3 \times 10^{-4}$ illustrates the very small (relative) anisotropy of the compound, which was ascribed to zero-field splitting competing with (i.e., opposed to) the dipolar anisotropy.

The analogous bromide, which is probably isostructural, exhibits a similar broad maximum in the susceptibility and is described by essentially the same value of the exchange constant, $J/k = -4.50\,\text{K}$. This is found despite the fact that the Mn–Mn intraplanar separation is 0.556 nm for the bromide, while it is 0.523 nm for the chloride. Thus, although the larger bromide ion separates the metals more than does the chloride, there is also a larger extension of the bromides' valence electron shell. Or, in other words, the bromide is larger but also more polarizable. The bromide is not a canted antiferromagnet, which is an odd result.

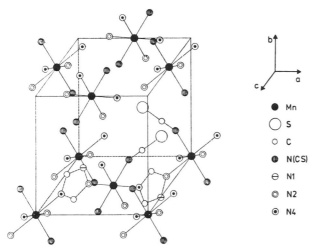

Fig. 7.59. The crystal structure of orthorhombic $Mn(trz)_2(NCS)_2$, where trz is 1,2,4-triazole. The top half of the unity cell has been omitted and for clarity only two triazole molecules (without hydrogen atoms) and two thiocyanate groups are shown. From Ref. [132]

The weak ferromagnetism in the chloride appears to be due to antisymmetric exchange, which is allowed because of the canted crystal structure. A more thorough study [129] on single crystals was able to explore the effects of domains on the ac susceptibility of the chloride.

An interesting series of layered complexes of 1,2,4-triazole with divalent metal thiocyanates has recently been investigated by Engelfriet [130–137]. A portion of the structure of these molecules is illustrated in Fig. 7.59; though the molecules are not isostructural the general features are all similar. The essential feature is that triazole-bridged layers are formed with tilted MN_6 octahedra. Each metal ion is coordinated by two N2 and two N4 nitrogen atoms originating from triazole rings and by two nitrogen atoms from the thiocyanate groups. The strongest exchange between metal ions takes place in the ac layer via the bridging triazole molecules; each metal ion is connected equivalently to four nearest magnetic neighbors. The superexchange path between the layers consists of an S atom in one layer hydrogen-bonded to a triazole ligand in the next layer, and is therefore quite weak.

The single-crystal susceptibilities of antiferromagnetic $Mn(trz)_2(NCS)_2$ are illustrated in Fig. 7.60, where it will be seen that there is the anticipated broad maximum, characteristic of a layered compound, at about 4.3 K. The data may be fit by the Heisenberg model with an exchange constant of about -0.25 K, and long-range order occurs at 3.29 K. Though the presence of canting is allowed because of the tilted octahedra which occur in the structure of this system, no canting is observed in the zero-field data.

The situation is different, however, with the iron(II) salt and the cobalt(II) salt, both of which are planar magnets with hidden canting. Broad featureless maxima characteristic of two-dimensional antiferromagnets are observed in the easy axis susceptibility of both compounds, but a sharp peak is observed in the zero-field data in one direction perpendicular to the preferred axis. This feature of the properties of the

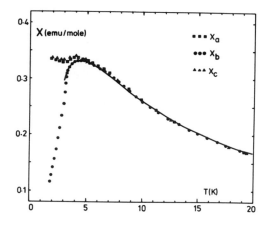

Fig. 7.60. Magnetic susceptibility of Mn(trz)$_2$(NCS)$_2$ measured along the three orthorhombic axes at a field strength of 5.57 kOe (0.557 T). The full line represents the fit to the series expansion for $\mathscr{S} = \frac{5}{2}$ with $J/k = -0.25$ K. From Ref. [132]

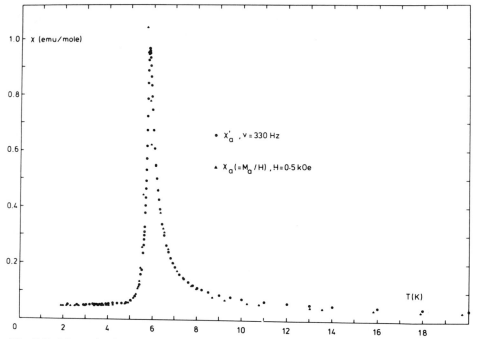

Fig. 7.61. Magnetization and *ac* susceptibility of Co(trz)$_2$(NCS)$_2$, as a function of temperature, measured along the *a*-axis. From Ref. [134]

cobalt compound is illustrated in Fig. 7.61; as T_c is approached from above, there is an enormous increase in the susceptibility, due to the development of the ordered, canted structure. At the temperature T_c (5.74 K) and below, a four-sublattice structure, perhaps as illustrated in Fig. 6.28b is locked in, the canted moments being arrayed in a fashion such that there is no net moment, and the susceptibility drops to zero.

Another substance which behaves in similar fashion is Co(urea)$_2$Cl$_2 \cdot 2$H$_2$O [138]. A view of the structure, Fig. 7.62, illustrates the planarity of the lattice. The zero-field

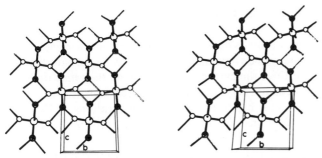

Fig. 7.62. Stereo view of $[CoCl_2(H_2O)_2]_\infty$ network in the *bc*-plane of $Co(urea)_2Cl_2 \cdot 2 H_2O$. The urea ligands, which lie above and below the plane, are omitted for clarity. The filled circles are the chlorine atoms. From Ref. [138]

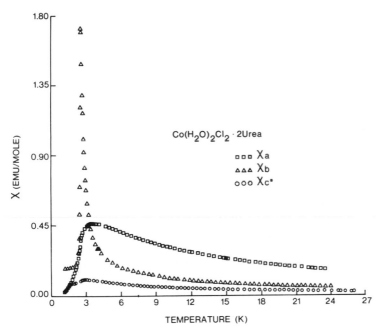

Fig. 7.63. The three orthogonal susceptibilities of monoclinic $Co(urea)_2Cl_2 \cdot 2 H_2O$. From Ref. [138]

susceptibilities, Fig. 7.63, resemble those of $Co(trz)_2(NCS)_2$; χ_a in this Ising system has the broad maximum anticipated for a measurement parallel to the preferred axis of spin alignment. Both χ_b and χ_{c*} (where $c*$ is perpendicular to a and b, a common choice of axes for a monoclinic system) level off below T_c, consistent with their being the perpendicular or hard axes, but note the sharp spike in χ_b near T_c. The magnitude of χ_b is too large for that of a simple antiferromagnet and too small to suggest ferromagnetism. No absorption (χ'') was noted in χ_b over the whole measured temperature range. Such data are characteristic of a system with hidden canting.

7.10 Some Ferromagnetic Linear Chains

The dominant interaction in linear chain systems, by definition, is the one-dimensional nearest neighbor one. This may be antiferromagnetic or ferromagnetic in nature; when the chains interact to form the three-dimensional ordered state, they may do so in several ways. Ferromagnetic chains may interact antiferromagnetically or ferromagnetically, for example. Several hypothetical spin configurations of the two-sublattice ordered state are illustrated in Fig. 7.64; helical and other unusual configurations are also possible.

The linear chain interaction remains, by and large, the most interesting one in a system of ordered linear chains. Antiferromagnetically ordered chains are the most common, but there have been a number of investigations recently of ferromagnetically aligned one-dimensional systems. We refer the reader to the article of Willett et al. for an extensive review [18]. Examples of such compounds discussed there include $(C_6H_{11}NH_3)CuCl_3$ (the cation is cyclohexylammonium and the compound is frequently referred to as CHAC), $CuCl_2 \cdot DMSO$ [the ligand is $(CH_3)_2SO$], $CsNiF_3$ and $[(CH_3)_3NH]CoCl_3 \cdot 2H_2O$. The last compound is one of a series, and is a metamagnet as discussed above. This extensive series [139] of compounds will now be discussed. For the most part, the compounds show a large degree of linear chain behavior, but a number of other features are of interest. In particular, they exhibit remarkably well a correlation of magnetic properties with structure.

The crystal structure of $[(CH_3)_3NH]MnCl_3 \cdot 2H_2O$ is illustrated in Fig. 6.36; this is the basic structure of all the molecules in this series, even for those which belong to a different space group. The Co/Cl [118] and Mn/Cl compounds are orthorhombic, while the Cu/Cl [140] and Mn/Br analogs are monoclinic, with angles β not far from $90°$. Chains of edge-sharing octahedra are obtained by means of μ-di-halo bridges; the coordination sphere is $trans$-$[MCl_4(H_2O)_2]$, and it will be noticed that the O–M–O axis on adjoining octahedra tilt in opposite directions. This has important implications for the magnetic properties, as it allows canting. Furthermore, the chains are linked together, weakly, into planes by another chloride ion, Cl_3, which is hydrogen-bonded to the water molecules in two adjoining chains. The trimethylammonium ions lie between the planes and act to separate them both structurally and magnetically.

It is convenient to begin this discussion of the magnetic properties of this series of compounds by turning first to the monoclinic copper salt, $[(CH_3)_3NH]CuCl_3 \cdot 2H_2O$. All the significant magnetic interactions are weak and occur below 1 K [140]. Both susceptibility [141] and specific heat [142, 143] studies show that long-range order sets in at only about 0.16 K. The principle interactions are ferromagnetic, and in fact the crystal is the first example of a Heisenberg linear chain in which ferromagnetic interactions were found to predominate. The magnetic specific heat is illustrated in

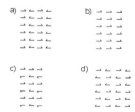

Fig. 7.64. Some hypothetical spin configurations by which ordered nearest-neighbor chains can interact. In each case, the chain axes are vertical. a) Ferromagnetic chains forming an antiferromagnetic 3D system. b) Ferromagnetic chains interacting ferromagnetically. c) Antiferromagnetic chains interacting ferromagnetically. d) Antiferromagnetic chains interacting antiferromagnetically

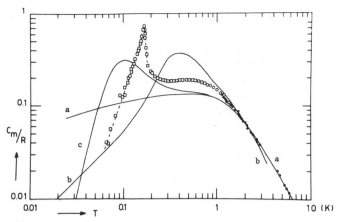

Fig. 7.65. The magnetic specific heat of $[(CH_3)_3NH]CuCl_3 \cdot 2H_2O$ compared with different theoretical model predictions. Curves a, b, and c are theoretical fits to the data. The broken curve is a guide to the eye. From Ref. [143]

Fig. 7.65, where curve a is the prediction for a Heisenberg ferromagnetic chain with $J/k=0.80\,K$; note in particular the flatness of the curve over a large temperature interval, and compare with the theoretical curves in Fig. 7.18. Curve b, which does not fit the data adequately, corresponds to the quadratic ferromagnet, while curve c, the best fit, is the prediction for a ferromagnetic chain with slight (Ising) anisotropy, $J_{\parallel}/k=0.85\,K$, $J_{\perp}/k=0.9\,J_{\parallel}/k$. These data have been used [143] for a discussion of lattice-dimensionality crossover. This phenomenon is concerned with the shape of the specific heat curve for a system which behaves as a linear chain at higher temperatures but whose dimensionality increases (from 1D to 2D or 3D) as the temperature is lowered. Broad maxima in the specific heat are expected in Heisenberg systems which are either 1D or 2D, but the transition to long-range, 3D order gives rise to a λ-anomaly. Deviations in the data for $[(CH_3)_3NH]CuCl_3 \cdot 2H_2O$ from ideal 1D behavior have been attributed to this phenomenon. The result of this analysis is that the three orthogonal exchange constants for this compound are of relative magnitudes $1:0.05:0.015$. Recent EPR experiments [144] are consistent with this interpretation.

It is interesting to compare the magnetic behavior observed for this compound with that of $CuCl_2 \cdot 2H_2O$ and $CuCl_2 \cdot 2C_5H_5N$. A variety of low-temperature techniques have been used to study the magnetic properties of $CuCl_2 \cdot 2H_2O$. These have revealed a typical antiferromagnet displaying a transition to an ordered state at 4.3 K. The crystal structure consists of chains similar to those found in $[(CH_3)_3NH]CuCl_3 \cdot 2H_2O$. These chains are connected by the hydrogen bonds formed between the water molecules of one chain and the chlorine atoms of the neighboring chains. It has been established by NMR [146] and neutron-diffraction [147] studies that the antiferromagnetic transition in $CuCl_2 \cdot 2H_2O$ results in sheets of ferromagnetically ordered spins in the ab-planes with antiparallel alignment in adjacent ab-planes. The susceptibility measured in the a-direction goes through a broad maximum at about 4.5–5.5 K [148]. Although there is no clear experimental proof of magnetic chain behavior along the c-axis other than the presence of short-range order, several calculations have estimated the exchange in this direction.

Measurements [149] of the heat capacity suggest substantial short-range order above the Néel temperature since only about 60% of the magnetic entropy has been acquired at the transition temperature.

In the case of $CuCl_2 \cdot 2 C_5H_5N$ [150], square coplanar units aggregate into nearly isolated antiferromagnetic chains by weaker Cu–Cl bonds which complete the distorted octahedron about each copper. The two nonequivalent Cu–Cl bond distances differ more in this structure than those in either $[(CH_3)_3NH]CuCl_3 \cdot 2 H_2O$ or $CuCl_2 \cdot 2 H_2O$. This compound exhibits [151–153] a broad maximum at 17.5 K in the susceptibility which is characteristic of a magnetic linear chain. Long-range order sets in at 1.13 K. The specific-heat measurements yield a value of $J/k = -13.4$ K.

An interesting feature of these three compounds is the relative magnitude of the exchange within the chemical chains. The values for the exchange parameters are as follows: $J/k < 1$ K for $[(CH_3)_3NH]CuCl_3 \cdot 2 H_2O$, $J/k \approx -7$ K for $CuCl_2 \cdot 2 H_2O$, and $J/k \approx -13$ K for $CuCl_2 \cdot 2 C_5H_5N$, where the values have all been derived from a Heisenberg model. It appears that the compound with the most asymmetry in the $(CuCl_2-)_n$ unit, i.e., $CuCl_2 \cdot 2 C_5H_5N$, has the largest exchange while the least asymmetric unit, in $[(CH_3)_3NH]CuCl_3 \cdot 2 H_2O$, displays clearly the least amount of exchange along the chemical chain. It is likely that the exchange interactions in $CuCl_2 \cdot 2 H_2O$ causing the transition itself will in some way influence the exchange along the chain. However, it is also perhaps likely that the nearest-neighbor interaction as reflected in the asymmetry will make the largest contribution in determining this exchange.

Now, $[(CH_3)_3NH]CuCl_3 \cdot 2 H_2O$ has a lower transition temperature than $CuCl_2 \cdot 2 H_2O$. If the transition in $CuCl_2 \cdot 2 H_2O$ is assumed to be largely determined by the exchange within the ab-plane then the lower transition temperature of the trimethylammonium compound can be hypothesized to be the result of the intervening anionic chloride ions between the chains, which reduce the predominant mode of such exchange. This exchange possibility has been all but eliminated in $CuCl_2 \cdot 2 C_5H_5N$ because of the insulating (diluting) effects of the pyridine molecules. Spence [154, 155], using NMR, compared the exchange interactions in $MnCl_2 \cdot 2 H_2O$ and $KMnCl_3 \cdot 2 H_2O$ and pointed out the extremely delicate balance there is between the geometry and exchange effects. The bond distances within the coordinating octahedron in $MnCl_2 \cdot 2 H_2O$ are $(Mn–Cl_1) = 0.2592$ nm, $(Mn–Cl'_1) = 0.2515$ nm, and $(Mn–O) = 0.2150$ nm, while in $KMnCl_3 \cdot 2 H_2O$ they are $(Mn–Cl_1) = 0.2594$ nm, $(Mn–Cl'_1) = 0.2570$ nm, and $(Mn–O) = 0.218$ nm. From the NMR studies, the spins in $MnCl_2 \cdot 2 H_2O$ were found to order antiferromagnetically in chains, while the spins in $KMnCl_3 \cdot 2 H_2O$ are ordered ferromagnetically within dimeric units even though both compounds have the same basic octahedral edge-sharing coordination. In any event, very small changes in the coordination geometry are associated with profound changes in the exchange to the point where even the sign of the exchange has been reversed.

It is then reasonable to suggest that small changes in the coordination geometry for this series of copper compounds cause the effects observed in the magnitude of the exchange along the chemical chain.

The specific heat of $[(CH_3)_3NH]CoCl_3 \cdot 2 H_2O$ is illustrated in Fig. 7.66, where it will be observed that T_c is about 4.18 K, and that the λ-shaped peak is quite sharp [118]. The solid curve is the lattice heat capacity, which was estimated by means of a corresponding states calculation using the measured heat capacity of the copper

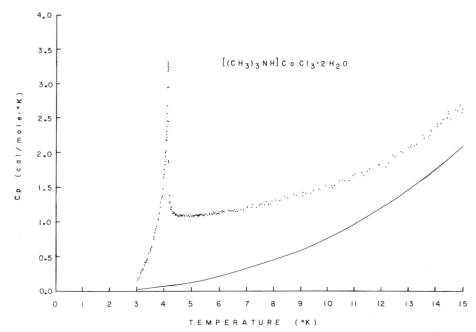

Fig. 7.66. Specific heat of $[(CH_3)_3NH]CoCl_3 \cdot 2H_2O$. The solid curve is the estimated lattice contribution. From Ref. [118]

analog, which is paramagnetic throughout this temperature region. The most significant fact about the measured specific heat is that a mere 8% of the theoretical maximum entropy for a mol of $\mathscr{S} = \frac{1}{2} Co^{2+}$ ions, $R \ln 2$, has been acquired at the transition temperature. This may be appreciated by looking at Fig. 7.67, a plot of the experimental magnetic entropy as a function of temperature. It is clear that substantial short-range order persists from T_c to well above 15 K in this compound. This short-range order is an indication of the lowered dimensionality of the magnetic spin system, and is consistent with the chain structure. Indeed, the magnetic heat capacity was successfully fitted to Onsager's complete solution for the rectangular Ising lattice, as illustrated in Fig. 7.68, with the very anisotropic parameters $|J/k| = 7.7$ K and $|J'/k| = 0.09$ K. Although the sign of the exchange constants is not provided by this analysis, the parameters are consistent with T_c as calculated independently from another equation of Onsager,

$$\sinh(2J/kT_c) \cdot \sinh(2J'/kT_c) = 1 .$$

(The parameter J/k reported above differs by approximately a factor of two from analyses by other authors because of a change in the definition of the Ising Hamiltonian.)

   These data have been analyzed slightly differently [143, 156], again from the point of view of lattice dimensionality crossover. The major difference between $[(CH_3)_3NH]CuCl_3 \cdot 2H_2O$ and $[(CH_3)_3NH]CoCl_3 \cdot 2H_2O$ is that while the former has about 10% spin anisotropy, the latter compound has about 75% spin anisotropy

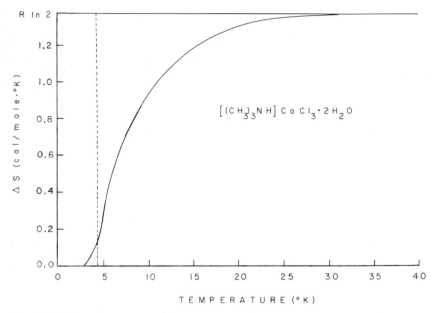

Fig. 7.67. Magnetic entropy of $[(CH_3)_3NH]CoCl_3 \cdot 2H_2O$ as a function of temperature. The dashed line indicates the ordering temperature. From Ref. [118]

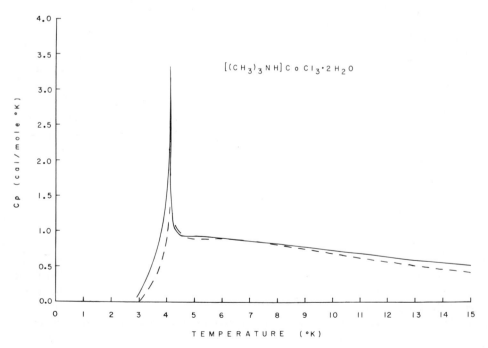

Fig. 7.68. Magnetic heat capacity (solid line) of $[(CH_3)_3NH]CoCl_3 \cdot 2H_2O$. The dashed line represents the fit to Onsager's equation. From Ref. [118]

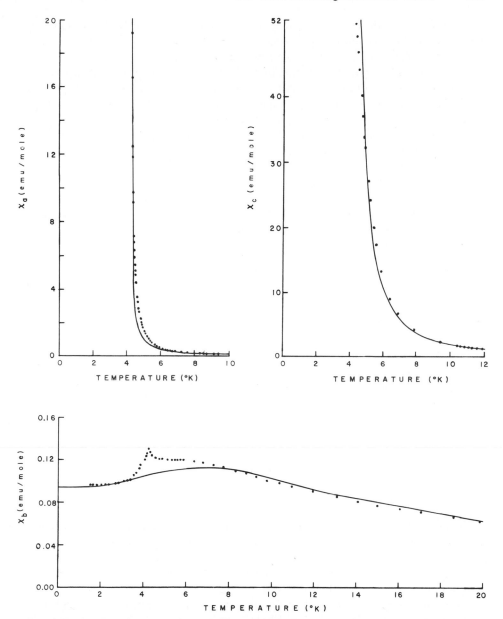

Fig. 7.69. Results of the best-fit analyses (solid lines) of the three principal axis magnetic susceptibilities of $[(CH_3)_3NH]CoCl_3 \cdot 2H_2O$. From Ref. [118]

and is therefore an Ising system. Since a two-dimensional Ising system can exhibit a singularity in the specific heat (in contrast to a Heisenberg system), the analysis in terms of crossover [156] showed that the system is an example of the $(d=1) \rightarrow (d=2)$ crossover. The system is a two-dimensional array of weakly coupled Ising chains, with the coupling in the third dimension as a small perturbation.

The field-dependence of the specific heat of $[(CH_3)_3NH]CoCl_3 \cdot 2H_2O$ has also been examined [157] and interpreted in terms of the Ising model.

The zero-field susceptibilities [118] provide not only the sign of the exchange constant but also a lot of other information about the spin-structure of this crystal. As illustrated in Fig. 7.69, the susceptibilities are quite anisotropic, and χ_b in particular can be fitted over the entire temperature region, both above and below T_c, quite well by Fisher's Ising chain equation, with a positive exchange constant.

The a-axis susceptibility retains a constant value below T_c, suggesting a weak ferromagnetic moment, and in fact χ_a could only be fitted by requiring the presence of a canted moment. The symmetry of the crystal allows this, for it will be recalled that a center of inversion symmetry is absent in this lattice because of the way the octahedra in a chain are successively tilted first one way and then the other. In contrast to the original analysis of the data [118], it has been pointed out [120] that the canting does not arise from an antisymmetric exchange term. Rather, it arises from the anisotropic g-values in combination with the tilted octahedra.

The susceptibilities of $[(CH_3)_3NH]CoCl_3 \cdot 2H_2O$ have recently been remeasured and analyzed [120]. The transition temperature is reported as 4.179(5)K, and the angle of canting is 22° towards the a-axis, in the ac-plane. The g-values are $g_c = 7.5$, $g_b = 3.82$, and $g_a = 3.13$, while the intrachain exchange constant is $J_b/k = 13.8(5)K$. The interchain interactions are given as $z_cJ_c/k = 0.28(5)K$ and $z_aJ_a/k = -0.0320(5)K$. There is thus a large degree of correlation between the crystal structure and the magnetic interactions. The ratio of the exchange constants is $1:1.1 \times 10^{-2}:1.2 \times 10^{-3}$; the two strongest interactions are ferromagnetic while the weakest (J_a) is antiferromagnetic. The arrangement of the spin-structure, showing the canting, is illustrated in Fig. 7.70.

The effect of deuterium substitution to form $[(CH_3)_3ND]CoCl_3 \cdot 2D_2O$ on the magnetic properties has also been examined [158]. Since J_b couples the cobalt ions by a

exchange path, deuterium substitution should have little effect on J_b. The exchange path involving J_c proceeds via hydrogen bonding between chains. The weakest interaction J_a is thought to be mainly of dipolar origin, and provides the three-dimensional magnetic coupling between the layers. When the crystal is deuterated, one therefore expects that only J_c will change.

Measuring both protonic and deuterated samples at the same time, the transition temperatures were found as $T_c^D = 4.124(1)K$ and $T_c^H = 4.182(1)K$, the ratio of the ordering temperatures being $T_c^D/T_c^H = 0.986$. The directly-measured change in all the exchange constants is small, and $J_c^D/J_c^H = 0.94$ was estimated from the transition temperatures. The change of only 6% upon deuteration is due in part of the dipolar contribution to the exchange constants.

The metamagnetic properties of $[(CH_3)_3NH]CoCl_3 \cdot 2H_2O$ have been discussed in Sect. 6.12. The tricritical point is found at 4.125 K and an (internal) magnetic field of 29 Oe (0.0029 T). Interestingly, T_c for the analogous bromide compound is the lower value of 3.86 K, and the external transition field increases to 120 Oe (0.012 T).

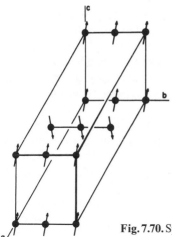

Fig. 7.70. Spin-structure of $[(CH_3)_3NH]CoCl_3 \cdot 2H_2O$. From Ref. [119]

The analogous spin $\mathscr{S}=1$ compound $[(CH_3)_3NH]NiCl_3 \cdot 2H_2O$ is isostructural with the cobalt compound [159] and, likewise, a ferromagnetic moment dominates the susceptibility. The transition temperature is 3.67 K. The susceptibilities are remarkably like those of the cobalt compound: the b-axis is the hard axis, the c-axis is the easy, ferromagnetic axis, and canting is observed parallel to the a-axis. The data analysis is complicated because of the inclusion of zero-field splitting terms, but the parameter D/k appears to be about -3.7 K, and E/k is non-zero. As expected, the material is a metamagnet, with tricritical temperature of 3.48 K.

The manganese analogs are similar in both structure and magnetic properties, except that the chains are antiferromagnetically ordered. Broad maxima are observed in the specific heat, which again are evidence of a high degree of short-range order. The susceptibilities indicate that canting also occurs in these molecules. It has been found [160–162] that g is isotropic, but that the zero-field splitting parameter D takes the rather large value of $-0.368\,\text{cm}^{-1}$ (-0.53 K) for $[(CH_3)_3NH]MnBr_3 \cdot 2H_2O$; it is about three times smaller for the chloride. It has been argued [160] that the anisotropy in the susceptibilities requires in addition to the usual bilinear term, an important biquadratic term, $\alpha(S_i \cdot S_j)^2$, with $\alpha = 0.3$. The phase diagrams of both the chloride and the bromide have been reported [162].

The iron analog, $[(CH_3)_3NH]FeCl_3 \cdot 2H_2O$, appears to be similar [163]. The substance is isomorphic to the others and also contains canted chains. The transition temperature is 3.2 K, and the substance is a metamagnet.

Paramagnetic resonance of $[(CH_3)_3NH]CoCl_3 \cdot 2H_2O$ doped separately with either Mn^{2+} or Cu^{2+} has been studied [164].

7.11 Solitons

While spin wave and most other elementary excitations are not discussed in this book, recent interest in the excitations called *solitons* prompts this section. This is in part because solitons are to be found primarily in one-dimensional magnets, and partly because they may in time come to be used to explain a number of phenomena concerning linear chain magnets. Solitons are non-linear solitary excitations [165, 166], subtle phenomena more easy for us to describe than to define. A soliton corresponds to a one-dimensional domain wall which separates, e.g., the spin-up and spin-down regions of a ferromagnetic chain, or the two degenerate configurations of an antiferromagnetic chain related by an interchange of the two (ferromagnetic) sublattices. Solitons were first sought in neutron diffraction studies of the ferromagnetic chain compound, $CsNiF_3$ [166].

The idea of a soliton is illustrated in Fig. 7.71. At the top, a ferromagnetically-oriented chain is illustrated, and a π-soliton corresponds to a moving domain wall between the two opposed configurations; it is important to realize that a soliton is a dynamic excitation, and is not related to a static configuration. A 2π-soliton, as illustrated, corresponds to a 2π rotation of the spins. In an ideal one-dimensional magnet, such walls will be excited thermally because of entropy considerations. These wall movements may only occur in the region above the 3-dimensional ordering temperature, since the interchain interactions will tend to slow down the wall movements and eventually freeze them below T_c, where the static 3D ordered magnetic structure is established.

In general, π-solitons are easier to detect than 2π-solitons. This is because, after the passage of a 2π-soliton, the spins have returned to their original orientation, while the π-soliton overturns the spins.

Evidence for the existence of solitons has been presented for several linear chain systems, and includes neutron scattering studies [166] on $CsNiF_3$ as well as on TMMC [167]. NMR measurements [167] as well as specific heat measurements in applied fields [168] on TMMC have also been interpreted in terms of the presence of solitons. A soliton contribution to the Mössbauer linewidth in the Ising compounds $RbFeCl_3 \cdot 2H_2O$, $CsFeCl_3 \cdot 2H_2O$, and $(N_2H_5)_2Fe(SO_4)_2$ has also been reported [169]. Relaxation behavior in the ac susceptibility of the nickel analog of TMMC, $[(CH_3)_4N]NiCl_3$, has yielded evidence for the existence of solitons [170], and experiments on the ferromagnetically-oriented chain in $(C_6H_{11}NH_3)CuBr_3$ (CHAB) have also been interpreted on this basis [171]. Nuclear relaxation in the Ising systems $CsFeCl_3 \cdot 2H_2O$ and $RbFeCl_3 \cdot 2H_2O$ has been interpreted in terms of diffusive solitons [172]. An extensive investigation of CHAC, CHAB, and $[(CH_3)_3NH]NiCl_3 \cdot 2H_2O$ appears in the thesis of Hoogerbeets [173].

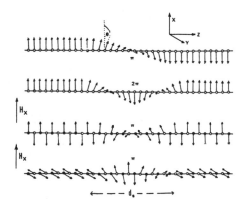

Fig. 7.71. Solitons in magnetic chains. From top to bottom are shown: a π- and 2π-soliton in the ferromagnetic chain and the antiferromagnetic π-soliton for $H < H_{SF}$ and $H > H_{SF}$, respectively

7.12 References

1. de Jongh L.J. and Miedema A.R., Adv. Phys. **23**, 1 (1974)
2. Hone D.W. and Richards P.M., Ann. Rev. Material Sci. **4**, 337 (1974)
3. Carlin R.L., Accts. Chem. Res. **9**, 67 (1976)
4. Lieb E.H. and Mattis D.C., (1966) Mathematical Physics in One Dimension, Academic Press, New York
5. Fisher M.E., (1972) Essays in Physics, Ed.: Conn G.K.T. and Fowler G.N., Vol. 4, p. 43, Academic Press, New York
6. A variety of papers on this subject are included in the symposium volume, "Extended Interactions between Metal Ions in Transition Metal Complexes," Ed.: Interrante L.V., (1974) Am. Chem. Soc. Symp. Ser. 5, Washington, D.C.
7. Bonner J.C., J. Appl. Phys. **49**, 1299 (1978)
8. de Jongh L.J., J. Appl. Phys. **49**, 1305 (1978)
9. O'Connor C.J., Prog. Inorg. Chem. **29**, 203 (1982)
10. Hatfield W.E., J. Appl. Phys. **52**, 1985 (1981)
11. Carlin R.L., J. Appl. Phys. **52**, 1993 (1981)
12. Willett R.D. and Landee C.P., J. Appl. Phys. **52**, 2004 (1981)
13. Bonner J.C., J. Appl. Phys. **52**, 2011 (1981)
14. de Jongh L.J., J. Appl. Phys. **53**, 8018 (1982)
15. Bonner J.C. and Müller G., Lower Dimensional Magnetism, to be published
16. A number of interesting articles appear in "Extended Linear Chain Compounds," Vol. 3, Ed.: Miller J.S., (1983) Plenum Publishing Co., New York, N.Y.
17. Hatfield W.E., Estes W.E., Marsh W.E., Pickens M.W., ter Haar L.W., and Weller R.R., in Ref. 16, p. 43
18. Willett R.D., Gaura R.M., and Landee C.P., in Ref. 16, p. 143
19. de Jongh L.J., (1981) in: Recent Developments in Condensed Matter Physics, Vol. 1, p. 343, (Ed.: Devreese J.T.,) Plenum Press, New York, N.Y.
20. See the articles by de Jongh L.J. and by Bonner J.C., in: Magneto-Structural Correlations in Exchange Coupled Systems, edited by Willett R.D., Gatteschi D., and Kahn O., NATO ASI Series, D. Reidel Publ. Co., Dordrecht, 1985
21. Ising E., Z. f. Physik **31**, 253 (1925)
22. Fritz J.J. and Pinch H.L., J. Am. Chem. Soc. **79**, 3644 (1957); Eisenstein J.C., J. Chem. Phys. **28**, 323 (1958); Watanabe T. and Haseda T., J. Chem. Phys. **29**, 1429 (1958); Haseda T. and Miedema A.R., Physica **27**, 1102 (1961)
23. Griffiths R.B., Phys. Rev. **135**, A659 (1964)
24. Bonner J.C. and Fisher M.E., Phys. Rev. **135**, A640 (1964)
25. Newell G.F. and Montroll E.W., Rev. Mod. Phys. **25**, 353 (1953)
26. McCoy B.M. and Wu T.T., (1973) The Two-Dimensional Ising Model, Harvard University Press, Cambridge, Mass., (USA)
27. Dunitz J.D., Acta Cryst. **10**, 307 (1957); Clarke P.J. and Milledge H.J., Acta Cryst. **B31**, 1543 (1975)
28. Takeda K., Matsukawa S., and Haseda T., J. Phys. Soc. Japan **30**, 1330 (1971); see also: de Jonge W.J.M., van Vlimmeren Q.A.G., Hijmans J.P.A.M., Swüste C.H.W., Buys J.A.H.M., and van Workum G.J.M., J. Chem. Phys. **67**, 751 (1977) and Pires A. and Hone D., J. Phys. Soc. Japan **44**, 43 (1978)
29. Morosin B. and Graeber E.J., Acta Cryst. **16**, 1176 (1963)
30. Fisher M.E., J. Math. Phys. **4**, 124 (1963)
31. Blöte H.W.J., Physica **78**, 302 (1974); de Neef T., Kuipers A.J.M., and Kopinga K., J. Phys. A **7**, L171 (1974)
32. Bhatia S.N., O'Connor C.J., Carlin R.L., Algra H.A., and de Jongh L.J., Chem. Phys. Lett. **50**, 353 (1977)
33. Morosin B. and Graeber E.J., Acta Cryst. **23**, 766 (1967)
34. Dingle R., Lines M.E., and Holt S.L., Phys. Rev. **187**, 643 (1969)

35. Dietz R.E., Walker L.R., Hsu F.S.L., Haemmerle W.H., Vis B., Chau C.K., and Weinstock H., Solid State Comm. **15**, 1185 (1974); de Jonge W.J.M., Swüste C.H.W., Kopinga K., and Takeda K., Phys. Rev. **B12**, 5858 (1975)

36. Vis B., Chau C.K., Weinstock H., and Dietz R.E., Solid State Comm. **15**, 1765 (1974)

37. Walker L.R., Dietz R.E., Andres K., and Darack S., Solid State Comm. **11**, 593 (1972)

38. Fisher M.E., Am. J. Phys. **32**, 343 (1964)

39. Smith T. and Friedberg S.A., Phys. Rev. **176**, 660 (1968); Kobayashi H., Tsujikawa I., and Friedberg S.A., J. Low Temp. Phys. **10**, 621 (1973). The compound $CsMnCl_3 \cdot 2 H_2O$ is one of the most extensively studied linear chain systems (1). Other references include: Kopinga K., de Neef T., and de Jonge W.J.M., Phys. Rev. **B11**, 2364 (1975); de Jonge W.J.M., Kopinga K., and Swüste C.H.W., Phys. Rev. **B14**, 2137 (1976); Iwashita T. and Uryû N., J. Chem. Phys. **65**, 2794 (1976)

40. Weng C.Y., Thesis, (1969) Carnegie-Mellon University

41. Blöte H.W.J., Physica **79B**, 427 (1975)

42. de Neef T. and de Jonge W.J.M., Phys. Rev. **B11**, 4402 (1975); de Neef T., Phys. Rev. **B13**, 4141 (1976)

43. Achiwa N., J. Phys. Soc. Japan **27**, 561 (1969)

44. Dupas C. and Renard J.-P., J. Chem. Phys. **61**, 3871 (1974); Kopinga K., de Neef T., de Jonge W.J.M., and Gerstein B.C., Phys. Rev. **B13**, 3953 (1975)

45. Ackerman J.F., Cole G.M., and Holt S.L., Inorg. Chim. Acta **8**, 323 (1974)

46. Caputo R.D. and Willett R.D., Phys. Rev. **B13**, 3956 (1976); Takeda K., Schouten J.C., Kopinga K., and de Jonge W.J.M., Phys. Rev. B **17**, 1285 (1978); Buys J.A.H.M., van Workum G.J.M., and de Jonge W.J.M., J. Chem. Phys. **70**, 1811 (1979)

47. Kopinga K., Phys. Rev. **B16**, 427 (1977)

48. Simizu S., Chen J.-Y., and Friedberg S.A., J. Appl. Phys. **55**, 2398 (1984). The magnetic phase diagram has been measured by: Paduan-Filho A. and Oliveira, Jr. N.F., J. Phys. **C 17**, L857 (1984). They find $H_{SF}(0) = 19$ kOe (1.9 T), $T_b = 3.22$ K, $H_b = 22.4$ kOe (2.24 T)

49. Boersma F., Tinus A.M.C., Kopinga K., Paduan-Filho A., and Carlin R.L., Physica **B114**, 231 (1982)

50. Merchant S., McElearney J.N., Shankle G.E., and Carlin R.L., Physica **78**, 380 (1974)

51. Richards P.M., Quinn R.K., and Morosin B., J. Chem. Phys. **59**, 4474 (1973); Caputo R., Willett R.D., and Morosin B., J. Chem. Phys. **69**, 4976 (1978)

52. Caputo R.E. and Willett R.D., Acta Cryst. **B37**, 1618 (1981)

53. McElearney J.N., Merchant S., and Carlin R.L., Inorg. Chem. **12**, 906 (1973)

54. Klaaijsen F.W., Blöte H.W.J., and Dokoupil Z., Physica **B81**, 1 (1976)

55. Carlin R.L., Burriel R., Rojo J.A., and Palacio F., Inorg. Chem. **23**, 2213 (1984)

56. Katsura S., Phys. Rev. **127**, 1508 (1962)

57. Witteveen H.T. and Reedijk J., J. Solid State Chem. **10**, 151 (1974); Klaaijsen F.W., den Adel H., Dokoupil Z., and Huiskamp W.J., Physica **79B**, 113 (1975)

58. Algra H.A., de Jongh L.J., Blöte H.W.J., Huiskamp W.J., and Carlin R.L., Physica **82B**, 239 (1976); McElearney J.N., Merchant S., Shankle G.E., and Carlin R.L., J. Chem. Phys. **66**, 450 (1977)

59. Smit J.J. and de Jongh L.J., Physica **B97**, 224 (1979)

60. Duxbury P.M., Oitmaa J., Barber M.N., van der Bilt A., Joung K.O., and Carlin R.L., Phys. Rev. **B24**, 5149 (1981)

61. Yoshizawa H., Shirane G., Shiba H., and Hirakawa H., Phys. Rev. **B28**, 3904 (1983)

62. Metselaar J.W., de Jongh L.J., and de Klerk D., Physica **79B**, 53 (1975)

63. Date M., Motokowa M., Hori H., Kuroda S., and Matsui K., J. Phys. Soc. Japan **39**, 257 (1975); Duffy Jr. W., Weinhaws F.M., and Strandburg D.L., Phys. Lett. **59A**, 491 (1977)

64. Henkens L.S.J.M., Diederix K.M., Klaassen T.O., and Poulis N.J., Physica **81B**, 259 (1976); Henkens L.S.J.M., van Tol M.W., Diederix K.M., Klaassen T.O., and Poulis N.J., Phys. Rev. Lett. **36**, 1252 (1976); Henkens L.S.J.M., Diederix K.M., Klaassen T.O., and Poulis N.J., Physica **83B**, 147 (1974); Henkens L.S.J.M., Thesis (1977), Leiden

65. McElearney J.N., Inorg. Chem. **15**, 823 (1976); Caputo R.E., Roberts S., Willett R.D., and Gerstein B.C., Inorg. Chem. **15**, 820 (1976)

66. Smit J.J., Mostafa M., and de Jongh L.J., J. Phys. (Paris) **39**, C6-725 (1978)
67. McElearney J.N., Inorg. Chem. **17**, 248 (1978)
68. Reinen D. and Friebel C., Structure and Bonding **37**, 1 (1979)
69. Algra H.A., de Jongh L.J., and Carlin R.L., Physica **93B**, 24 (1978)
70. Blöte H.W.J., J. Appl. Phys. **50**, 1825 (1979)
71. Reinen D. and Krause S., Solid State Comm. **29**, 691 (1979)
72. Steiner M., Villain J., and Windsor C.G., Adv. Phys. **25**, 87 (1976)
73. Butterworth G.J. and Woollam J.A., Phys. Lett. **A29**, 259 (1969)
74. de Jonge W.J.M., Hijmans J.P.A.M., Boersma F., Schouten J.C., and Kopinga K., Phys. Rev. **B17**, 2922 (1978); Hijmans J.P.A.M., Kopinga K., Boersma F., and de Jonge W.J.M., Phys. Rev. Lett. **40**, 1108 (1978)
75. Oguchi T., Phys. Rev. **133**, A1098 (1964)
76. Diederix K.M., Blöte H.W.J., Groen J.P., Klaassen T.O., and Poulis N.J., Phys. Rev. **B19**, 420 (1979)
77. Bonner J.C., Blöte H.W.J., Bray J.W., and Jacobs I.S., J. Appl. Phys. **50**, 1810 (1979)
78. Smit J.J., de Jongh L.J., van Ooijen J.A.C., Reedijk J., and Bonner J.C., Physica **B97**, 229 (1979)
79. Hall J.W., Marsh W.E., Weller R.R., and Hatfield W.E., Inorg. Chem. **20**, 1033 (1981); Olmstead M.M., Musker W.K., ter Haar L.W., and Hatfield W.E., J. Am. Chem. Soc. **104**, 6627 (1982)
80. Mohan M. and Bonner J.C., J. Appl. Phys. **53**, 8035 (1982)
81. de Groot H.J.M., de Jongh L.J., Willett R.D., and Reedijk J., J. Appl. Phys. **53**, 8038 (1982)
82. Bonner J.C., Friedberg S.A., Kobayashi H., Meier D.L., and Blöte H.W.J., Phys. Rev. **B27**, 248 (1983)
83. Hiller W., Strähle J., Datz A., Hanaek M., Hatfield W.E., ter Haar L.W., and Gütlich P., J. Am. Chem. Soc. **106**, 329 (1984)
84. Benelli C., Gatteschi D., Carnegie, Jr. D.W., and Carlin R.L., J. Am. Chem. Soc. **107**, 2560 (1985)
85. Jacobs I.S., Bray J.W., Hart, Jr. H.R., Interrante L.V., Kasper J.S., Watkins G.D., Prober D.E., and Bonner J.C., Phys. Rev. **B14**, 3036 (1976)
86. Northby J.A., Greidanus F.J.A.M., Huiskamp W.J., de Jongh L.J., Jacobs I.S., and Interrante L.V., J. Appl. Phys. **53**, 8032 (1982)
87. Northby J.A., Groenendijk H.A., de Jongh L.J., Bonner J.C., Jacobs I.S., and Interrante L.V., Phys. Rev. **B25**, 3215 (1982)
88. Bray J.W., Interrante L.V., Jacobs I.S., and Bonner J.C., in: Ref. 16, p. 353
89. Onsager L., Phys. Rev. **65**, 117 (1944)
90. Mess K.W., Lagendijk E., Curtis D.A., and Huiskamp W.J., Physica **34**, 126 (1967); Wielinga R.F., Blöte H.W.J., Roest J.A., and Huiskamp W.J., Physica **34**, 223 (1967)
91. Tanaka Y. and Uryû N., Phys. Rev. **B21**, 1994 (1980)
92. Carlin R.L., Chirico R.D., Sinn E., Mennenga G., and de Jongh L.J., Inorg. Chem. **21**, 2218 (1982)
93. Colpa J.H.P., Physica **57**, 347 (1972); de Jongh L.J., van Amstel W.D., and Miedema A.R., Physica **58**, 277 (1972); Bloembergen P. and Miedema A.R., Physica **75**, 205 (1974); Bloembergen P., Physica **79B**, 467 (1975); **81B**, 205 (1976); de Jongh L.J., Physica **82B**, 247 (1976)
94. Steijger J.J.M., Frikkee E., de Jongh L.J., and Huiskamp W.J., Physica **B123**, 271, 284 (1984)
95. von Känel H., Physica **B96**, 167 (1979)
96. Block R. and Jansen L., Phys. Rev. **B26**, 148 (1982); Straatman P., Block R., and Jansen L., Phys. Rev. **B29**, 1415 (1984)
97. Breed D.J., Physica **37**, 35 (1967)
98. Navarro R., Smit J.J., de Jongh L.J., Crama W.J., and Ijdo D.J.W., Physica **83B**, 97 (1976); Arts A.F.M., van Uyen C.M.J., van Luijk J.A., de Wijn H.W., and Beers C.J., Solid State Comm. **21**, 13 (1977)
99. Gurewitz E., Makovsky J., and Shaked H., Phys. Rev. **B14**, 2071 (1976)
100. Newman P.R., Imes J.L., and Cowen J.A., Phys. Rev. **B13**, 4093 (1976)

101. McElearney J.N., Forstat H., Bailey P.T., and Ricks J.R., Phys. Rev. **B13**, 1277 (1976)
102. Mennenga G., Bartolome J., de Jongh L.J., and Willett R.D., Chem. Phys. Lett. **105**, 351 (1984)
103. Mulder C.A.M., Stipdonk H.L., Kes P.H., van Duyneveldt A.J., and de Jongh L.J., Physica **B113**, 380 (1982); de Jongh L.J., Regnault L.P., Rossat-Mignod J., and Henry J.Y., J. Appl. Phys. **53**, 7963 (1982)
104. Paduan-Filho A., Phys. Lett. **A104**, 477 (1984)
105. Hijmans J.P.A.M., de Jonge W.J.M., van der Leeden P., and Steenland M.J., Physica **69**, 76 (1973); Kopinga K., Borm P.W.M., and de Jonge W.J.M., Phys. Rev. **B10**, 4690 (1974); Hijmans J.P.A.M., van Vlimmeren Q.A.G., and de Jonge W.J.M., Phys. Rev. **B12**, 3859 (1975)
106. White J.J. and Bhatia S.N., J. Phys. **C8**, 1227 (1975)
107. Basten J.A.J. and Bongaarts A.L.M., Phys. Rev. **B14**, 2119 (1976)
108. Mizuno J., J. Phys. Soc. Japan **15**, 1412 (1960)
109. Langs D.A. and Hare C.R., Chem. Commun. **1967**, 890; Klop E.A., Duisenberg A.J.M., and Spec A.L., Acta Cryst. **C39**, 1342 (1983)
110. Gregson A.K. and Mitra S., J. Chem. Phys. **50**, 2021 (1969)
111. McElearney J.N., Losee D.B., Merchant S., and Carlin R.L., J. Chem. Phys. **54**, 4585 (1971); van Santen, J.A., Thesis (1978), Leiden
112. van Santen J.A., van Duyneveldt A.J., and Carlin R.L., Physica **79B**, 91 (1975)
113. Numasawa Y., Kitaguchi H., and Watanabe T., J. Phys. Soc. Japan **38**, 1415 (1975)
114. Numasawa Y. and Watanabe T., J. Phys. Soc. Japan **41**, 1903 (1976)
115. Dokoupil Z., den Adel H., and Algra H.A., unpublished
116. Klaaijsen F.W., Suga H., and Dokoupil Z., Physica **51**, 630 (1971)
117. Stryjewski E. and Giordano, N., Adv. Phys. **26**, 487 (1977)
118. Losee D.B., McElearney J.N., Shankle G.E., Carlin R.L., Cresswell P.J., and Robinson W.T., Phys. Rev. **B8**, 2185 (1973)
119. Spence R.D. and Botterman A.C., Phys. Rev. **B9**, 2993 (1974)
120. Groenendijk H.A. and van Duyneveldt A.J., Physica **B115**, 41 (1982)
121. Griffin J.A., Schnatterly S.E., Farge Y., Regis M., and Fontana M.P., Phys. Rev. **B10**, 1960 (1974); Griffin J.A. and Schnatterly S.E., Phys. Rev. Lett. **33**, 1576 (1974)
122. Dillon, Jr. J.F., Chen E.Y., and Guggenheim H.J., Phys. Rev. **B18**, 377 (1978)
123. Bongaarts A.L.M. and de Jonge W.J.M., Phys. Rev. **B15**, 3424 (1977)
124. Spence R.D., J. Chem. Phys. **62**, 3659 (1975)
125. Vermeulen A.J.W.A., Frikkee E., Flokstra J., and de Jonge W.J.M., Physica **B98**, 205 (1980)
126. Basten J.A.J., van Vlimmeren Q.A.G., and de Jonge W.J.M., Phys. Rev. **B18**, 2179 (1978); van Vlimmeren Q.A.G. and de Jonge W.J.M., Phys. Rev. **B19**, 1503 (1979)
127. LeFever H.Th., Thiel R.C., Huiskamp W.J., and de Jonge W.J.M., Physica **B111**, 190 (1981); LeFever H.Th., Thiel R.C., Huiskamp W.J., de Jonge W.J.M., and van der Kraan A.M., Physica **B111**, 209 (1981)
128. Groenendijk H.A., van Duyneveldt A.J., and Willett R.D., Physica **B98**, 53 (1979)
129. Groenendijk H.A., van Duyneveldt A.J., and Willett R.D., Physica **B101**, 320 (1980)
130. Engelfriet D.W., Thesis (1980), Leiden. See: van der Griendt B., Haasnoot J.G., Reedijk J., and de Jongh L.J., Chem. Phys. Lett. **111**, 161 (1984) for a susceptibility study of $Fe(trz)_2(NCS)_2$ as it is diluted by the isomorphous zinc analog
131. Engelfriet D.W., Haasnoot J.G., and Groeneveld W.L., Z. Naturforsch. **32a**, 783 (1977)
132. Engelfriet D.W. and Groeneveld W.L., Z. Naturforsch. **33a**, 848 (1978)
133. Engelfriet D.W., den Brinker W., Verschoor G.C., and Gorter S., Acta Cryst. **B35**, 2922 (1979)
134. Engelfriet D.W., Groeneveld W.L., Groenendijk H.A., Smit J.J., and Nap G.M., Z. Naturforsch. **35a**, 115 (1980)
135. Engelfriet D.W., Groeneveld W.L., and Nap G.M., Z. Naturforsch. **35a**, 852 (1980)
136. Engelfriet D.W., Groeneveld W.L., and Nap G.M., Z. Naturforsch. **35a**, 1382 (1980)
137. Engelfriet D.W., Groeneveld W.L., and Nap G.M., Z. Naturforsch. **35a**, 1387 (1980)
138. Carlin R.L., Joung K.O., van der Bilt A., den Adel H., O'Connor C.J., and Sinn E., J. Chem. Phys. **75**, 431 (1981)
139. McElearney J.N., Shankle G.E., Losee D.B., Merchant S., and Carlin R.L., (1974) Am. Chem.

Soc. Symp. Ser. 5, Extended Interactions Between Metal Ions in Transition Metal Complexes, p. 194 (Ed.: Interrante L.V.) Am. Chem. Soc., Washington, D.C.

140. Losee D.B., McElearney J.N., Siegel A., Carlin R.L., Khan A.A., Roux J.P., and James W.J., Phys. Rev. **B6**, 4342 (1972)
141. Stirrat C.R., Dudzinski S., Owens A.H., and Cowen J.A., Phys. Rev. **B9**, 2183 (1974)
142. Algra H.A., de Jongh L.J., Blöte H.W.J., Huiskamp W.J., and Carlin R.L., Physica **78**, 314 (1974)
143. Algra H.A., de Jongh L.J., Huiskamp W.J., and Carlin R.L., Physica **92B**, 187 (1977)
144. Ritter M.B., Drumheller J.E., Kite T.M., Snively L.O., Emerson K., Phys. Rev. **B28**, 4949 (1983)
145. Engberg A., Acta Chem. Scand. **24**, 3510 (1970)
146. Poulis N.J. and Hardeman G.E.G., Physica **18**, 201 (1952); **18**, 315 (1952); **19**, 391 (1953)
147. Shirane G., Frazer B.C., and Friedberg S.A., Phys. Lett. **17**, 95 (1965)
148. van der Marel L.C., van den Broek J., Wasscher J.D., and Gorter C.J., Physica **21**, 685 (1955)
149. Friedberg S.A., Physica **18**, 714 (1952)
150. Dunitz J.D., Acta Cryst. **10**, 307 (1957)
151. Takeda K., Matsukawa S., and Haseda T., J. Phys. Soc. Japan **30**, 1330 (1971)
152. Duffy, Jr. W., Venneman J.E., Strandburg D.L., and Richards P.M., Phys. Rev. **B9**, 2220 (1974)
153. Andres K., Darack S., and Holt S.L., Solid State Comm. **15**, 1087 (1974); Hughes R.C., Morosin B., Richards P.M., and Duffy, Jr. W., Phys. Rev. **B11**, 1795 (1975); Tinus A.M.C., Boersma F., and de Jonge W.J.M., Phys. Lett. **86A**, 300 (1981)
154. Spence R.D. and Rama Rao K.V.S., J. Chem. Phys. **52**, 2740 (1970)
155. Spence R.D., de Jonge W.J.M., and Rama Rao K.V.S., J. Chem. Phys. **54**, 3438 (1971)
156. Navarro R. and de Jongh L.J., Physica **94B**, 67 (1978)
157. Takeda K. and Wada M., J. Phys. Soc. Japan. **50**, 3603 (1981)
158. Groenendijk H.A., van Duyneveldt A.J., and Carlin R.L., Physica **115B**, 63 (1982)
159. O'Brien S., Gaura R.M., Landee C.P., and Willett R.D., Solid State Comm., **39**, 1333 (1981); Hoogerbeets R., Wiegers S.A.J., van Duyneveldt A.J., Willett R.D., and Geiser U., Physica B **125**, 135 (1984)
160. Yamamoto I. and Nagata K., J. Phys. Soc. Japan **43**, 1581 (1977); Yamamoto I., Iio K., and Nagata K., J. Phys. Soc. Japan **49**, 1756 (1980)
161. Chikazawa S., Sato T., Miyako Y., Iio K., and Nagata K., J. Mag. Mag. Mat. **15–18**, 749 (1980)
162. Takeda K., Koike T., Harada I., and Tonegawa T., J. Phys. Soc. Japan **51**, 85 (1982)
163. Landee C.P., Greeney R.E., Reiff W.M., Zhang J.H., Chalupa J., and Novotny M.A., J. Appl. Phys. **57**, 3343 (1985)
164. Phaff A.C., Swüste C.H.W., and de Jonge W.J.M., Phys. Rev. **B25**, 6570 (1982)
165. de Jongh L.J., J. Appl. Phys. **53**, 8018 (1982). See also the recent application of soliton theory to phase diagrams in de Jongh L.J. and de Groot H.J.M., Sol. St. Comm. **53**, 731, 737 (1985)
166. Steiner M., J. Mag. Mag. Mat. **14**, 142 (1979); ibid., **31–34**, 1277 (1983)
167. Boucher J.P., Regnault L.P., Rossat-Mignod J., Renard J.P., Bouillot J., and Stirling W.G., Solid State Commun. **33**, 171 (1980); Boucher J.P. and Renard J.P., Phys. Rev. Lett. **45**, 486 (1980)
168. Borsa F., Phys. Rev. **B25**, 3430 (1982); Borsa F., Pini M.G., Rettori A., and Tognetti V., Phys. Rev. **B28**, 5173 (1983)
169. Thiel R.C., de Graaf H., and de Jongh L.J., Phys. Rev. Lett. **47**, 1415 (1981)
170. de Jongh L.J., Mulder C.A.M., Cornelisse R.M., van Duyneveldt A.J., and Renard J.P., Phys. Rev. Lett **47**, 1672 (1981). The metamagnetic phase diagram has been determined by: Hijmans T.W., van Duyneveldt A.J., and de Jongh L.J., Physica **125B**, 21 (1984)
171. Kopinga K., Tinus A.M.C., and de Jonge W.J.M., Phys. Rev. **B29**, 2868 (1984)
172. Tinus A.M.C., Denissen C.J.M., Nishihara H., and de Jonge W.J.M., J. Phys. **C15**, L791 (1982)
173. Hoogerbeets R., Thesis (1984), Leiden

8. The Heavy Transition Metals

8.1 Introduction

We group together under the above title both the 4d and 5d transition metal ions. The reasons for this are several: the similarities among these elements, at least for our purposes, are much greater than their differences. They also behave distinctly differently from the 3d or iron series ions. Finally, because of expense, synthetic problems and a general lack of an interesting variety of materials to study, there simply has been much less recent research on the magnetochemistry of these elements compared to that on the iron series compounds. Yet, as we shall discuss below, the work at Oxford on the IrX_6^{2-} ions, for example, has had a profound influence on the subject.

These elements have a rich chemistry, much different from that of the iron series ions, but much of it is of little interest to us in this book. The number of magnetic ML_6 compounds is relatively small, and there are few aquo ions. The available oxidation states differ appreciably from those of the 3d series; for example, the divalent oxidation state of cobalt, cobalt(II), is of great interest and importance to us, but as one goes below cobalt in the periodic chart, we find that only a few Rh(II) compounds are known, and Ir(II) is unknown. In contrast to the important place of copper(II) in so many discussions in this book, there is scarcely any magnetic information available on discrete coordination compounds of either silver(II) or gold(II).

Because of the lanthanide contraction, the size of the 4d and 5d atoms (or ions) is almost the same. Thus, the chemistries of Zr(IV) and Hf(IV) are virtually identical. These are non-magnetic atoms, but this illustrates the properties of the other atoms as well.

Higher oxidation states are generally more readily available with the heavier elements. There are no 3d analogs of such molecules as RuO_4, WCl_6, and PtF_6. The heavy metals have a particularly rich carbonyl and organometallic chemistry, but since most of the compounds are not magnetic, they are of no interest to us here. Metal-metal bonding is also frequently found with the heavier elements, but likewise there is little for us there.

The divalent and trivalent aquo ions are of little importance and the aqueous chemistry is complicated.

The distinguishing feature for us of the 4d and 5d metals is the importance of spin-orbit coupling. This complicates the magnetic properties of these ions and, in concert with the stronger ligand field found with the heavier metals, causes essentially all the magnetic compounds to be spin-paired (or low spin). Thus, virtually all octahedral nickel salts are paramagnetic, while virtually all octahedral salts of its congener, platinum, are diamagnetic.

We may compare the spin-orbit coupling constant, λ, for several ions. Considering only the d^4 configuration, the values of λ for, respectively, Cr(II), Mn(III), and Os(IV) are about 90, 170, and 3600 cm^{-1}. Since spin-orbit coupling tends to cause energy level degeneracies to be resolved, greater energy savings are to be had by the pairing of spins. Furthermore, the more important the spin-orbit coupling becomes, the more that Russell-Saunders coupling breaks down. One moves into the regime called intermediate coupling.

By and large, the μ_{eff} of the iron series ions is temperature independent, except when collective phenomena occur at low temperatures. But, this is not true for the heavier metals.

There seem to be no examples as yet of the fitting of data on these heavier metals to any of the magnetic models discussed earlier. No examples of linear chain or planar magnets appear to have been reported either; the expense of a synthetic program to find such materials has probably been too great.

8.2 Molybdenum(III)

This species is one of the better known examples of the 4d metals, yet it would be an exaggeration to say that it is "well-known". Molybdenum(III), but not tungsten(III), forms complexes of the type $[MX_6]^{3-}$. The red salts K_3MoCl_6, $(NH_4)_3MoCl_6$, and $(NH_4)_2[MoCl_5(H_2O)]$ are known, but have limited stability in air. The acetylacetonate of Mo(III) is air-sensitive, hindering its investigation; the analogous Cr(acac)$_3$ has long-term air stability.

Since the spin-orbit coupling constant λ for the free ion is 273 cm^{-1}, one would expect through Eq. (4.6) that magnetic moments would be slightly reduced from the spin-only value of 3.88 μ_B, and this seems to be true. Room temperature values of 3.7–3.8 μ_B are common for such compounds as has been mentioned above.

Spin-orbit coupling seems to have a much more dramatic effect on the zero-field splitting of the $^4A_{2g}$ ground state of Mo(III). As with Cr(III), this splitting is called 2D, but the small amount of data available suggests that this ZFS is more than an order of magnitude larger with Mo(III). Susceptibility measurements [1] on Mo(acac)$_3$ suggested that D is either $+7$ or -6.3 cm^{-1}; the ambiguity arises because single crystal measurements are not available. The $|\pm\frac{3}{2}\rangle$ EPR transition has been observed [2] and found to increase in intensity as the temperature is decreased. This result establishes the ZFS as negative in sign. The doping of CsMgCl$_3$ and CsMgBr$_3$ with molybdenum leads to Mo(III) pairs, joined by a vacancy for charge compensation. The exchange constant and the zero-field splitting, as measured by EPR [3], are both larger than for the corresponding chromium(III) pairs.

One of the few coordination compounds of molybdenum(III) which has been shown to undergo magnetic ordering appears to be K_3MoCl_6. Static susceptibility measurements [4] suggested that there are two transition temperatures, 6.49 and 4.53 K, and that the substance is a weak ferromagnet. One set of specific heat data [5] indicated magnetic transitions at 4.65 and 6.55 K, but another data set, while indicating a λ-anomaly at 6.8 K, differs appreciably [6]. Unfortunately, all measurements have been carried out on polycrystalline materials that were obtained commercially; several of the samples were estimated to be only 95% pure. Since a small amount of ferromagnetic impurity could overwhelm the true susceptibility of the materials, most of these results must be viewed as uncertain. The measurements should be repeated on pure materials.

The crystal structures of $(NH_4)_2[MoCl_5(H_2O)]$ and $(NH_4)_2[MoBr_5(H_2O)]$ have recently been reported [7] and the susceptibility measured [8]. The substances are isomorphic with the iron analogs, which are discussed at length in Chap. 10. The susceptibility measurements show that the bromide orders at about 12–13 K, but the substance appears to have suffered a crystallographic phase transition on cooling. Single crystal measurements are not available as yet.

8.3 Ruthenium(III)

This $4d^5$ ion is of interest because the only configuration yet observed is the strong-field or spin-paired one, t_{2g}^5, with spin $\mathscr{S} = \frac{1}{2}$. This is because of the combined effect of strong ligand fields and spin orbit coupling, which is of the order of $1000\,cm^{-1}$ for this oxidation state and thus is very important. By contrast, iron(III), which lies just above ruthenium in the periodic chart, with five 3d electrons, exhibits both spin-free and spin-paired configurations.

The theory of the paramagnetic properties of t_{2g}^5 is due originally to Stevens [9] and to Bleaney and O'Brien [10]. The results are expressed in terms of the axial crystal field parameter Δ and spin orbit coupling constant ζ. The axial field separates the orbital states into a singlet and a doublet, split by Δ. The two perturbations acting together yield three Kramers doublets of which the lowest has g-values (for Δ either positive or negative),

$$g_{\parallel} = 2|(1+k)\cos^2\alpha - \sin^2\alpha|$$
$$g_{\perp} = 2|2^{1/2}\,k\cos\alpha\,\sin\alpha + \sin^2\alpha|,$$

(8.1)

where $\tan 2\alpha = 2^{1/2}(\frac{1}{2} - \Delta/\zeta)^{-1}$ and $0 < 2\alpha < \pi$. When Δ is zero, $g_{\parallel} = g_{\perp} = 2(1+2k)/3$. The orbital reduction factor k of Stevens is introduced to account for the overlap of the d functions onto the ligands. One can see from the relationships for the g-values that they can be quite anisotropic and differ widely in value, depending on the local crystalline field. This is well-illustrated by the range of values listed in Table 8.1; there is a reciprocal relationship between the g-values reminiscent of the situation with octahedral cobalt(II). The g-values are plotted in Fig. 8.1, the variable parameter being Δ/ζ. These g-values are affected by covalency [9], resulting in slightly smaller values than those predicted by an ionic model alone. The molecular orbital theory of covalency in metal complexes has been reviewed by Owen and Thornley [11].

The EPR spectra of a number of concentrated ruthenium(III) compounds have also been reported [12–15]. The compounds have ligands such as phosphines and arsines, as well as a variety of organic chelating ligands. A number of the compounds are of relatively low symmetry, and the work has been carried out with the purpose of parameterizing the molecular orbital coefficients. The g-values vary widely, and the parameters such as Δ have been evaluated. While the parameters derived may have limited significance, an important point of all this work is the continued validation of the theoretical work behind such relationships as given in Eq. (8.1).

Table 8.1. Magnetic parameters of Ru(III).

Compound or Host	$g_\parallel$	$g_\perp$	Method	Ref.
Al_2O_3	≤ 0.06	2.43	EPR	a
YGa garnet*	3.113	1.148	EPR	b
LuGa garnet	3.161	1.136	EPR	c
YAl garnet	2.88	1.300	EPR	b
LuAl garnet	2.907	1.286	EPR	c
Al(acac)$_3$	2.82	1.28, 1.74	EPR	d
Ru(acac)$_3$	2.48	1.52	susc.	e
	2.45	2.16, 1.45	EPR	f
Ru(dtc)$_3$	1.979	2.156, 2.109	EPR	f
[Ru(en)$_3$]$^{3+}$	0.330	2.640	EPR	g
[Ru(C$_2$O$_4$)$_3$]$^{3-}$	1.76	2.30, 2.02	EPR	f
Ru(sacsac)$_3$	1.992	2.109, 2.031	EPR	f
[Ru(mnt)$_3$]$^{3-}$	1.968	2.120, 2.026	EPR	f
Ru(dtp)$_3$	1.982	2.085, 2.055	EPR	f

* The garnets have the general formula $A_3B_5O_{12}$.
a Geschwind S and Remeika J.P., J. Appl. Phys. **33**, 370 S (1962)
b Miller I.A. and Offenbacher E.L., Phys. Rev. **166**, 269 (1968)
c Offenbacher E.L., Miller I.A., and Kemmerer G., J. Chem. Phys. **51**, 3082 (1969)
d Jarret H.S., J. Chem. Phys. **27**, 1298 (1957)
e Gregson A.K. and Mitra S., Chem. Phys. Lett. **3**, 392 (1969)
f DeSimone R.E., J. Am. Chem. Soc. **95**, 6238 (1973)
g Stanko J.A., Peresie H.J., Bernheim R.A., Wang R., and Wang P.S., Inorg. Chem. **12**, 634 (1973)

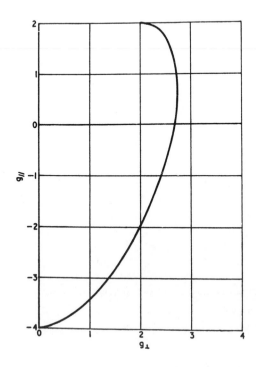

Fig. 8.1 Theoretical g-values for the configuration $4d^5$ (and $5d^5$) in strong crystalline fields of octahedral symmetry

Because of the large anisotropies expected, susceptibility measurements on powders, especially at high temperatures [16], cannot be expected to be vary informative. This was proved by the single-crystal measurements on paramagnetic Ru(acac)$_3$ [17]. in which it was shown that fits to the single crystal data were much more sensitive to the fitting parameters than was the fit to the average or powder data. Furthermore, measurements at high temperatures (above 80 K) will be affected by the thermally-populated excited states, which can then result in very large Weiss constants [18]. These of course are then not necessarily related to magnetic exchange interactions.

A careful study of [Ru(H$_2$O)$_6$]$^{3+}$ in alum crystals has recently appeared [19]. The susceptibility of polycrystalline CsRu(SO$_4$)$_2 \cdot 12$H$_2$O, apparently a β-alum, between 2.6 and 200 K yielded a temperature-independent moment of 1.92 μ_B with an average $\langle g \rangle = 2.22$. EPR spectra of doped CsGa(SO$_4$)$_2 \cdot 12$H$_2$O yielded typical values of $|g_\parallel| = 1.494$ and $|g_\perp| = 2.517$, along with hyperfine constants. These results require a strong crystal field distortion.

When CsMgCl$_3$ is doped with Ru(III), no EPR is observed down to 77 K [20]. This may be due to very strong exchange between pairs, which appears unlikely, or else to fast relaxation. When Li$^+$, Na$^+$ or Cu(I) are added for charge compensation, then g$_\parallel$ is found to be about 2.6–2.9, and g$_\perp$ is 1.2–1.5. With (diamagnetic) In(III) or Ir(III), a vacancy is formed, and the g-values become g$_\parallel \simeq 2.35$ and g$_\perp \simeq 1.62$. Interestingly, when Gd(III) is added, weakly-coupled Gd–Ru pairs (with a vacancy between them) are formed.

Several Mössbauer experiments have been reported [21] on ruthenium(III) compounds, but this is a difficult experiment to perform. Several attempts to observe magnetic interactions between pairs of Ru(III) atoms bridged by pyrazines and related molecules were unsuccessful [22].

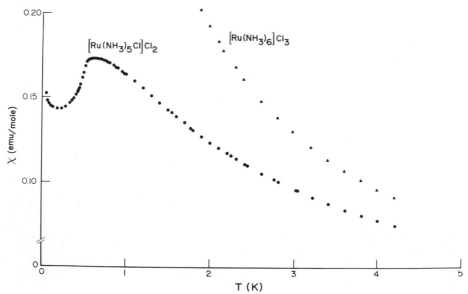

Fig. 8.2. The susceptibility of [Ru(NH$_3$)$_5$Cl]Cl$_2$, compared with that of paramagnetic [Ru(NH$_3$)$_6$]Cl$_3$. From Ref. [25]

The first discrete coordination compound of ruthenium(III) in which magnetic ordering was reported was $[Ru(NH_3)_5Cl]Cl_2$ [23]. The $\langle g \rangle$ value for the powder, as measured by susceptibilities in the 1.2–4.2 K interval, is 2.11, and $\theta = -1.36$ K. EPR measurements at 77 K of the compound doped into $[Co(NH_3)_5Cl]Cl_2$ have yielded [24] the results that $g_x = 0.987$, $g_y = 1.513$, and $g_z = 2.983$, or an average value of 2.013. Antiferromagnetic ordering was observed at 0.525 K. For comparison, the compound $[Ru(NH_3)_6]Cl_3$ has $\langle g \rangle = 2.04$ and a θ of only -0.036 K. Single crystal susceptibilities of $[Ru(NH_3)_5Cl]Cl_2$ have also been measured [25]; see Fig. 8.2. Several other compounds recently studied [26] are:

	T_c, K
$Cs_2[RuCl_5(H_2O)]$	0.58
$Rb_2[RuCl_5(H_2O)]$	2.5
$K_2[RuCl_5(H_2O)]$	4

These ordering temperatures may be compared with those for the analogous iron(III) compounds, which appear to be isomorphous and have spin $\mathscr{S} = \frac{5}{2}$:

	T_c, K
$Cs_2[FeCl_5(H_2O)]$	6.54
$Rb_2[FeCl_5(H_2O)]$	10.00
$K_2[FeCl_5(H_2O)]$	14.06

The latter compounds are discussed in Chap. 10.

8.4 Rhenium(IV)

The compound K_2ReCl_6 stands out uniquely for the amount of work done on the compounds of this $5d^3$ metal ion. That is because it has become the prototype for the study of certain structural phase transitions, and a rigid ion model description of this crystal has been proposed [27]. The spectra of K_2ReCl_6 were reviewed some time ago [28].

Tetravalent rhenium has a $^4A_2$ ground state in a cubic crystalline field, but spin orbit coupling is so large (ζ is some 3000 cm^{-1}), that it is more properly labeled as a Γ_8 spin orbit state. Calculations of the magnetic properties of this state [29] are complicated, and the simple spin-Hamiltonian formalism applied earlier to $3d^3$ chromium(III) is not directly applicable. There appear to be no reports on the zero field splitting of this state, but in any case most of the detailed studies of this metal ion have concerned the cubic compound K_2ReCl_6.

Thus the EPR spectra of K_2PtCl_6 doped with Re(IV) have been reported to consist of a cubic site with $g = 1.815$. The most recent EPR studies, on Re/$(NH_4)_2PtCl_6$ [30], have likewise found octahedral sites with $g = 1.817$ and other, axial sites with $g = 1.815$. Any trigonal distortion is therefore quite small. Only the $|\pm\frac{1}{2}\rangle$ transition has been observed, and the ZFS is indeterminate. The reduction in the g-value from 2 is due both to covalency and strong spin orbit coupling.

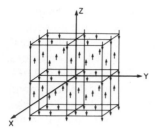

Fig. 8.3. Magnetic structure of K_2ReCl_6

Studies on these systems have been hindered because of the reasons cited earlier. There are no single crystal susceptiblity studies of K_2ReCl_6 and K_2ReBr_6, which are cubic at room temperature, and indeed the only susceptibility studies at low temperatures [31] were carried out by the Faraday method in an applied field of 20 kOe (2 T); the calibration procedure used was also not straightforward. In any case, the data show that K_2ReCl_6 and K_2ReBr_6 order antiferromagnetically at, respectively, 11.9 and 15.2 K. These results are confirmed by specific heat measurements [32–34]. Neutron diffraction studies [35] show that the order is of the kind called Type 1, in which the magnetic moments are ferromagnetically aligned in the (001) planes and the magnetic moments in adjacent planes are oriented antiparallel to each other. This is illustrated in Fig. 8.3. The magnetic susceptibility of polycrystalline $(CH_3NH_3)_2ReCl_6$ and of $[(CH_3)_2NH_2]ReCl_6$ has been measured [36] over a wide temperature interval. The substances order antiferromagnetically at temperatures, respectively, of about 3.8 K and 9.8 K. The crystal structure of neither substance is known, but the decrease in ordering temperatures as the alkali ions are replaced by the ammonium ions will be noted. It was also found that $[N(CH_3)_4]_2ReCl_6$ did not order down to 1.5 K. Aside from some susceptibility studies at high temperatures [37], that is all there is to say about this magnetic ion!

8.5 Osmium(III)

There has scarcely been any magnetic work on this $5d^5$ ion, despite the availability of quite a few suitable compounds. Perhaps this has occurred because it is isoelectronic with iridium(IV) which, as we shall see in the next section, has been studied so extensively. Aside from some susceptibility data on a few compounds at high temperatures [16b], most of the published data concern EPR spectra [12, 13, 38]. We defer the discussion of the theory of the electronic structure of the $5d^5$ configuration until the next section.

The only single crystal study of an osmium(III) compound appears to be the EPR study by Hill [38] on monoclinic $OsCl_3[P(C_2H_5)_2(C_6H_5)]_3$ doped into the isomorphous rhodium(III) compound. Large g-anisotropy was observed at 77 K, the g-values being 3.32, 1.44, and 0.32. The effective value of the spin-orbit coupling constant in the molecule was estimated as about 2650 cm^{-1} and the orbital reduction factor as 0.80. The inclusion of configuration interaction (mixing metal ion excited states into the ground state wave function) was shown to lead to g-value shifts of as much as 0.2 units.

The compound $[Os(NH_3)_5Cl]Cl_2$ has recently been studied [39].

8.6 Iridium(IV)

Iridium is the heaviest element that we shall deal with in this book, and it has only one oxidation state that interests us. That is the tetravalent state, which has five d electrons. Because of the strength of the crystalline field and of the very large spin-orbit coupling constant, only spin-paired compounds are known, and they all have spin $\mathscr{S}=\frac{1}{2}$. The most thoroughly studied complexes of Ir(IV) are the hexahalides, IrX_6^{2-}. These are a classic series of molecules because they gave rise to the first studies of ligand hyperfine structure in the EPR spectra and therefore to the most direct evidence for covalency in transition metal complexes [40].

The theory of $5d^5$ is essentially the same as that of ruthenium(III), a $4d^5$ ion. The only difference is that the spin-orbit coupling constant (some $2000\,cm^{-1}$) is larger (Table 1.2). All the compounds are strong-field. The cubic field ground state is thus a $^2T_2$. The available EPR results are listed in Table 8.2.

The g-values are as plotted in Fig. 8.1, the variable parameter being Δ/ζ. These g-values are affected by covalency [2], resulting in slightly smaller values than those predicted by an ionic model alone. Most of the iridium salts which have been studied are cubic and the lowest doublet appears to be well-isolated from the higher ones. There are relatively few susceptibility studies, the major one [41] being devoted to measurements on powders.

Both K_2IrCl_6 and $(NH_4)_2IrCl_6$ behave as antiferromagnets at low temperatures, the potassium salt exhibiting a λ-anomaly at 3.05 K and the ammonium salt at 2.15 K [42]. These ordering temperatures agree with those measured by means of susceptibilities [41] but single crystal measurements are lacking. A field of 1 kOe (0.1 T) did not affect the susceptibility, but one would expect that much larger fields would have to be applied in order to observe any effect. The rubidium salt, which is also an antiferromagnet, orders at 1.85 K, while the cesium salt orders at 0.5 K [43–45].

The interesting thing about a face-centered cubic lattice, which is the one to which both K_2IrCl_6 and $(NH_4)_2IrCl_6$ belong, is that it can be shown that nearest-neighbor (nn) interactions alone cannot produce a magnetic phase transition. That is, there is "frustration" in that all nn spins cannot be antiparallel in the fcc lattice. There must be next-nearest neighbor (nnn) interactions as well. The beauty of these two systems is that they have allowed all these interactions to be measured, primarily by EPR.

Substituting the value $\mathscr{S}=\frac{1}{2}$ in the molecular field relation Eq. (6.16), one has $\theta=3J/k$, when the number of nn is 12. Using the experimental values for θ, one finds $2J/k$

Table 8.2. Paramagnetic resonance data on $[IrX_6]^{-2}$ compounds.[a]

Host Lattice	
Cs_2GeF_6 (cubic)	$|g_x|=|g_y|=|g_z|=1.853$
K_2PtF_6 (trigonal)	$|g_x|=|g_y|=2.042, |g_z|=1.509$
$(NH_4)_2PtCl_6$ (cubic)	$|g_x|=|g_y|=|g_z|=1.786$
K_2PtCl_6 (cubic)	$|g|=1.79$
$Na_2PtCl_6\cdot6H_2O$ (triclinic)	$|g_x|=2.168, |g_y|=2.078, |g_z|=1.050$
$Na_2PtBr_6\cdot6H_2O$ (triclinic)	$|g_x|=2.25, |g_y|=2.21, |g_z|=0.75$
K_2PtBr_6 (cubic)	$|g_x|=|g_y|=1.87, |g_z|=1.60$

[a] Compiled by Lustig C.D., Owen J., and Thornley J.H.M., in Ref. [49].

$= -6.7 \, \text{K}$ for $(NH_4)_2IrCl_6$ and $-10.7 \, \text{K}$ for K_2IrCl_6, with our usual definition of the exchange term. Likewise, $2J/k = -4.7 \, \text{K}$ for Rb_2IrCl_6. These parameters are close to the values found by the EPR studies, which were carried out on the crystals K_2PtCl_6 and $(NH_4)_2PtCl_6$ doped with iridium. This shows nicely that the nn exchange constant does not change with concentration in these systems. The exchange constants were evaluated by the following procedure.

When an isomorphic host crystal such as K_2PtCl_6 is doped with iridium, there will be a large number of isolated ions as well as a small concentration of pairs. Because of the crystal structure of the system, the structure of the pairs will be as illustrated:

With an antiferromagnetic exchange interaction between the metal ions, then the electronic structure of the pair is just like that described for other $\mathscr{S} = \frac{1}{2}$ dimers in Chap. 5. That is, there is a singlet ground state with a triplet $-2J/k$ higher in energy. Since the EPR transitions occur within the (excited) triplet state, a measurement of the temperature dependence of the relative intensity of the pair spectrum allows an evaluation of the exchange constant [46]. This was found to be $-7.5 \, \text{K}$ for iridium in $(NH_4)_2PtCl_6$ and $-11.5 \, \text{K}$ for doped K_2PtCl_6, which is in excellent agreement with the results mentioned above.

The nnn interactions were also evaluated by EPR [47]. In this case, the spectra of triads were analyzed; this was an impressive achievment, for there are six different triads to consider; linear, right-angled, equilateral, and so on. The spectra of each of these triples was observed and assigned, and led to an evaluation of $2J(nnn)/k = -0.55 \, \text{K}$ in K_2PtCl_6 and $-0.39 \, \text{K}$ in $(NH_4)_2PtCl_6$. The exchange constant for the nearest-neighbor ions was found not be changed when a triad was formed. All of these results show that, at least for these systems, the exchange constant does not change with concentration.

One of the remarkable results that have come out of this work is the observation of a high degree of covalency in the IrX_6^{2-} ions. There are two independent measurements of this covalency, the value of the g-value and the observation of ligand hyperfine structure [48]. These results require that the magnetic electron spend about 5% of its time on each of the six ligand atoms, or only about 70% of its time on the metal ion.

In earlier sections of this book, we have emphasized the fact that deviations of the g-value from 2 are due to spin-orbit coupling. The g-values in Eq. (8.1) above however include the Stevens covalency factor, k, which has a value less than one. A more thorough theoretical study [49] showed that the g-value contains contributions from the t_4^2e excited state, and π-bonding covalency. This allows an explanation for the variation of g-value of the IrX_6^{2-} ions with halide, this being 1.853 (F), 1.786 (Cl), and 1.76 (Br).

The iridium nucleus has spin $\mathscr{I} = \frac{3}{2}$ and should therefore exhibit four EPR lines. The EPR spectra also exhibited ligand hyperfine structure (hfs). The chloride ion has nuclear spin $\frac{3}{2}$, and interaction of the nucleus with the electron spin (the (hyperfine structure, hfs) causes each line to split into four. For $IrCl_6^{2-}$ there are six equivalent chloride ions and so the binomial expression must be used [48] in order to count the

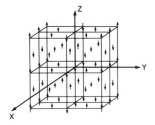

Fig. 8.4. Magnetic structure of K_2IrCl_6

number of anticipated EPR hyperfine lines. All the calculated fine and hyperfine structure was in fact resolved experimentally.

The occurrence of magnetic ordering in these substances at low temperatures has been demonstrated by specific heat [42], susceptibility [41] and neutron diffraction [35, 50] measurements. The λ-anomaly found at 3.01 K in the most recent specific heat measurements [34] on K_2IrCl_6 is accompanied by a shoulder at 2.83 K which is due to a displacive phase transition [27]. There are as yet no single-crystal nor field-dependent studies of these materials in the ordered state. Mössbauer studies [51] have been reported on K_2IrF_6, which orders at 0.460 K, Rb_2IrCl_6 (at 1.85 K), and Cs_2IrCl_6. The neutron work has shown that K_2IrCl_6 [and probably also $(NH_4)_2IrCl_6$] has the Type 3A magnetic structure, as illustrated in Fig. 8.4. There are four parallel and eight anti-parallel magnetic neighbors for each iridium ion. The magnetic structure of Rb_2IrCl_6 and Cs_2IrCl_6 may be different [43–45], because of a superexchange path through the alkali ion which increases in importance with the increase in size of the alkali metal. The theory of the exchange interaction in the $IrCl_6^{2-}$ systems has been explored by Judd [52] and others [53]. Using data for the three salts A_2IrCl_6, $A = NH_4$, K and Rb, a proportionality of T_c with the -32 power of the room temperature Ir–Ir separation has been observed [43].

Other studies on these systems include further analyses of hyperfine interactions [54] and of spin-lattice relaxation [55]. Some measurements on substances such as triclinic $Na_2IrCl_6 \cdot 6H_2O$ have been reported in the papers referred to, but the structure of this compound is unknown and the material seems to be incompletely characterized.

8.7 References

1. Gregson A.K. and Anker M., Aust. J. Chem. **32**, 503 (1979). The crystal structure is reported by: Raston C.L. and White A.H., Aust. J. Chem. **32**, 507 (1979)
2. Averill B.A. and Orme-Johnson W.H., Inorg. Chem. **19**, 1702 (1980)
3. McPherson G.L., Varga J.A., and Nodine M.H., Inorg. Chem. **18**, 2189 (1979)
4. Meijer H.C., Pimmelaar L.M.W.A., Brouwer S.R., and van den Handel J., Physica **51**, 588 (1971) and references therein
5. Herweijer A. and Gijsman H.M., Physica **36**, 269 (1967)
6. van Dalen P.A., Gijsman H.M., Love N., and Forstat, H., Proc. LT 9; Chem. Abst. **65**, 11516a (1966)
7. Cavell R.G. and Quail J.W., Inorg. Chem. **22**, 2597 (1983)
8. Carlin R.L., Carnegie, Jr. D.W., unpublished
9. Stevens K.W.H., Proc. Roy. Soc. (London) A **219**, 542 (1953)
10. Bleaney B. and O'Brien M.C.M., Proc. Phys. Soc. (London) B **69**, 1216 (1956)
11. Owen J. and Thornley J.H.M., Repts. Prog. Phys. **29**, 675 (1966)

12. Hudson A. and Kennedy M.J., J. Chem. Soc. **A 1969**, 1116
13. Sakaki S., Hagiwara N., Yanase Y., and Ohyoshi A., J. Phys. Chem. **82**, 1917 (1978)
14. Medhi O.K. and Agarwala U., Inorg. Chem. **19**, 1381 (1980)
15. Sakaki S., Yanase Y., Hagiwara N., Takeshita T., Naganuma H., Ohyoshi A., and Ohkubo K., J. Phys. Chem. **86**, 1038 (1982)
16. a) Figgis B.N., Lewis J., Mabbs F.E., and Webb G.A., J. Chem. Soc. **A 1966**, 422; b) Lewis J., Mabbs F.E., and Walton R.A., J. Chem. Soc. **A 1967**, 1366
17. Gregson A.K. and Mitra S., Chem. Phys. Lett. **3**, 392 (1968)
18. Peresie H.J., Stanko J.A., and Mulay L.N., Chem. Phys. Lett. **31**, 392 (1975)
19. Bernhard P., Stebler A., and Ludi A., Inorg. Chem. **23**, 2151 (1984)
20. McPherson G.L. and Martin L.A., J. Am. Chem. Soc. **106**, 6884 (1984)
21. Foyt D.C., Siddall III, T.H., Alexander C.J., and Good M.L., Inorg. Chem. **13**, 1793 (1974); Bouchard R.J., Weiher J.F., and Gillson J.L., J. Solid State Chem. **21**, 135 (1977); DaCosta F.M., Greatrex R., and Greenwood N.N., J. Solid State Chem. **20**, 381 (1977)
22. Bunker B.C., Drago R.S., Hendrickson D.N., Richman R.M., and Kessell S.L., J. Am. Chem. Soc. **100**, 3805 (1978); Johnson E.C., Callahan R.W., Eckberg R.P., Hatfield W.E., and Meyer T.J., Inorg. Chem. **18**, 618 (1979)
23. Carlin R.L., Burriel R., Seddon K.R., and Crisp R.I., Inorg. Chem. **21**, 4337 (1982)
24. Kaplan D. and Navon G., J. Phys. Chem. **78**, 700 (1974)
25. Carlin R.L. and Burriel R., to be published
26. Carlin R.L., Burriel R., Mennenga G., and de Jongh L.J., to be published
27. Armstrong R.L., Phys. Reports **57**, #6, 343 (1980)
28. Dorain P.B., (1968) Transition Metal Chemistry, Vol. 4, (Ed. by: Carlin R.L., Dekker M., New York)
29. Kamimura H., Koide S., Sekiyama H., and Sugano S., J. Phys. Soc. Japan **15**, 1264 (1960)
30. Bufaical R.F. and Harris E.A., Solid State Comm. **39**, 1143 (1981)
31. Busey R.H. and Sonder E., J. Chem. Phys. **36**, 93 (1962)
32. Busey R.H., Dearman H., and Bevan R.B., J. Phys. Chem. **66**, 82 (1962)
33. Busey R.H., Bevan R.B., and Gilbert R.A., J. Phys. Chem. **69**, 3471 (1965)
34. Moses D., Sutton M., Armstrong R.L., and Meicke P.P.M., J. Low Temp. Phys. **36**, 587 (1979)
35. Minkiewicz V.J., Shirane G., Frazer B.C., Wheeler R.G., and Dorain P.B., J. Phys. Chem. Solids **29**, 881 (1968)
36. Mrozinski J., Bull. Acad. Pol. Sci. Ser. Sci. Chim. **26**, 789 (1978)
37. Figgis B.N., Lewis J., and Mabbs F.E., J. Chem. Soc. **1961**, 3138
38. Hill N.J., J. Chem. Soc. Faraday Trans. II, **68**, 427 (1972)
39. Carlin R.L., Burriel R., unpublished
40. Griffiths J.H.E., Owen J., and Ward I.M., Proc. Roy. Soc. (London) A **219**, 526 (1953)
41. Cooke A.H., Lazenby R., McKim F.R., Owen J., and Wolf W.P., Proc. Roy. Soc. (London) A **250**, 97 (1959)
42. Bailey C.A. and Smith P.L., Phys. Rev. **114**, 1010 (1959)
43. Svare I. and Raaen A.M., J. Phys. C **15**, 1329 (1982)
44. Raaen A.M., Svare I., and Pedersen B., Physica **121 B**, 89 (1983)
45. Raaen A.M. and Svare I., Physica **121 B**, 95 (1983)
46. Griffiths J.H.E., Owen J., Park J.G., and Partridge M.F., Proc. Roy. Soc. (London) A **250**, 84 (1959)
47. Harris E.A. and Owen J., Proc. Roy. Soc. (London) A **289**, 122 (1965)
48. Griffiths J.H.E. and Owen J., Proc. Roy. Soc. (London) A **226**, 96 (1954)
49. Thornley J.H.M., J. Phys. C **1**, 1024 (1968)
50. Hutchings M.T. and Windsor C.G., Proc. Phys. Soc. **91**, 928 (1967)
51. Potzel W., Wagner F.E., Gierisch W., Gebauer E., and Kalvius G.M., Physica B **86–88**, 899 (1977)
52. Judd B.R., Proc. Roy. Soc. (London) A **250**, 110 (1959)
53. Stedman G.E. and Newman D.J., J. Phys. C **4**, 884 (1971)
54. Davies J.J. and Owen J., J. Phys. C **2**, 1405 (1969)
55. Harris E.A. and Kngvesson K.S., J. Phys. C **1**, 990, 1011 (1968)

9. The Rare Earths or Lanthanides

9.1 Introduction

The paramagnetic properties of the rare earth ions have been investigated so extensively that a book could be written on the subject, never mind a chapter. This is due in large part to the many studies of the EPR spectra of the lanthanides when doped into diamagnetic hosts, and the attendant theoretical development in terms of crystal field theory. Many of the current ideas in magnetism have followed from such studies, especially with regard to relaxation phenomena.

Relatively few coordination compounds of the lanthanides have been studied in detail. Superexchange is relatively unimportant, and dipole-dipole interactions therefore tend to dominate the magnetic ordering phenomena; this is enhanced by both the relatively large spin and g-values of many of the lanthanides. The compounds which order do so largely at temperatures of 4 K and below as a result. These effects are enhanced by the high coordination number of most rare earth compounds, which works to reduce the strength of any superexchange paths. Also, as a result, there appear to be no lower-dimensional rare earth compounds, at least in the sense described in Chap. 7.

It is difficult to generalize about each ion, since the crystal field splitting of any ion, though small, can change from one system to another.

The chemistry of the lanthanides has been reviewed [1–3]. The distinguishing features for our purposes are:

a) The metal ions are found primarily in the trivalent state.

b) High coordination numbers (eight appears to be the commonest) and a variety of geometries are found. Six-coordination and lower are rare. Several typical coordination polyhedra are illustrated in Fig. 9.1.

c) Oxygen and nitrogen are the favored donor atoms, but a number of halides have also been studied.

d) Spin-orbit coupling is important, the more so energetically than crystal field splittings, and increases with atomic number. This is summarized in Table 9.1. The letter $\mathcal{J}$ is once again used as the total angular momentum quantum number.

e) The magnetic properties are determined by the 4f electrons, which are well-shielded by the occupied outer shells of 5s and 5p electrons. The 4f electrons are little-involved with chemical bonding, which is why superexchange is relatively unimportant. This is well-illustrated in Fig. 9.2, which shows the radial extension of the wave functions [6]. The interactions which dominate the cooperative phenomena with rare earth compounds are dipolar in nature. This is because of the important orbital contributions to the magnetic moment. Thus, since dipole-dipole interactions are long-

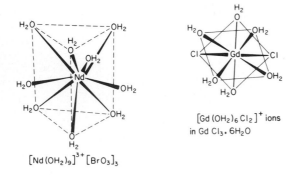

$[Nd(OH_2)_9]^{3+}[BrO_3]_3$

$[Gd(OH_2)_6 Cl_2]^+$ ions
in Gd Cl$_3$ · 6H$_2$O

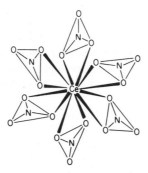

$[Ce(NO_3)_6]^{3-}$ in Ce$_2$Mg$_3$(NO$_3$)$_{12}$· 24 H$_2$O

Fig. 9.1. Coordination polyhedra of several simple lanthanide complexes; from Ref. [1]

Table 9.1. Summary of experimental spin-orbit interaction parameters ζ for the tripositive lanthanide ions together with the electronic configuration and spectroscopic terms of the ground state and first excited state and their separation.

Tripositive lanthanide ion	Ionic con-figuration	Ground-state term	Excited state term	Energy separation (cm^{-1})	Spin-orbit, ζ experimental (cm^{-1})
Ce	$4f^1$	$^2F_{5/2}$	$^2F_{7/2}$	2 200	643
Pr	$4f^2$	$^3H_4$	$^3H_5$	2 100	800
Nd	$4f^3$	$^4I_{9/2}$	$^4I_{11/2}$	1 900	900
Pm	$4f^4$	$^5I_4$	$^5I_5$	1 600	–
Sm	$4f^5$	$^6H_{5/2}$	$^6H_{7/2}$	1 000	1200
Eu	$4f^6$	$^7F_0$	$^7F_1$	400	1415
Gd	$4f^7$	$^8S_{7/2}$	^6P	30 000	–
Tb	$4f^8$	$^7F_6$	$^7F_5$	2 000	1620
Dy	$4f^9$	$^6H_{15/2}$	$^6H_{13/2}$	–	1820
Ho	$4f^{10}$	$^5I_8$	$^5I_7$	–	2080
Er	$4f^{11}$	$^4I_{15/2}$	$^4I_{13/2}$	6 500	2360
Tm	$4f^{12}$	$^3H_6$	$^3H_5$	–	2800
Yb	$4f^{13}$	$^2F_{7/2}$	$^2F_{5/2}$	10 000	2940

From Refs. [4] and [5].

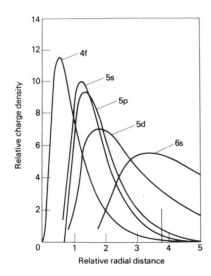

Relative charge density

Relative radial distance

4f
5s
5p
5d
6s

Fig. 9.2. The distribution of the radial part of the f wave function. From Refs. [5] and [6]

range, behavior most closely resembling mean or molecular field theories is found here. (Calculations for such models as the Heisenberg and Ising Hamiltonians are based primarily on short-range, nearest neighbor interactions.)

f) The metal ion radius decreases regularly as atomic number increases from La^{3+} (0.106 nm) to Lu^{3+} (0.085 nm). Though there are great similarities among all the lanthanide compounds of a given formula, ligand-ligand repulsion increases as the metal ion gets smaller. This is why the compounds of the heavier ions are sometimes different from those of the lighter ones. The rhombohedral LaMN lattice, where LaMN is $La_2Mg_3(NO_3)_{12} \cdot 24 H_2O$, is not formed for metals heavier than Sm or Eu, for example.

g) Most of the low lying energy levels are mixed by both spin-orbit coupling and the crystalline field, and so the ground state can rarely be characterized by a single value of $\mathscr{J}_z$. The g values are generally quite anisotropic and deviate substantially from 2. EPR usually can be observed only at temperatures of 20 K or lower. Furthermore, several of the ions are non-Kramers ions, Eu^{3+} ($^7F_0$) being an example. With a $\mathscr{J}=0$ ground state, no EPR can be observed.

h) The ionic radius of Y^{3+}, which is not a lanthanide but which has chemical properties similar to lanthanum, is intermediate between the radii of Ho^{3+} and Er^{3+}. Its coordination behavior is normally identical with these ions, and yttrium salts often furnish a colorless, diamagnetic host for the heavier rare earth ions.

The closely-related subject of the optical spectra of rare earth ions in crystals has been reviewed by Dieke [7]. Though we shall mention a number of EPR results, we shall not review the paramagnetic relaxation results of the Leiden group (e.g., [8,9]) nor of the Berkeley group [10–13]. Many of these results have been reviewed by Orton [14] and Abragam and Bleaney [15]. Surprisingly, most of the compounds studied at low temperatures contain primarily water as a ligand. The bulk of the chapter is devoted to these substances. Unlike many other tripositive ions, however, these cations do not form alums.

9.2 Cerium

We begin with the $4f^1$ ion cerium(III) because it should be the most straightforward. The ground state, with $\mathscr{J} = \frac{5}{2}$, is the $^2F_{5/2}$ state, and the $^2F_{7/2}$ state is some $2200\,\mathrm{cm}^{-1}$ higher in energy. Since the latter state is unpopulated at low temperatures, it does not contribute directly to the magnetic properties of cerium. However, it does alter the properties to a certain extent because of quantum mechanical mixing of its wave function with that of the lower state. The theory of the magnetic properties of cerium(III), developed largely by Elliott and Stevens, arose because of the EPR experiments at Oxford and is closely tied historically with the development of crystal field theory. The results have been described in many texts [4, 5, 16], especially those of Orton [14] and Abragam and Bleaney [15].

The orbital angular momentum of the lanthanide ions is generally not quenched by the crystal field. The Lande factor g_J is of the form

$$g_J = \frac{3\mathscr{J}(\mathscr{J}+1) + \mathscr{S}(\mathscr{S}+1) - \mathscr{L}(\mathscr{L}+1)}{2\mathscr{J}(\mathscr{J}+1)}. \tag{9.1}$$

The $\mathscr{J} = \frac{5}{2}$ state of cerium is composed from the $\mathscr{L} = 3$, $\mathscr{S} = \frac{1}{2}$ microstate, so $g_J = \frac{6}{7}$. With the introduction of this parameter one may deal directly with the $|m_{\mathscr{J}}\rangle$ states by using the Zeeman operator $g_J \mu_B \boldsymbol{H} \cdot \boldsymbol{J}$.

The $\mathscr{J} = \frac{5}{2}$ state has six-fold degeneracy consisting of the $\mathscr{J}_z = \pm\frac{1}{2}$, $\pm\frac{3}{2}$, and $\pm\frac{5}{2}$ components. The degeneracy of these states is not resolved by spin-orbit coupling but by the crystalline field supplied by the ligands. One of the major problems in the study of the magnetic properties of cerium(III) has concerned the determination of the relative order of these levels and their separation. The overall separation in the compounds which have been studied is less than $300\,\mathrm{K}$.

To first order, the g-values of the three doublets are found, by using the operator introduced above:

$$|\pm\tfrac{1}{2}\rangle; \quad g_{\parallel} = \tfrac{6}{7} = 0.86, \quad g_{\perp} = \tfrac{18}{7} = 2.57$$
$$|\pm\tfrac{3}{2}\rangle; \quad g_{\parallel} = \tfrac{18}{7} = 2.57, \quad g_{\perp} = 0$$
$$|\pm\tfrac{5}{2}\rangle; \quad g_{\parallel} = \tfrac{30}{7} = 4.29, \quad g_{\perp} = 0.$$

Lanthanum ethyl sulfate, $La(EtOSO_3)_3 \cdot 9\,H_2O$ (LaES) is a common host crystal used for studies of the rare earths because it grows easily and is uniaxial. The crystal belongs to the space group $P6_3/m$, and the point symmetry of the lanthanide ion is C_{3h}. The metal ions in this lattice have nine water molecules as ligands; six form a triangular prism with three above and three below the symmetry plane containing the other three water molecules and the lanthanide ion [17]. The shortest metal-metal distance in this lattice is about $0.7\,\mathrm{nm}$, along the symmetry axis. Since the most direct superexchange path would go through two water molecules, and therefore be quite weak, magnetic dipole-dipole interactions are expected to be the most important interaction in these salts. The ground state for LaES doped with Ce^{3+} exhibits an EPR spectrum with $g_{\parallel} = 0.955$ and $g_{\perp} = 2.185$, which implies that the $\mathscr{J}_z = \pm\frac{1}{2}$ state is the ground state. The next higher state is only about $5\,\mathrm{K}$ higher and exhibits $g_{\parallel} = 3.72$ and $g_{\perp} = 0.2$, so this must be the $|\pm\frac{5}{2}\rangle$ state. Comparison of these results with the theoretical values stated above shows that a more complete theory is required. This can be formulated [18] by writing the wave-functions of the $|\mathscr{I}_x\rangle = |\pm\frac{1}{2}\rangle$ state as

$$|\tfrac{1}{2}\rangle = \cos\theta|1, -\tfrac{1}{2}\rangle - \sin\theta|0, \tfrac{1}{2}\rangle$$
$$|-\tfrac{1}{2}\rangle = -\cos\theta|-1, +\tfrac{1}{2}\rangle + \sin\theta|0, -\tfrac{1}{2}\rangle,$$

where the notation used on the right side is $|m, m_s\rangle$ and θ is simply an empirical parameter. The g-values are then found as

$$g_\parallel = 2\sin^2\theta$$
$$g_\perp = 2|\sqrt{3}\sin 2\theta - \sin^2\theta|.$$

The experimental values are still not fit exactly by these results, the best value for θ being near 50°. This yields $g_\parallel = 1.17$ and $g_\perp = 2.24$.

The calculation of the properties of the $|\mathscr{I}_x = \pm\frac{5}{2}\rangle$ state is more complicated because it mixes with the $|\mathscr{I}_x = \mp\frac{7}{2}\rangle$ (excited) state. The state may be written as

$$|\pm\tfrac{5}{2}\rangle = \pm p|\pm 2, \pm\tfrac{1}{2}\rangle \pm q|\pm 3, \mp\tfrac{1}{2}\rangle \pm r|\mp 3, \mp\tfrac{1}{2}\rangle$$

with $p^2 + q^2 + r^2 = 1$. The g-values are

$$g_\parallel = 2|3p^2 + 2q^2 - 4r^2|$$

and

$$g_\perp = 2|r(2q + \sqrt{6}p)|$$

and p, q, r may be determined by comparison with experiment. The parameters depend on the radial crystal field parameters. The sensitivity of the ion to the local crystalline field may be illustrated by the fact that the $|\pm\frac{5}{2}\rangle$ state becomes the ground state in pure CeES, with $|\pm\frac{1}{2}\rangle$ being at 6.7 K. The position of the $|\pm\frac{3}{2}\rangle$ state has not been located exactly but it is some 200 K above the ground state in both dilute and concentrated salts.

Specific heat measurements [19–21] of CeES show that the lattice contribution is not T^3 even at 1 K. This was one of the first salts cooled below 1 K by adiabatic demagnetization; it becomes antiferromagnetic at approximately 0.025 K. The quantity cT^2/R takes the value 6.7×10^{-4} K^2. A long-standing problem has concerned the nature of the nondipolar interactions which are especially important in this salt [22]. The best available estimate of the ground state of CeS may be written as

$$\cos\alpha|\pm\tfrac{5}{2}\rangle \mp \sin\alpha|\mp\tfrac{1}{2}\rangle,$$

with the low-lying excited state as

$$\cos\alpha|\pm\tfrac{1}{2}\rangle \mp \sin\alpha|\mp\tfrac{5}{2}\rangle,$$

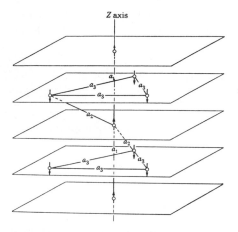

Z axis

Fig. 9.3. Relative position of the cerium ions in the ethyl sulfate lattice. $a_1 = 0.711$ nm, $a_2 = 0.855$ nm, and $a_3 = 1.400$ nm. The arrows indicate the assumed direction of the magnetic spins. At temperatures below the Néel point, the spins on alternate planes are assumed to point in the same direction. From Ref. [20]

and $\sin^2 \alpha = 0.036$ to 0.039. Thus the states are mixed a small amount, and a change in the mixing can result from small changes in the local crystalline field.

Incidentally, cerium is the only rare earth of which all its naturally occuring isotopes have no nuclear spin. This means there cannot be any electron spin-nuclear spin hyperfine interaction in the EPR spectra, nor can such an interaction contribute to the low-temperature specific heat.

A likely structure for the antiferromagnetic state is illustrated in Fig. 9.3.

Another salt which has been investigated extensively [23–26] is cerous magnesium nitrate, $Ce_2Mg_3(NO_3)_{12} \cdot 24 H_2O$, and generally referred to as either CMN or CeMN. It has a rhombohedral lattice, with all the cerium ions equivalent [27]. The coordination of the cerium ion by six nitrate groups was illustrated in Fig. 9.1. The $|\mathscr{I}_z = \pm\frac{1}{2}\rangle$ state is the largely unmixed ground state, with $g_\parallel = 0.25$ and $g_\perp = 1.84$. The parameters do not change upon dilution with the diamagnetic La analog. The salt obeys the Curie law [24] down to at least 0.006 K! One calculates that the specific heat, cT^2/R, from dipole-dipole interaction alone is 6.6×10^{-6} K^2, and early measurements gave the result $cT^2/R = 7.5 \times 10^{-6}$ K^2. Thus, dipole-dipole interaction accounts for nearly the whole specific heat term. More recent results [26a] may be expressed as $c/R \times 10^3 = 0.012\, T^{-2} + 0.026\, T^{-1.6}$, where the latter term is used only as an empirical (fitting) parameter, while other investigators [26b] have reported $cT^2/R = (5.99 \pm 0.02) \times 10^{-6}$, i.e., a value less than the calculated dipolar contribution. The specific heat is therefore some 2300 times smaller than that of such another good Curie law salt as potassium chrome alum. The next higher level is known, rather precisely, to lie at 36.00 K.

The magnetic interactions between cerium ions in CMN therefore are smaller than in any other known salt. With its large magnetic anisotropy, it is ideal for conducting experiments below 1 K, as both a cooling salt and as a magnetic thermometer. The parallel susceptibility, $\chi_\parallel$, has the added feature of a broad maximum at around 25 K, which is due to the small $g_\parallel$ of the ground state. This is a susceptibility analog of a Schottky anomaly, for the salt is practically non-magnetic in the parallel direction at very low temperatures. Finally, this is the salt in which the Orbach process of spin-lattice relaxation was first tested. The relevant state, which is primarily the $|\pm\frac{3}{2}\rangle$ state, is of the form

$$a|\mathscr{I}=\tfrac{5}{2}, \mathscr{I}_z=\pm\tfrac{3}{2}\rangle+b|\tfrac{7}{2}, \pm\tfrac{3}{2}\rangle+c|\tfrac{7}{2}, \mp\tfrac{3}{2}\rangle,$$

with $g_\perp=0$. The only real practical problem with using CMN is that, as with many hydrated salts, it tends to lose the water of hydration.

Magnetic ordering in CMN has been found to occur at 1.9 mK [28] from specific heat measurements. The ordering has variously been described as ferromagnetic or antiferromagnetic [28–31]. Calculations of the ground state energies confirm the dipole-dipole nature of the interaction [32].

A similar salt that has been little-studied is hydrated cerium zinc nitrate, CZN, in which the slightly larger zinc ion replaces the magnesium ion. The two salts, CMN and CZN, are isomorphous. The susceptibilities of CZN have been measured in the ^{4}He region [33], and the g-values change: $g_\parallel=0.38$ and $g_\perp=1.94$. The order of the energy levels appears not to change from that in CMN, but the first excited state in CZN is estimated to be only at 30 K, about 83% of the separation in CMN.

Several attempts [34–36] have been made to find salts more suitable than CMN for magnetic cooling but, while there are several promising candidates, none so far has been found to be as useful as CMN. For example, $Na_3[Ce(C_4H_4O_5)_3]\cdot 2\,NaClO_4 \cdot 6\,H_2O$, where the ligand is diglycollate, is more dilute magnetically than CMN, but large single crystals of the compound have not yet been prepared.

Another substance of interest is anhydrous $CeCl_3$ [37]. This salt also attains a relatively simple, trigonal structure, as illustrated in Fig. 9.4. In this case the ground state is the almost-pure $|\mathscr{I}_z=\pm\tfrac{5}{2}\rangle$, which is well-isolated (by 59 K) from the next higher state; when diluted by $LaCl_3$, the EPR parameters are $g_\parallel=4.037$, $g_\perp=0.20$. The cerium therefore is in an effective $\mathscr{S}=\tfrac{1}{2}$ state at low temperatures. Since $\Delta\mathscr{I}_z$ of the ground state doublet is 5, one expects highly anisotropic interactions. The specific heat exhibits a rounded peak near 0.115 K, and the substance, which appears to exhibit a large amount of short range order, seems to be an antiferromagnet.

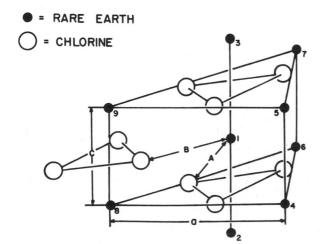

● = RARE EARTH

○ = CHLORINE

Fig. 9.4. Structure of the rare-earth trichlorides for rare earths between lanthanum and gadolinium. For $LaCl_3$, $a=0.7483$ nm, $c=0.4375$ nm; for $CeCl_3$, $a=0.7450$ nm, $c=0.4315$ nm. From Ref. [37]

9.3 Praseodymium

This trivalent ion, with a $4f^2$ electron configuration, is an even-electron or non-Kramers ion. This means that a crystal field alone can resolve degeneracies to give a non-magnetic ground state. Thus, the $^3H_4$ ground state of the free ion is split by the crystalline field into three doublets and three singlets. In the ethylsulfate, $Pr(EtOSO_3)_3$ $\cdot 9\,H_2O$ [19, 38] the lowest state is a doublet, but the specific heat shows that it is split into two singlets separated by a small energy, Δ. The analysis of even-electron systems was described by Baker and Bleaney [39]. The spin-Hamiltonian (neglecting nuclear terms) may be written as

$$H = g_{\parallel}\mu_B S_z H_z + \Delta_x S_x + \Delta_y S_y, \tag{9.2}$$

where $\Delta = (\Delta_x^2 + \Delta_y^2)^{1/2}$ represents the effect of random local departures in the crystal from the normal symmetry.

The ground state non-Kramers doublet may be written as

$$|\pm\rangle = \cos\theta|\mathscr{J} = 4, \mathscr{J}_z = \pm 2\rangle - \sin\theta|\mathscr{J} = 4, \mathscr{J}_z = \mp 4\rangle$$

for which $g_{\parallel} = g_J(4\cos^2\theta - 8\sin^2\theta)$ and $g_J = \frac{4}{5}$. EPR has been observed between these levels in Pr/YES, with $g_{\parallel} = 1.52$; $g_{\perp}$ is necessarily zero in this situation. The results require $\theta \simeq 25°$ but admixture of $\mathscr{J} = 5$ states is probably also likely.

The final result of the analysis of the ground state leads to the wave function

$$|+\rangle = \cos\theta|4, 2\rangle - \sin\theta|4, -4\rangle - 0.22|5, 2\rangle + 0.056|5, -4\rangle$$

and

$$g_{\parallel} = g_J(4\cos^2\theta - 8\sin^2\theta) - 0.09,$$

and $\theta = 24°$. The parameter $\Delta = 0.11\,\mathrm{cm}^{-1}$ (0.16 K).

This analysis is representative of the magnetochemistry of the lanthanides: there are a large number of levels and they can mix to a varying extent. Small variations in the local crystalline fields cause changes in the mixing, along with a concomitant change in the observables. Specific heat measurements below 1 K on pure PrES show that Δ is larger, 0.53 K, while measurements at higher temperatures revealed a Schottky anomaly corresponding to a singlet at 16.7 K. There are no other levels below 200 K. The single crystal susceptibilities of PrES have been measured over a wide temperature interval [40].

For Pr^{3+} in LaMN, $g_{\parallel}$ by EPR also is 1.55, and $g_{\perp} = 0$. Susceptibility measurements at 1.1–4.2 K show that $g_{\parallel}$ is 1.56 in PrMN [41] but that $g_{\perp}$ differs from zero, being 0.18 ± 0.03. Similarly, measurements on PrZN [36] show that $g_{\parallel} = 1.60$ and $g_{\perp} = 0.16$. There is as yet no suggestion why these values disagree with both theory and the EPR results. The quantity Δ has been measured as $0.18\,\mathrm{cm}^{-1}$ in pure PrMN [42].

9.4 Neodymium

This $4f^3$ ion has a 10-fold degenerate $^4I_{9/2}$ ground state, with the next higher state ($^4I_{11/2}$) some 300 K higher. As with the other examples discussed so far, the crystalline field resolves this state into five doublets, only one of which is populated at temperatures of 20 K or below. Thus, as with most of the rare earth ions (the major exceptions being Sm^{3+} and Eu^{3+}, as we shall see) Nd^{3+} has an effective doublet ground state; fast spin-lattice relaxation prevents EPR spectra from being observed at temperatures greater than 20 K.

Since we are neglecting most of the nuclear interactions, the spin-Hamiltonian for Nd^{3+} is relatively simple, being

$$H = \mu_B[g_\parallel H_z S_z + g_\perp(H_x S_x + H_y S_y)]$$

When diluted by LaES, the parameters are

$$g_\parallel = 3.535, \quad g_\perp = 2.072$$

while, in LaMN, the parameters are

$$g_\parallel = 0.45, \quad g_\perp = 2.72.$$

Notice the dramatic difference between the two sets of values. The theoretical development for Nd^{3+} is by Elliott and Stevens [43]. In Nd/LaES, the ground state appears, to first order, to be the doublet

$$\cos\theta|\pm\tfrac{7}{2}\rangle + \sin\theta|\mp\tfrac{5}{2}\rangle$$

and $\theta = 24°$. A second-order calculation leads to good agreement with experiment, and to the following calculated energy levels:

State	Energy, cm^{-1}	$g_\parallel$	$g_\perp$
$0.92|\pm\tfrac{7}{2}\rangle + 0.38|\mp\tfrac{5}{2}\rangle$	0	3.56	2.12
$|\pm\tfrac{1}{2}\rangle$	130	0.73	3.64
$0.75|\pm\tfrac{9}{2}\rangle + 0.66|\mp\tfrac{3}{2}\rangle$	170	2.73	0
$0.38|\pm\tfrac{7}{2}\rangle - 0.92|\mp\tfrac{5}{2}\rangle$	340	2.34	2.03
$0.66|\pm\tfrac{9}{2}\rangle - 0.75|\mp\tfrac{3}{2}\rangle$	350	1.62	0

The susceptibility follows the Curie law up to 100 K, which is consistent with only one doublet being populated until this temperature is reached. The susceptibilities to 300 K as calculated from the above energy levels agree very well with the measurements of van den Handel and Hupse [44].

The specific heat [45] is featureless down to 0.24 K, as expected (i.e., there is no Schottky term). The Curie-Weiss law is obeyed down to 0.15 K, and the low temperature specific heat, $cT^2/R = 1.14 \times 10^{-3} \, K^2$, can be entirely ascribed to dipole-dipole interaction and nuclear hyperfine interaction.

The Nd/LaMN system requires that the first-order wavefunction be of the form

$$\pm \cos\theta \cos\phi |\tfrac{9}{2}, \pm\tfrac{7}{2}\rangle + \sin\theta \cos\phi |\tfrac{9}{2}, \mp\tfrac{5}{2}\rangle + \sin\phi |\tfrac{9}{2}, \pm\tfrac{1}{2}\rangle,$$

but that this be corrected by four extra admixed states from $\mathcal{J} = \tfrac{11}{2}$, with

$$\mathcal{J}_z = \pm\tfrac{7}{2}, \pm\tfrac{1}{2}, \mp\tfrac{5}{2}, \quad \text{and} \quad \mp\tfrac{1}{2}.$$

Susceptibility measurements in the ^{4}He temperature region [36, 41] yield the following results:

	$g_\parallel$	$g_\perp$
NdZN	0.42	2.75
NdMN	0.39	2.70

One of the few neodymium salts which has been found to order [46] is $Nd(OH)_3$, at 265 mK.

9.5 Samarium

The reason that this $4f^5$ ion, with a $^6H_{5/2}$ ground state, is interesting may be seen in Fig. 9.5. With a $|\mathcal{J}\rangle = |\tfrac{5}{2}\rangle$ ground state, the $|\tfrac{7}{2}\rangle$ state is at an energy comparable with kT

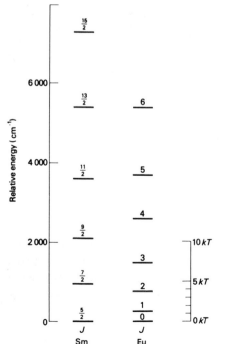

Fig. 9.5. Multiplet spacings of Sm^{3+} and Eu^{3+}. The scale showing thermal energy is for a temperature of 300 K

at room temperature, and is therefore thermally populated. The population of this state will change with temperature, and therefore this statistical occupation must be taken account of in any susceptibility calculation. Thus, the susceptibility will not follow a simple Curie-Weiss law. Furthermore, the $|\frac{7}{2}\rangle$ state (or its components) are at a low enough energy to mix strongly with the components of the $|\frac{5}{2}\rangle$ ground state, and thereby modify substantially the g-values, for example.

Figure 9.5 also illustrates a similar situation for europium(III). This will be discussed further below.

The g-values of Sm/LaES, by EPR, are $g_{\parallel} = 0.596$ and $g_{\perp} = 0.604$. The ground state deduced is

$$\cos\theta|\tfrac{5}{2}, \pm\tfrac{1}{2}\rangle \pm \sin\theta|\tfrac{7}{2}, \pm\tfrac{1}{2}\rangle,$$

with $\theta = 0.07$ rad. Susceptibility measurements [36, 41] on the pure salts provide the following results:

	$g_{\parallel}$	$g_{\perp}$
SmMN	0.736	0.406
SmZN	0.714	0.412

9.6 Europium

There is relatively little that can be said about the magnetochemistry of this trivalent $4f^6$ ion. Being a non-Kramers ion, it can and does have a singlet (non-magnetic) ground state. There is no electron paramagnetic resonance for this ion, for example. As mentioned above and illustrated in Fig. 9.5, the different $\mathscr{J}$ states are close together, and their occupation will change substantially with temperature. The Curie-Weiss law will not be obeyed, and the temperature-dependent susceptibility should go to zero at low temperatures. There should, however, be an important temperature-independent susceptibility, because of the close proximity of the low lying excited states.

The divalent ion, Eu^{2+}, is more important and interesting. This is because it has a half-filled 4f shell, electron configuration $4f^7$, and this has a well-isolated $^8S_{7/2}$ ground state. EPR is readily observed and, since the orbital angular momentum is nearly completely quenched, g-values of 1.99 are commonly observed. Europium is also a useful nucleus for Mössbauer studies, and the magnetic ordering in a number of compounds has been reported [47]. The results include the following transition temperatures: $EuSO_4$, 0.43 K; EuF_2, 1.0 K; $EuCO_3$, 1.05 K; $EuCl_2$, 1.1 K, and $EuC_2O_4 \cdot H_2O$, 2.85 K.

Perhaps the most interesting and important europium compounds are the chalcogenides, EuX, where X is oxygen, sulfur, selenium and even tellurium. They are important because they are magnetic semiconductors; the high ordering temperatures also show that these are some of the few rare earth compounds in which superexchange interaction is clearly important [48]. The transition temperatures in zero applied field are: EuO, 69.2 K; EuS, 16.2 K; EuSe, 4.6 K; and EuTe, 9.6 K. The magnetic structures of the compounds are complicated: EuO and EuS are both ferromagnets, EuSe has a complicated phase diagram [49], while EuTe is an antiferromagnet. Because of the complex magnetic structure, based on a magnetic unit cell that is 8 times larger than the chemical unit cell, EuTe is not an easy-axis antiferromagnet. In the four compounds, the factor determining the macroscopic magnetic properties appears to be the relative sign and magnitude of the nearest neighbor and next-nearest neighbor exchange constants.

There has been considerable interest lately in the properties of EuS as it is diluted by diamagnetic SrS [50].

9.7 Gadolinium

With a half-filled 4f shell, trivalent gadolinium has an $^8S_{7/2}$ ground state; the orbital contribution is almost entirely quenched and isotropic g-values of 1.99 are universal. Due to the long spin-lattice relaxation times, EPR spectra may be observed at room temperature, and there is a host of reports on the paramagnetic properties of gadolinium [51]. Its magnetochemistry is therefore straightforward and perhaps the best known of all the rare earth ions. The ground state is usually resolved (by a Kelvin or less) by zero-field splitting effects.

There are many studies of the EPR spectra of Gd(III) in the ethylsulfate host lattice, but there seem to be no modern reports on the magnetic properties of pure $Gd(EtOSO_3)_3 \cdot 9\,H_2O$. The gadolinium compound analogous to CMN, GdMN, does not appear to exist. Salts such as $Gd_2(SO_4)_3 \cdot 8\,H_2O$ have long been investigated for adiabatic demagnetization purposes because of the high spin of the ion.

Several salts have been shown to order magnetically. Susceptibility studies [52] show that both $Gd_2(SO_4)_3 \cdot 8\,H_2O$ and $GdCl_3 \cdot 6\,H_2O$ are antiferromagnets, with transition temperatures, respectively, of 0.182 and 0.185 K. The ordering is dipolar in both cases, even though the chloride contains $[Gd(OH_2)_6Cl_2]^+$ units. Zero-field splittings of the ground state make a large Schottky contribution to the specific heat, although the overall splittings are only about 1 K. Interestingly, the $|\pm\frac{1}{2}\rangle$ level is the lowest for the sulfate while the order of the levels is reversed for the chloride, the $|\pm\frac{7}{2}\rangle$ state being the lowest. These different behaviors may be observed from an examination of the specific heat data, which are illustrated in Figs. 9.6 and 9.7. The exchange constant in $GdCl_3 \cdot 6\,H_2O$ is 10 times smaller than it is in ferromagnetic $GdCl_3$, where J/k is -0.08 K for nearest neighbor exchange.

One of the most thoroughly studied [53] salts is gadolinium hydroxide, $Gd(OH)_3$. This material is an antiferromagnet with predominantly (but not exclusively) nearest-neighbor interactions. Since the nearest neighbors are aligned along the c-axis, the compound is accidentally a linear chain system. While isomorphic $GdCl_3$ is a ferromagnet ($T_c = 2.20$ K) with the spins aligned parallel to the c-axis, the spins have a perpendicular arrangement in $Gd(OH)_3$. The critical temperature is 0.94 K. All the $R(OH)_3$ salts exist with the same structure as $LaCl_3$, with two magnetically-equivalent ions per unit cell. The local point symmetry is C_{3h}. The lattice parameters are 14% less than for the corresponding RCl_3, and so cooperative magnetic effects are expected at reasonable temperatures.

The magnetic ions lie on identical chains parallel to the c-axis with the two nearest neighbors in the same chain separated by ± 0.36 nm. The g-value is isotropic, at 1.992.

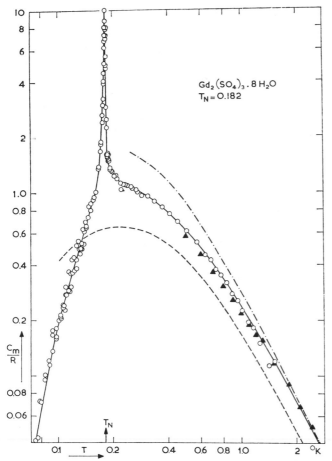

Fig. 9.6. Heat capacity c/R of $Gd_2(SO_4)_3 \cdot 8\,H_2O$ as a function of temperature on a logarithmic scale. A singularity occurs at $T_c = 0.182 \pm 0.001$ K. The dashed line represents the Schottky specific heat for $m_d = \pm\frac{1}{2}$ lowest and the dot-dashed line gives the sum of Schottky and dipolar heat capacities. From Ref. [52]

Small crystal field splittings (of order 0.02 K) have been observed by EPR, but do not affect data taken in the relevant temperature range for this salt.

The specific heat (Fig. 9.8) exhibits an anomaly at T_c, but also exhibits a broad peak to higher temperatures. Since this cannot arise from the small Schottky contribution, it must arise, unexpectedly, from short range order. Indeed the critical entropy S_c is found as only 45% of the total. Analysis of a wealth of data yielded J_1/k, the exchange constant along the c-axis, as 0.18 K. This is the largest interaction, but the dipole-dipole interaction turns out to be only a bit smaller for nn pairs and remains appreciable for many of the more distant neighbors. It provides the only source of anisotropy. Since the strongest interaction lies along the c-axis, the system then behaves as a magnetic linear chain, with spins lying antiparallel and perpendicular to the c-axis. What is not clear is how the chains are arranged. They could be as illustrated in Fig. 9.9 with 2 parallel and

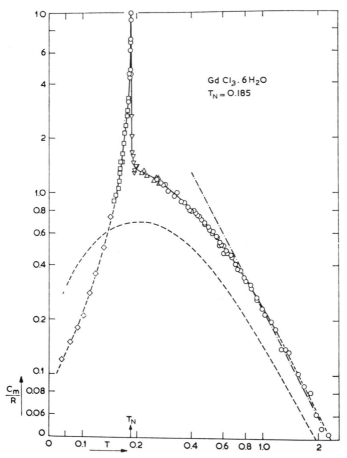

Fig. 9.7. Heat capacity c/R of $GdCl_3 \cdot 6H_2O$ as a function of temperature T on a logarithmic scale. A singularity is found at $T_c = 0.185 \pm 0.001$ K. The dashed line indicates the estimated Schottky specific heat. The dash-dotted line corresponds to the sum of Schottky specific heat and dipolar specific heat. From Ref. [52]

4 antiparallel next nearest neighbors, in a spiral state (helix) or some other such arrangement. The long-range dipole-dipole interaction helps to determine the structure of the ordered state.

Thus, $Gd(OH)_3$ behaves to a certain extent as a linear chain antiferromagnet, not because of a chemical linking as described so frequently in Chap. 7, but because of the structure and relative strength of the interactions.

Two substances, parts of a larger series, that have been extensively studied are gadolinium orthoferrite [54], $GdFeO_3$, and gadolinium orthoaluminate [55], $GdAlO_3$. The first compound is complicated because of the two magnetic systems, the iron ions (which order antiferromagnetically at 650 K) and the gadolinium ions, which order at 1.5 K, under the influence of the magnetic iron lattice. The isomorphous aluminate is simpler because of the absence of any other magnetic ions.

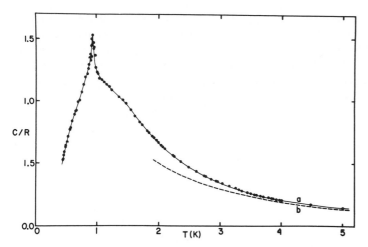

Fig. 9.8. Total specific heat of $Gd(OH)_3$ from calorimetric measurements, curve a, which is very close to the magnetic specific heat in that the lattice specific heat contributes a maximum of about 3% at 5 K. The peak corresponds to $T_c = 0.94 \pm 0.02$ K and is interpreted as the onset of a complex long-range antiferromagnetic order. Curve b is the high-temperature asymptote $c_M/R = c_2/T^2 + c_3/T^3$ as calculated from the final values of the interactions. From Ref. [53]

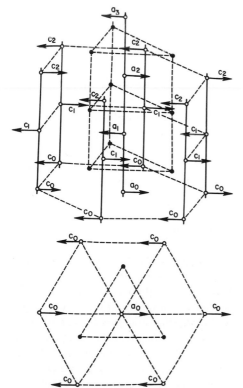

Fig. 9.9. One possible arrangement of the spins in $Gd(OH)_3$. From Ref. [53]

In contrast to so many rare earth compounds, the principal interaction between the ions in $GdFeO_3$ and $GdAlO_3$ is exchange interaction between nearest neighbors. This happens because, although the substances are orthorhombic, the gadolinium ions form a pseudocubic unit cell which is only slightly distorted. This is important, since for a cubic lattice the magnetic dipole interaction field on a particular ion due to all the ions in a spherical sample will vanish both for the ferromagnetic case and for the simple antiferromagnetic case (all nearest neighbors antiparallel). In the antiferromagnetic state of $GdAlO_3$, in which the Gd ions are displaced from the ideal cubic positions, the dipole interaction field amounts to only 300 Oe (0.03 T). It must play little part in determining the direction of antiferromagnetic alignment.

The ordering temperature of $GdAlO_3$ is 3.87 K. Each magnetic ion on one sublattice appears to be surrounded by six nearest neighbors belonging to the opposite sublattice. This is the ordering pattern to be expected when the predominant mechanism of interaction is isotropic (Heisenberg) exchange between nearest neighbors. The evaluation of the exchange field from the saturation magnetization yields an exchange constant $J/k = -0.067$ K.

In this salt the crystalline field splitting, in terms of the spin-Hamiltonian parameters D and E, is the source of the anisotropy field.

9.8 Terbium

Terbium, $4f^8$, has a $^7F_6$ ground state, $g_J = \frac{3}{2}$ and is a non-Kramers ion. A set of doublets, $|\pm \mathscr{J}_z\rangle$ and a singlet $|\mathscr{J}_z = 0\rangle$ arise from the $\mathscr{J} = 6$ state. The value of g for any doublet is $3|\mathscr{J}_z|$, with a maximum value of 18 for $|\mathscr{J}_z| = 6$. For Tb^{3+} in YES [56] $g_{\parallel} = 17.72$ is observed, suggesting that the $|\mathscr{J}_z = \pm 6\rangle$ state is indeed the major contributor to the ground state. These levels are split by a small amount, $\Delta = 0.387\,cm^{-1}$ (0.56 K). Pure TbES orders ferromagnetically [57, 58] at 240 mK, largely by dipole-dipole interaction. The parameter Δ is now about 0.63 K.

The nuclear hyperfine splitting is strong relative to the crystal field splitting.

The monoclinic compound $Tb_2(SO_4)_3 \cdot 8\,H_2O$ also orders ferromagnetically, at 150 mK [59]. Systems of non-Kramers ions with two singlets separated by an amount Δ can exhibit long-range order of induced electronic moments only if the interionic magnetic coupling exceeds a critical fraction of Δ. In this case, Δ was estimated as 1.4 K. The nuclear spins may play an important role in inducing magnetic ordering in situations such as this.

The compound $Tb(OH)_3$ has also been studied extensively [60]. In contrast to isotropic $Gd(OH)_3$, it is strongly anisotropic, with $g_{\parallel} \sim 17.9$ and $g_{\perp} \sim 0$. It thus behaves like a three-dimensional Ising system, and is a ferromagnet. The transition temperature is 3.82 K, and the spins are constrained to lie parallel to the c-axis.

Though the interaction is Ising-like, the large $g_{\parallel}$ causes dipolar interactions to be strong, and we do not imply that the usual Ising superexchange Hamiltonian is in effect here. Indeed, there is a competition between the dipolar and nondipolar (largely exchange) terms. In particular, the strong nearest-neighbor magnetic dipole interaction is partly canceled by the corresponding nondipolar energy, while the relatively weak next-nearest neighbor dipolar term is dominated by a somewhat stronger nondipolar term which is ferromagnetic in sign. Further neighbors interact largely by the dipole term. Thus, while $Tb(OH)_3$ is an almost ideal Ising system from the point of view of the $(S_{zi}S_{zj})$ form of the interactions, it is quite different from the usual Ising models in the range dependence of its individual pair interactions.

9.9 Dysprosium

This $4f^9$ ion has a $^6H_{15/2}$ ground state, with $g_J = \frac{4}{3}$ [Eq. (9.1)], and is of interest because of the large magnetic moment of the ground state. No EPR has been observed in the ethylsulfate, indicating that $g_\perp = 0$. This is consistent with susceptibility measurements [61, 62] on the pure material, which yield $g_\parallel = 10.8$ and $g_\perp = 0$.

The $^6H_{15/2}$ state is split into 8 doublets by the crystalline field of point symmetry C_{3h}, which allows states differing by a change of 6 in $\mathscr{J}_z$ to mix. The ground doublet in DyES is a combination of the $|\pm\frac{15}{2}\rangle, |\pm\frac{3}{2}\rangle$, and $|\mp\frac{9}{2}\rangle$ states, and the position of the next higher doublet is only at some 23 K.

All the magnetic properties of DyES are determined by dipole-dipole interactions [32, 61, 62]. A large deviation from Curie law behavior even at temperatures in the ^4He region is due to dipole-dipole coupling. The experimental specific heat follows the relationship $cT^2/R = 130.5 \times 10^{-4} K^2$, and 93% of this value is attributed to the dipole-dipole term, the remainder to nuclear hyperfine terms. The ordering temperature is 0.13 K, and ferromagnetic domains which are long and thin are established parallel to the c-axis. Because of the large anisotropy in g, DyES can be treated as an Ising system, and since the nearest neighbors (0.7 nm) are along the c-axis and have the strongest interaction with a given reference ion, the system can actually be treated as a linear chain magnet. The interactions with next nearest neighbors along the chain are comparable in strength with any other nnn interaction not on the same axis. The low critical entropy is consistent with this interpretation.

The monoclinic compound $DyCl_3 \cdot 6H_2O$ is another that orders exclusively by dipole-dipole interaction [32, 63]. This effective spin $\mathscr{S} = \frac{1}{2}$ system has $g_\parallel = 16.52$ and an average $g_\perp$ of 1.76, and this is consistent with the ground state being predominantly $|\pm\frac{15}{2}\rangle$. With a $T_c = 0.289$ K, the compound is again Ising-like. Only about half the critical entropy is acquired below T_c, which indicates, as with many other dipole-dipole systems, that there is a large amount of short-range order. The specific heat is illustrated in Fig. 9.10.

Ising-like behavior is frequently exhibited by compounds of dysprosium, other examples being $Dy(OH)_3$ [64], $DyPO_4$ [65], and $Dy_3Al_5O_{12}$ [66], the latter compound being a garnet and usually referred to as DAG. The zero-field ordering temperatures are, respectively, 3.48, 3.39, and 2.53 K, all relatively high for rare earth compounds. Dipolar coupling is the predominant phenomenon, and the large moment of Dy^{3+} is responsible for the strength of the interaction. Both $DyPO_4$ and DAG are metamagnets [65–67]. DAG is interesting (in part) because the strong and highly anisotropic crystal field constrains the moment of each Dy^{3+} ion to point along one of the three cubic crystal axes, and when the material orders, it does so as a six-sublattice antiferromagnet. In the presence of an applied magnetic field along the [111] axis these sublattices become equivalent in threes, $(+x, +y, +z)$ and $(-x, -y, -z)$, and one may treat the system as a simple two-sublattice antiferromagnet.

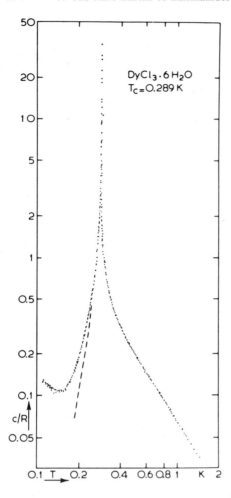

Fig. 9.10. Specific-heat data of $DyCl_3 \cdot 6H_2O$ versus temperature at zero applied field. Note the very high maximum value: $c/R_{max} \approx 40$. Dashed curve is obtained after subtraction of the hyperfine specific heat. From Ref. [63]

9.10 Holmium

Holmium(III), $4f^{10}$, is another non-Kramers ion, with a $^5I_8$ ground state. EPR of Ho^{3+} in yttrium ethyl sulfate (YES) (yttrium is closer in radius to Ho^{3+} than is La^{3+}) shows that the lowest levels are a doublet, the admixed $|\mathscr{I}_z = \pm 7\rangle, |\pm 1\rangle$ and $|\mp 5\rangle$ states, and a singlet, $|\mathscr{I}_z = \pm 6,0\rangle$, with transitions between all three levels. The singlet was estimated as being $5.5\,cm^{-1}$ (7.9 K) in energy above the doublet. The system then is a pseudo-triplet, remarkably like a nickel(II) ion with a zero-field splitting. Using an effective spin of $\mathscr{S} = 1$, $g_\parallel = 7.705$ is found.

The principal magnetic susceptibilities of pure HoES have been measured [68] over the temperature range 1.5–300 K and the specific heat near 1 K [69]. The ZFS was taken, from independent spectroscopic analyses, as $6.01\,cm^{-1}$ (8.66 K), and the resolution of the ground state non-Kramers doublet is assumed to be about 0.14 K. The magnetic properties can all be fit by means of the crystal field calculation of the single-ion electronic states and by dipolar interactions between the ions. The specific heat in the temperature range 1 to 2 K is approximately $cT^2/R \sim 0.3\,K^2$, a value much larger than given above for related salts. Indeed, c/R is not even exactly $1/T^2$ in its dependence. This is not due to the Schottky contribution from the singlet state mentioned above, but rather to the nuclear hyperfine terms which are of the same order of magnitude as kT at this temperature. The dipole-dipole term, the only other one of importance, contributes only about 16% of c/R.

Holmium hydroxide, $Ho(OH)_3$, has also been investigated [64]. This salt orders at 2.54 K, and like so many other salts discussed above, the ordered state is ferromagnetic and determined by dipolar interactions. The ground doublet, which appears to be split by about $1.6\,cm^{-1}$ (1 K) at 4.2 K, may be written as

$$|\pm\rangle = 0.94|\pm 7\rangle + 0.31|\pm 1\rangle + 0.15|\mp 5\rangle$$

with the singlet

$$0.59|+6\rangle + 0.55|0\rangle + 0.59|-6\rangle$$

lying at $11.1\,cm^{-1}$ (16 K). With $g_\parallel = 15.5$ and $g_\perp = 0$, Ising-like behavior is found again.

9.11 Erbium

This Kramers ion has eleven f electrons and a $^4I_{15/2}$ ground state; the first excited state $\mathcal{J} = \frac{13}{2}$ is some $8000\,cm^{-1}$ higher and therefore of little importance for magnetochemistry. In LaES, the g-values are $g_\parallel = 1.47$ and $g_\perp = 8.85$; the susceptiblity values [68] for pure ErES are $g_\parallel = 1.58$ and $g_\perp = 8.51$. Heat capacity data [70] over the range 12–300 K exhibit a broad maximum at about 40 K and decrease in value at higher temperatures, but do not reach a $1/T^2$ region even at 250 K. In other words, the data resemble one large Schottky curve, consistent with a set of doublets at 0, 44, 75, 110, 173, 216, 255, and 304 K, all of which arise from the $^4I_{15/2}$ state. At low temperatures [69], the specific heat rises again, due to dipolar interactions and nuclear splittings. The quantity $cT^2/R = 7.3 \times 10^{-3}\,K^2$ is reported, with the dipolar contribution being about 88% of the total. Since the lowest excited state is higher than that of holmium in the ethylsulfate, the analysis of the susceptibility data of ErES is more straightforward, and can be fit by combined crystal field and dipolar terms. The transition temperature of ErES is below 40 mK [32].

The monoclinic compound $ErCl_3 \cdot 6\,H_2O$, which is isostructural with $GdCl_3 \cdot 6\,H_2O$, orders [32, 63, 71] by dipolar interaction at 0.356 K; the specific heat is illustrated in Fig. 9.11. The lowest excited doublet appears to be at 24.5 K, and the magnetic parameters are $g_\parallel = 13.74$ and $g_\perp = 0.75$. The compound, like its dysprosium analog, has predominantly a $|\pm\frac{15}{2}\rangle$ ground state and is therefore also Ising-like.

The varieties of magnetic behavior exhibited by the rare earths may be observed by comparing the g-values quoted above with those reported [72] for Er/LaMN, $g_\parallel = 4.21$ and $g_\perp = 7.99$, and with those reported [73] for $Er(C_2O_4)(C_2O_4H) \cdot 3\,H_2O$, $g_\parallel = 12.97$ and $g_\perp = 2.98$. Recall that the g-value reflects the nature of the ground state wave function and how it is mixed with the low-lying excited states.

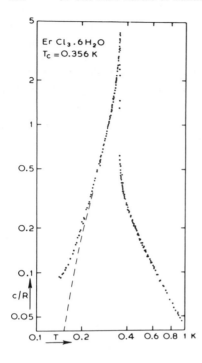

Fig. 9.11. Specific-heat data of $ErCl_3 \cdot 6H_2O$ versus temperature at zero applied field. Dashed curve is obtained after subtraction of the hyperfine specific heat. From Ref. [63]

9.12 Thulium

This $4f^{12}$ ion is a non-Kramers ion. No resonance has been observed in either the ethylsulfate or the chloride, $Tm/LaCl_3$, and the crystal field parameters predict that the ground state should be a singlet. The specific heat and susceptibility of TmES have been measured [74] over a wide temperature range and are consistent with a singlet ground state and a doublet at 32 K.

EPR spectra of Tm^{3+} in two monoclinic salts have been reported [75]. The ground state in both cases consists of a doublet with small zero-field splitting. For $Tm/YCl_3 \cdot 6H_2O$, $g_{\parallel} = 14$ and Δ, the ZFS, is $1.12\,cm^{-1}$ (1.61 K); for $Tm/Y_2(SO_4)_3 \cdot 8H_2O$, $g_{\parallel} = 13.4$ and $\Delta = 0.61\,cm^{-1}$ (0.88 K). Pure $Tm_2(SO_4)_3 \cdot 8H_2O$ exhibits a sharp peak in the heat capacity at 0.30 K, ascribed to magnetic ordering [76]. Since a ZFS is present, ordered magnetic moments can only appear in zero external field if the ratio of exchange interaction to crystal field splitting exceeds a critical value. A broad peak at 0.5 K above the λ-feature is assigned as a Schottky contribution to the specific heat, with Δ being closer to 1 K. On the other hand, experiments in which the $Tm_2(SO_4)_3 \cdot 8H_2O$ is diluted by the isomorphous yttrium salt suggest [77, 78] that Δ may be only 0.75 K.

An unusual discovery, reminiscent more of transition metal chemistry than of that of the rare earth ions, has recently been reported [79]. A variety of experiments on thulium nicotinate dihydrate, $[Tm(C_5H_4NCO_2)_3 \cdot (H_2O)_2]_2$, show it to be a dimer with magnetic interaction between the two ions. The Tm^{3+} ions occur in relatively isolated pairs, the two Tm^{3+} sites in each dimer being separated by 0.4346 nm. The ground state of each ion remains a degenerate doublet, so $\mathscr{S}_1 = \mathscr{S}_2 = \frac{1}{2}$ and the interaction is described by the Hamiltonian $\boldsymbol{H} = -2JS_{1z}S_{2z}$, with $-2J = 0.86\,\text{cm}^{-1}$ (1.24 K). The values $g_\parallel = 13.3(9)$ and $g_\perp \leq 0.4$ were found, and note the Ising-like character of the exchange term. The ions in pairs interact as described earlier (Chap. 5), with the $\mathscr{S} = \pm 1$ state the ground state. The only difference from the work on copper nitrate is in the sign and the anisotropy of the interaction.

9.13 Ytterbium

This is the heaviest magnetic rare earth, the trivalent ion having thirteen 4f electrons or one hole in the f shell. As with cerium(III) there are two multiplets $^2F_{7/2}$ and $^2F_{5/2}$, but the $^2F_{7/2}$ state becomes the lowest for ytterbium. The $^2F_{5/2}$ is at some $10\,300\,\text{cm}^{-1}$ and has little effect on the magnetic properties. The ion has been studied as pure YbES [80] and diluted by YES [81].

The ground doublet is $\mathscr{J}_x = \pm\frac{3}{2}$ from the $\mathscr{J} = \frac{7}{2}$ state, with the next higher doublet at $44\,\text{cm}^{-1}$. Because of the C_{3h} crystal field symmetry, the only states that can admix are those for which $\mathscr{J}_x$ differs by 6. Thus, the $|\pm\frac{1}{2}\rangle$ and $|\pm\frac{3}{2}\rangle$ states will be pure, except as they mix with the corresponding $|\mathscr{J} = \frac{5}{2}, \mathscr{J}_x\rangle$ states. Normal EPR spectra are not observed for Yb/YES because there are no allowed transitions within the $\pm\frac{3}{2}$ state for either parallel or perpendicular orientation of an external field. Electric dipolar transitions have been observed, however [81]. The parameter $g_\parallel = 3.40$ for pure YbES and 3.328 for Yb/YES, and $g_\perp = 0$ (theory) and ~ 0.01 (expt.). With $g_J = \frac{8}{7}$, one calculates $g_\parallel = 3g_J = 3.43$, in remarkable agreement with experiment. The specific heat result is $cT^2/R - 4.34 \times 10^{-4}\,\text{K}^2$, which agrees completely with that calculated for dipole-dipole interaction and nuclear hyperfine terms alone. It is clear that there is no exchange interaction in any of the ethylsulfates.

9.14 Some Other Systems

The results discussed thus far in this chapter concern some of the most thoroughly examined systems, though one can only marvel how little is known about so many of them. In this section a variety of other materials, many of them of only recent interest, will be discussed.

The anhydrous chlorides, RCl_3, have C_{3h} point symmetry at the rare earth ion, and the EPR results in the two salts (the chloride and the ethylsulfate) are remarkably similar. The major paper is that of Hutchison and Wong [82] which is discussed in detail by Abragam and Bleaney [15]. The authors measured the EPR spectra of most of the rare earth ions as diluted by $LaCl_3$. One of the results of interest is that the ground doublet of $Ce^{3+}/LaCl_3$ is the doublet which is found lowest in the undiluted CeES.

A useful publication is that of Lea, Leask and Wolf [83], who calculated the effects of a cubic crystalline field on a lanthanide ion. They predict the order and type of level for each $\mathscr{J}$ value and for all ratios of the fourth- to sixth-order crystal field parameters. These results have been used extensively in a variety of investigations [84–87] on lanthanide ions in the elpasolite lattice, which is of the formula A_2BRX_6, with A and B alkali metals, R the lanthanide and X either fluoride or, more commonly, chloride. The rare earth ion resides at a site of cubic symmetry in these compounds. From measurements of the susceptibility [84, 85], attempts have been made to assign the ground states and fix the crystal field parameters [86].

The specific heat and susceptibility of a series of powdered samples of compounds of formula R_2ZrS_5 have recently been published [88]. The coordination polyhedron of the lanthanide ions is a trigonal prism which is capped at two rectangular faces. In the (representative) dysposium compound, each Dy has five neighbors at about 0.4 nm and six at about 0.55 nm; the sulfide ions intervene, and provide a strong superexchange path. The crystals are orthorhombic. Both Ce_2ZrS_5 and Pr_2ZrS_5 remain paramagnetic down to 2 K, the Pr compound exhibiting a non-magnetic singlet ground state.

Antiferromagnetic ordering was found in several salts, as follows:

	T_c, K
Sm_2ZrS_5	4.5
Gd_2ZrS_5	12.74
Tb_2ZrS_5	9.27
Dy_2ZrS_5	5.24
Er_2ZrS_5	2.52

The remarkably high ordering temperatures provide good evidence for the importance of superexchange interactions in these compounds. The critical entropies, the amount of entropy acquired below T_c, are in good accord in each case for that calculated for three-dimensional magnetic lattices; as expected, the gadolinium salt follows the spin $\mathscr{S} = \frac{7}{2}$ Heisenberg model, while the Dy and Tb compounds follow the $\mathscr{S} = \frac{1}{2}$ Ising model. The low-lying excited states contribute a high-temperature Schottky maximum.

Another system of recent interest is the $LiRF_4$ series of materials, where R is a rare earth ion heavier than samarium. They possess a tetragonal structure with four magnetically equivalent rare earth ions per unit cell. The compounds have practical use as laser materials and can be obtained as single crystals. Though there is evidence for the presence of some superexchange interaction, dipole-dipole interactions predominate and control the nature of the magnetic ordering. Both susceptibility [89] and specific heat [90] data are available, at least for some of the compounds, as well as some EPR data [91].

A number of studies have resulted in measurements of the g-values of the compounds, with the following results: $LiTbF_4$ has $g_{\parallel} \sim 17.9$ and $g_{\perp} \sim 0$, while $LiHoF_4$ has $g_{\parallel} \sim 13.5$ and $g_{\perp} \sim 0$. Ising-type of anisotropy is therefore expected (and found). Both compounds are ferromagnets, with transition temperatures of 2.885 and 1.530 K, respectively.

On the other hand, $LiDyF_4$ has $g_{\parallel} = 3.52$ and $g_{\perp} = 10.85$ while $LiErF_4$ has $g_{\parallel} \sim 3.6$ and $g_{\perp} \sim 9$, so both compounds should exhibit the XY-type of anisotropy; they are antiferromagnets. The transition temperatures are respectively 0.581 and 0.383 K.

All these materials have effective spin $\mathscr{S} = \frac{1}{2}$ at low temperatures and so the shape of the specific heat curve should be mainly determined by the type of interaction (from the g-anisotropy) and the ratio of dipolar to exchange interactions.

Mennenga [90] has also measured the field- and concentration-dependent specific heats of several of these compounds.

A series of hydrated crystals $[R(H_2O)_9](BrO_3)_3$ has long been known and is currently under investigation [92]. The counter-ion is bromate and the hexagonal structure is similar to that of the RES compounds. Ferromagnetic transitions were observed at $0.125\,K$ for the Tb salt and at $0.170\,K$ for the Dy salt; as with the ethylsulfates, exchange interaction is negligible compared with dipolar interaction.

Another series of molecules in which dipolar interactions appear to predominate is based on the ligand pyridine N-oxide. The compounds are $[R(C_5H_5NO)_8]X_3$, where the lanthanide ion is at the center of a polyhedron that is close to a square antiprism [93]. The counterion may be perchlorate, nitrate or iodide. Unfortunately, the materials crystallize from the same solution in two slightly different monoclinic space groups. The gadolinium salts all order antiferromagnetically at $70\,mK$ [94].

9.15 References

1. Cotton S.A. and Hart F.A., (1975) The Heavy Transition Elements, Chap. 10. J. Wiley & Sons, New York
2. Cotton F.A. and Wilkinson G., (1980) Advanced Inorganic Chemistry, Chap. 23, 4th Ed.: J. Wiley & Sons, New York
3. Thompson L.C., (1979) in: Handbook on the Physics and Chemistry of Rare Earths, Vol. 3, Chap. 25 (Eds. Gschneidner, Jr. K.A. and Eyring L.) North-Holland, Amsterdam
4. Barnes R.G., (1979) in: Ref. 3, Vol. 2, Chap. 18
5. Siddall III T.H., Casey A.T., and Mitra S., (1976) in: Theory and Applications of Molecular Paramagnetism, (Ed.: by Boudreaux E.A. and Mulay L.N.) J. Wiley & Sons, New York
6. Freeman A.J. and Watson R.E., Phys. Rev. **127**, 2058 (1962)
7. Dieke G.H., (1968) Spectra and Energy Levels of Rare Earth Ions in Crystals, Interscience, New York
8. Tromp H.R.C. and van Duyneveldt A.J., Physica **45**, 445 (1969)
9. Soeteman J., Bevaart L., and van Duyneveldt A.J., Physica **74**, 126 (1974)
10. Scott P.L. and Jeffries C.D., Phys. Rev. **127**, 32 (1962)
11. Ruby R.H., Benoit H., and Jeffries C.D., Phys. Rev. **127**, 51 (1962)
12. Larson G.H. and Jeffries C.D., Phys. Rev. **141**, 461 (1966)
13. Larson G.H. and Jeffries C.D., Phys. Rev. **145**, 311 (1966)
14. Orton J.W., (1968) Electron Paramagnetic Resonance, Iliffe, London
15. Abragam A. and Bleaney B., (1970) Electron Paramagnetic Resonance of Transition Ions, Oxford
16. Taylor K.N.R. and Darby M.I., (1972) Physics of Rare Earth Solids, Chapman and Hall, London
17. Fitzwater D.R. and Rundle R.E., Z. f. Krist. **112**, 362 (1959); Gerkin R.E. and Reppart W.J., Acta Cryst. C **40**, 781 (1984)
18. Elliott R.J. and Stevens K.W.H., Proc. Roy. Soc. (London) A **215**, 437 (1952)
19. Meyer H. and Smith P.L., J. Phys. Chem. Solids **9**, 285 (1959)
20. Johnson C.E. and Meyer H., Proc. Roy. Soc. (London) A **253**, 199 (1959)
21. Blöte H.W.J., Physica **61**, 361 (1972)
22. Anderson R.J., Baker J.M., and Birgeneau R.J., J. Phys. C **4**, 1618 (1971); Baker J.M., J. Phys. C **4**, 1631 (1971)
23. Cooke A.H., Duffus H.J., and Wolf W.P., Phil. Mag. **44**, 623 (1953)
24. Daniels J.M. and Robinson F.N.H., Phil. Mag. **44**, 630 (1953)
25. Leask M.J.M., Orbach R., Powell M.J.D., and Wolf W.P., Proc. Roy. Soc. (London) A **272**, 371 (1963)
26. a) Colwell J.H., J. Low Temp. Phys. **14**, 53 (1974); b) Hudson R.P. and Pfeiffer E.R., J. Low Temp. Phys. **16**, 309 (1974)

27. Zalkin A., Forrester J.D., and Templeton D.H., J. Chem. Phys. **39**, 2881 (1963)
28. Mess K.W., Lubbers J., Niessen L., and Huiskamp W.J., Physica **41**, 260 (1969)
29. Abeshouse D.J., Zimmerman G.O., Kell D.R., and Maxwell E., Phys. Rev. Lett. **23**, 308 (1969)
30. Mess K.W., Niessen L., and Huiskamp W.J., Physica **45**, 626 (1970)
31. Zimmerman G.O., Abeshouse D.J., Maxwell E., and Kelland D., Physica **51**, 623 (1971)
32. Lagendijk E., Blöte H.W.J., and Huiskamp W.J., Physica **61**, 220 (1973)
33. Commander R.J. and Finn C.B.P., J. Phys. C **11**, 1169 (1978)
34. Roach P.R., Abraham B.M., Ketterson J.B., Greiner R., and Van Antwerp W., J. Low Temp. Phys. **13**, 59 (1973)
35. Imes J.L., Neiheisel G.L., and Pratt, Jr. W.P., J. Low Temp. Phys. **21**, 1 (1975)
36. Albertsson J., Chen P.Y., and Wolf W.P., Phys. Rev. **B 11**, 1943 (1975)
37. Landau D.P., Doran J.C., and Keen B.E., Phys. Rev. **B 7**, 4961 (1973); Skjeltorp A.T., J. Appl. Phys. **49**, 1567 (1978)
38. Meyer H., J. Phys. Chem. Solids **9**, 296 (1959)
39. Baker J.M. and Bleaney B., Proc. Roy. Soc. (London) **A 245**, 156 (1958)
40. Hellwege K.H., Schembs W., and Schneider B., Z. f. Physik **167**, 477 (1962)
41. Earney J.J., Finn C.B.P., and Najafabadi B.M., J. Phys. C **4**, 1013 (1971)
42. Sells V.E. and Bloor D., J. Phys. C **9**, 379 (1976)
43. Elliott R.J. and Stevens K.W.H., Proc. Roy. Soc. (London) **A 219**, 387 (1953)
44. van den Handel J. and Hupse J.C., Physica **9**, 225 (1940)
45. Meyer H., Phil. Mag. **2**, 521 (1957); see also, Clover R.B. and Skjeltorp A.T., Physica **53**, 132 (1971)
46. Ellingsen O.S., Bratsberg H., Mroczkowski S., and Skjeltorp A.T., J. Appl. Phys. **53**, 7948 (1982)
47. Ehnholm G.J., Katila T.E., Lounasmaa O.V., Reivari P., Kalvius G.M., and Shenoy G.K., Z. f. Physik **235**, 289 (1970). See also the recent paper on EuX_2, X = Cl, Br, I: Sanchez J.P., Friedt J.M., Bärnighausen H., and van Duyneveldt A.J., Inorg. Chem. **24**, 408 (1985)
48. ter Maten G. and Jansen L., Physica **95 B**, 11 (1978)
49. Shapira Y., Foner S., Oliveira, Jr. N.F., and Reed T.B., Phys. Rev. **B 5**, 2647 (1972); Shapira Y., Foner S., Oliveira, Jr. N.F., and Reed T.B., Phys. Rev. **B 10**, 4765 (1974); Shapira Y., Yacovitch R.D., Becerra C.C., Foner S., McNiff, Jr. E.J., Nelson D.R., and Gunther L., Phys. Rev. **B 14**, 3007 (1976); Silberstein R.P., Tekippe V.J., and Dresselhaus M.S., Phys. Rev. **B 16**, 2728 (1977)
50. Maletta H., J. Phys. (Paris) Colloq. **41**, C 5–115 (1980); Hüser D., Wenger L., van Duyneveldt A.J., and Myosh J.A., Phys. Rev. **B 27**, 3100 (1983)
51. Buckmaster H.A. and Shing Y.H., Phys. Status Solid **12a**, 325 (1972)
52. Wielinga R.F., Lubbers J., and Huiskamp W.J., Physica **37**, 375 (1967)
53. Skjeltorp A.T., Catanese C.A., Meissner H.E., and Wolf W.P., Phys. Rev. **B 7**, 2062 (1973)
54. Cashion J.D., Cooke A.H., Martin D.M., and Wells M.R., J. Phys. C **3**, 1612 (1970)
55. Cashion J.D., Cooke A.H., Thorp T.L., and Wells M.R., Proc. Roy. Soc. (London) **A 318**, 473 (1970)
56. Baker J.M. and Bleaney B., Proc. Roy. Soc. (London) **A 245**, 156 (1958)
57. Daniels J.M., Hirvonen M.T., Jauho A.P., Katila T.E., and Riski K.J., Phys. Rev. **B 11**, 4409 (1975)
58. Hirvonen M.T., Katila T.E., Riski K.J., Teplov M.A., Malkin B.Z., Phillips N.E., and Wun, M., Phys. Rev. **B 11**, 4652 (1975)
59. Simizu S. and Friedberg S.A., Physica **B 108**, 1099 (1981)
60. Catanese C.A., Skjeltorp A.T., Meissner H.E., and Wolf W.P., Phys. Rev. **B 8**, 4223 (1973)
61. Cooke A.H., Edmonds D.T., McKim F.R., and Wolf W.P., Proc. Roy. Soc. (London) **A 252**, 246 (1959)
62. Cooke A.H., Edmonds D.T., Finn C.B.P., and Wolf W.P., Proc. Roy. Soc. (London) **A 306**, 313, 335 (1968)
63. Lagendijk E. and Huiskamp W.J., Physica **65**, 118 (1973)
64. Catanese C.A. and Meissner H.E., Phys. Rev. **B 8**, 2060 (1973)
65. Wright J.C., Moos H.W., Colwell J.H., Mangum B.W., and Thornton D.D., Phys. Rev. **B 3**, 843 (1971)

66. Landau D.P., Keen B.E., Schneider B., and Wolf W.P., Phys. Rev. **B 3**, 2310 (1971)
67. Stryjewski E. and Giordano N., Adv. Phys. **26**, 487 (1977)
68. Cooke A.H., Lazenby R., and Leask M.J.M., Proc. Phys. Soc. **85**, 767 (1965)
69. Cooke A.H. and Finn C.B.P., J. Phys. C **1**, 694 (1968)
70. Gerstein B.C., Penney C.J., and Spedding F.H., J. Chem. Phys. **37**, 2610 (1962)
71. Beauvillain P., Dupas C., and Renard J.P., Phys. Lett. **54 A**, 436 (1975)
72. Judd B.R. and Wong E., J. Chem. Phys. **28**, 1097 (1958)
73. O'Connor C.J. and Carlin R.L., Chem. Phys. Lett. **49**, 574 (1977)
74. Gerstein B.C., Jennings L.D., and Spedding F.H., J. Chem. Phys. **37**, 1496 (1962)
75. Gruber J.B., Karlow E.A., Olsen D., and Ranon U., Phys. Rev. B **2**, 49 (1970)
76. Katila T.E., Phillips N.E., Veuro M.C., and Triplett B.B., Phys. Rev. B **6**, 1827 (1972)
77. Simizu S. and Friedberg S.A., J. Appl. Phys. **53**, 1885 (1982)
78. Simizu S. and Friedberg S.A., J. Mag. Mag. Mat. **31–34**, 1065 (1983)
79. Baker J.M., Bleaney B., Brown J.S., Hutchison Jr. C.A., Leask M.J.M., Martineau P.M., Wells M.R., Marin J.M., and Prout K., J. Mag. Mag. Mat. **31–34**, 657 (1983)
80. Cooke A.H., McKim F.R., Meyer H., and Wolf W.P., Phil. Mag. **2**, 928 (1957)
81. Wolfe J.P. and Jeffries C.D., Phys. Rev. **B 4**, 731 (1971)
82. Hutchison, Jr. C.A. and Wong E., J. Chem. Phys. **29**, 754 (1958)
83. Lea K.R., Leask M.J.M., and Wolf W.P., J. Phys. Chem. Solids **23**, 1381 (1962)
84. Bucher E., Guggenheim H.J., Andres K., Hull, Jr. G.W., and Cooper A.S., Phys. Rev. B **10**, 2945 (1974)
85. Hoehn M.V. and Karraker D.G., J. Chem. Phys. **60**, 393 (1974)
86. Dunlap B.D. and Shenoy G.K., Phys. Rev. B **12**, 2716 (1975)
87. Schwartz R.W., Watkins S.F., O'Connor C.J., and Carlin R.L., J. Chem. Soc. Faraday Trans. II **72**, 565 (1976); O'Connor C.J., Carlin R.L., and Schwartz R.W., J. Chem. Soc. Faraday Trans. II **73**, 361 (1977); North M.H. and Stapleton H.J., J. Chem. Phys. **66**, 4133 (1977); Schwartz R.W., Faulkner T.R., and Richardson F.S., Mol. Phys. **38**, 1767 (1979); Faulkner T.R., Morley J.P., Richardson F.S., and Schwartz R.W., Mol. Phys. **40**, 1481 (1980); Morley J.P., Faulkner T.R., Richardson F.S., and Schwartz R.W., J. Chem. Phys. **77**, 1734 (1982); Bill H., Magne G., Güdel H.U., and Neuenschwander K., Chem. Phys. Lett. **104**, 258 (1984)
88. Nap G.M. and Plug C.M., Physica **93 B**, 1 (1978)
89. Cooke A.H., Jones D.A., Silva J.F.A., and Wells M.R., J. Phys. C **8**, 4083 (1975); Beauvillain P., Renard J.-P., and Hansen P.-E., J. Phys. C **10**, L 709 (1977); Beauvillain P., Renard J.-P., Laursen I., and Walker P.J., Phys. Rev. B **18**, 3360 (1978)
90. Mennenga G., Thesis (1983) Leiden
91. Magariño J., Tuchendler J., Beauvillain P., and Laursen I., Phys. Rev. B **21**, 18 (1980)
92. Simizu S., Bellesis G.H., and Friedberg S.A., J. Appl. Phys. **55**, 2333 (1984)
93. Al-Karaghouli A.R. and Wood J.S., Inorg. Chem. **18**, 1177 (1979)
94. Carlin R.L., Burriel R., Mennenga G., and de Jongh L.J., to be published

10. Selected Examples

10.1 Introduction

The discussion in this chapter concerns some additional examples of interesting magnetic compounds that have been examined recently. It will be seen that all of the principles discussed earlier in this book come together here. No attempt is made at a complete literature survey, or even a summary of all the work reported on a given compound. Consonant with the point of view of this book will be the emphasis on results based on specific heat and magnetic susceptibility data, but especially we are trying to lay a firm foundation for a structural basis of magnetochemistry. The examples chosen are some of the best understood or most important ones, a knowledge of which should be had by everyone interested in this subject. Further information and references may be found in earlier sections of the book, as well as in Refs. [1–6].

10.2 Hydrated Nickel Halides

The hydrated halides of the iron-series ions are among the salts most thoroughly investigated at low temperatures. For example, $MnCl_2 \cdot 4H_2O$ with $T_c = 1.62$ K, is perhaps the most famous antiferromagnet. Fame is a matter of taste, however, and some would argue that $CuCl_2 \cdot 2H_2O$, $T_c = 4.3$ K, is the most important antiferromagnet. While there are other important candidates for the position as well, these two salts are of the class described. Since it would be difficult to discuss all of these salts properly, and since the hydrated halides of nickel have continued to be of interest and also offer some nice pedagogical examples, the discussion here will be limited to that class of salts. The limitation will be seen to be hardly restrictive.

We begin with the chemical phase diagram [7] of nickel chloride in water, which is illustrated in Fig. 10.1. The search for new magnetic materials depends heavily on the

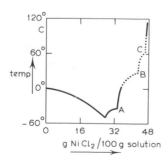

Fig. 10.1. Phase diagram of $NiCl_2$–H_2O system. Point A corresponds to $-33\,°C$, B to $28\,°C$, and C to $64\,°C$. From Ref. [7]

application of the information contained in such diagrams, and a careful exploration of such phase studies can be quite fruitful. In this simple example, four hydrates of nickel chloride will be seen to exist, although the heptahydrate, which must be obtained below $-33\,°C$, is little known. The hexahydrate, $NiCl_2 \cdot 6\,H_2O$ is the best known salt in this series and may be obtained from an aqueous solution over the wide temperature interval of -33 to $+28\,°C$. The tetrahydrate is found between 28 and $64\,°C$, and the dihydrate above $64\,°C$; though less well known than the hexahydrate, both of these salts have been studied at low temperatures recently, and are of some interest. The critical temperatures for each of these salts are given in Table 10.1.

Table 10.1. Critical temperatures of hydrated nickel chlorides.

	T_c, K	Ref.
$NiCl_2 \cdot 6\,H_2O$	5.34	[8]
$NiCl_2 \cdot 4\,H_2O$	2.99	[9]
$NiCl_2 \cdot 2\,H_2O$	7.258	[10]
$NiCl_2$	52	[11]

The T_c for $NiCl_2$ has been included in the table for comparison, because it shows the effect of concentration on superexchange interactions. Without any knowledge of the specific structures of the materials, one would guess that the metal ions will be further apart in the hydrates than in the anhydrous materials; furthermore, since chloride is more polarizable than water, one expects that increasing the amount of water present will lead to less effective superexchange paths, and therefore lower transition temperatures.

On the other hand, this is not a monotonic trend, as can be seen by observing that the tetrahydrate has a lower T_c than the hexahydrate. It is likely that the reason for this lies less with the structural features of the two systems than with the fact that the zero-field splitting appears to be larger in the tetrahydrate.

Turning to the bromides, only the data shown in Table 10.2 are available.

Table 10.2. Critical temperatures of hydrated nickel bromides.

	T_c, K	Ref.
$NiBr_2 \cdot 6\,H_2O$	8.30	[12]
$NiBr_2 \cdot 2\,H_2O$	6.23	[13]
$NiBr_2$	60	[14]

Upon comparison with the analogous (and isostructural) chlorides, these compounds illustrate the useful rule of thumb that bromides often order at higher temperatures than chlorides (the hexahydrates), as well as the violation of that rule (the dihydrates).

Two of the factors that influence the value of T_c are competing, for one expects the larger polarizability of bromide (over chloride) to lead to stronger superexchange interaction and thus a higher T_c, while the larger size of bromide should cause the metal ions to be separated further which leads, in turn, to a lower T_c. A graphical correlation of these trends has been observed [14].

Aside from the anhydrous materials, there are not enough data available on the fluorides and iodides of nickel to discuss here.

It is convenient to separate the discussion of the dihydrates from that of the other hydrates, because the structural and magnetic behaviors of the two sets of compounds are substantially different.

Nickel chloride hexahydrate is one of the classical antiferromagnets. Both the specific heat [8] and susceptibility [15] indicate a typical antiferromagnetic transition at 5.34 K. The crystal structure [16] shows that the monoclinic material consists of distorted $trans$-[$Ni(OH_2)_4Cl_2$] units, hydrogen bonded together by means of an additional two molecules of water. The salt is isomorphous to $CoCl_2 \cdot 6H_2O$ yet the preferred axes of magnetic alignment in the two salts are not the same [15]. There is relatively little magnetic anisotropy throughout the high-temperature (10–20 K) region [15, 17], and, a point of some interest, neutron diffraction studies [18] show that the crystal structure is unchanged upon cooling the substance from room temperature to 4.2 K, which is below the critical temperature. As illustrated in Fig. 10.2, additional neutron diffraction studies [19] have provided the magnetic structure, which consists of antiferromagnetic [001] planes with AF coupling between the planes. The result is that the magnetic unit cell is twice the size of the chemical cell, caused by a doubling along the c-axis. The arrangement of the spins, or magnetic ordering, is the same as in $CoCl_2 \cdot 6H_2O$, except for crystallographic orientation. The reason for the differing orientation probably lies with the differing single-ion anisotropies, for the nickel system is an anisotropic Heisenberg system, while the cobalt compound, as discussed earlier, is an XY magnet.

What is interesting about $NiCl_2 \cdot 6H_2O$ from a chemist's view-point is that the anisotropies are low and nearly uniaxial. The g value is normal, being isotropic at 2.22, but $D/k = -1.5 \pm 0.5$ K and $E/k = 0.26 \pm 0.40$ K, from the susceptibilities; these results may be contrasted with those [9] for the very similar molecule, $NiCl_2 \cdot 4H_2O$. In this case, cis-[$Ni(OH_2)_4Cl_2$]octahedra, which are quite distorted, are found. The compound is isomorphous with $MnCl_2 \cdot 4H_2O$, and in this case the easy axes of magnetization of the two materials are the same. With the nickel salt, large paramagnetic anisotropy persists throughout the high-temperature region, as illustrated in Fig. 10.3. With g again isotropic at 2.28, the zero-field splitting parameters

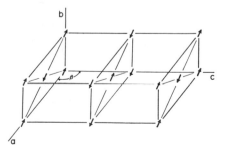

Fig. 10.2. Arrangement of magnetic moments in $NiCl_2 \cdot 6H_2O$ for the magnetic space group I_c2/c. The angle between the magnetic moment and the a-axis is approximately 10°. From Ref. [19]

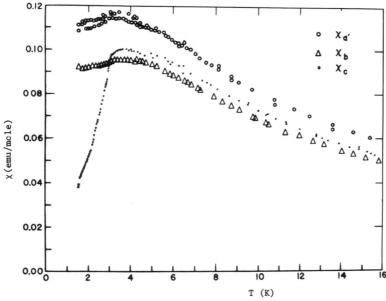

Fig. 10.3. The magnetic susceptibility of $NiCl_2 \cdot 4\,H_2O$ from 1 to 20 K along the a'-, b-, and c-axes. From Ref. [9]

reported in this case are $D/k = -11.5 \pm 0.1$ K and $E/k = 0.1 \pm 0.1$ K. This apparent increased single-ion anisotropy is perhaps the governing factor in reducing T_c for $NiCl_2 \cdot 4\,H_2O$ below that of $NiCl_2 \cdot 6\,H_2O$; the antiferromagnetic exchange constants in the two crystals have been evaluated as almost equal. (On the other hand, studies of the magnetic phase diagram of $NiCl_2 \cdot 4\,H_2O$ [20, 21] have suggested that D/k is only -2 K. The reason for the discrepancy in the values of these parameters is not known, though it may be due to systematic errors related to the crystallographic phase transition which the salt appears to undergo as it is cooled.) Short-range order effects have been observed [22] in $NiCl_2 \cdot 4\,H_2O$ (as well as several other salts) from measurements of the field-dependence of the susceptibility above the magnetic transition temperature.

The magnetic phase diagram of $NiCl_2 \cdot 6\,H_2O$ has been studied in great detail [23]. Although the study of the critical behavior of the phase boundaries very near the bicritical point (H_b, T_b) was a major impetus of the work, the boundaries with the applied field both parallel and perpendicular to the easy axis were determined. The diagram is illustrated in Fig. 10.4, where the high fields required for this study will be noted. The exchange and anisotropy parameters could be obtained from extrapolations to $T = 0$ of the spin-flop and C–P boundaries. The values obtained were $H_{SF}(0) = 39.45 \pm 0.30$ kOe (3.94 T) and $H_c(0) = 143.9 \pm 1.5$ kOe (4.4 T). Using these values, the phenomenological exchange and anisotropy fields could be calculated as $H_E = 77.4 \pm 0.9$ kOe (7.74 ± 0.09 T) and $H_A = 10.8 \pm 0.3$ kOe (1.08 ± 0.03 T). With $g = 2.22 \pm 0.01$, the corresponding exchange and anisotropy parameters of the single-ion Hamiltonian are $2z|J|/k = 11.5 \pm 0.3$ K and $|D|/k = 1.61 \pm 0.07$ K. These are probably the best available values of these parameters. The bicritical point was found at 3.94 K and 44.22 kOe (4.422 T).

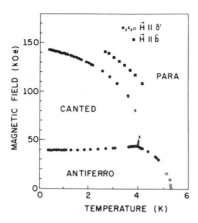

Fig. 10.4. Experimentally determined magnetic phase diagram of $NiCl_2 \cdot 6H_2O$ for H parallel to the easy axis a', and for H parallel to the b-axis. The open circles represent points from the calorimetric work of Johnson and Reese. All other points are from Ref. [23]. Crosses represent points obtained from plots of (dM/dH) vs. T at constant H. All other points were obtained from plots of dM/dH vs. H at constant T. From Ref. [23]

The compound $NiBr_2 \cdot 6H_2O$ has also been investigated [12], the specific heat, zero-field susceptibility and the phase diagram being reported. As was discussed above, a higher transition temperature than that of the chloride is observed, and the single-ion anisotropies are again small. All in all, the compound is much like $NiCl_2 \cdot 6H_2O$, with which it is isostructural.

Estimates of the internal fields can be made in several ways. An anisotropy constant K can be related to the zero-field splitting parameter D by the relationship $K(T=0) = |D|N\mathscr{S}(\mathscr{S}-\frac{1}{2})$ [17], where N is Avogadro's number and $\mathscr{S}$ is the spin of the ion. The anisotropy field can be defined as $H_A = K/M_S$, where $M_S = (\frac{1}{2})Ng\mu_B\mathscr{S}$. The exchange field is described by molecular field theory as $H_E = 2z|J|\mathscr{S}/g\mu_B$. These fields have been estimated for both $NiCl_2 \cdot 6H_2O$ [17] and $NiBr_2 \cdot 6H_2O$ [12].

The compounds $NiCl_2 \cdot 2H_2O$ and $NiBr_2 \cdot 2H_2O$ differ appreciably from the other hydrates, primarily because they have the characteristic linear chain structure of *trans*-$[NiX_4(OH_2)_2]$ units. The specific heats are of some interest because, as illustrated in Fig. 10.5, the magnetic phase transition is accompanied by two sharp maxima for each

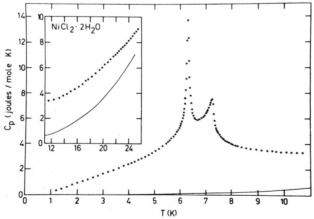

Fig. 10.5. Measured specific heat of $NiCl_2 \cdot 2H_2O$. Note that the insert which shows the high-temperature results has an abscissa four times as coarse as the main graph. The solid line denotes the lattice estimate, $4.5 \times 10^{-4} T^3$ J/mol-K. From Ref. [10]

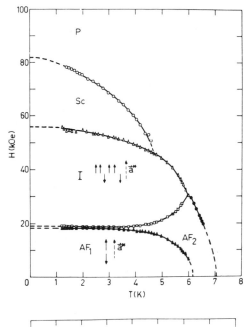

Fig. 10.6. Magnetic phase diagram of $NiCl_2 \cdot 2H_2O$ for $H \| a^*$. AF_1 and AF_2 denote different antiferromagnetic phases. I is a metamagnetic phase; Sc is a screw phase; P is the paramagnetic phase. From Ref. [24]

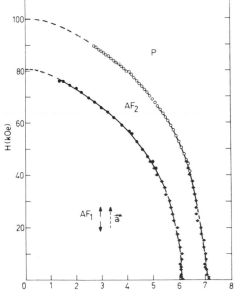

Fig. 10.7. Magnetic phase diagram of $NiCl_2 \cdot 2H_2O$ for $H \| b$. For symbols see caption of Fig. 10.6. Open and black circles represent data obtained by isothermal field sweeps. The other data are obtained by scanning in constant field. From Ref. [24]

compound. For the chloride, they occur at 6.309 and 7.258 K, while they occur at 5.79 and 6.23 K for the bromide: The implication is that there are two phase boundaries in the $H–T$ plane, even at $H = 0$. In fact, $NiCl_2 \cdot 2H_2O$ has a very complicated magnetic phase diagram for the external field applied along the easy axis, with five different regions appearing [24]. This is illustrated in Fig. 10.6, while the phase diagram with H applied perpendicular to the easy axis (Fig. 10.7) is somewhat simpler. Neutron diffraction studies in zero-field as well as magnetization and susceptibility measurements show that at the lowest temperatures the system consists of ferromagnetic chains; neighboring chains in one crystallographic direction also have their moments pointing in the same direction, thus forming a ferromagnetic layer parallel to the *ab*-plane. Successive layers in the *c*-direction have opposite moments. As a result, although the predominant interactions are ferromagnetic in sign, the weak antiferromagnetic interaction between the planes leads to a net antiferromagnetic configuration. The area of the phase diagram, Fig. 10.6, in which the magnetic moments are parallel to the easy axis a^* is denoted by AF_1. In the area denoted by I the magnetization is independent of the magnetic field strength and amounts to one-third of the saturation magnetization. After the phase boundary denoted by the open triangles has been crossed, the susceptibility remains constant until the transition to the paramagnetic state (open circles) occurs. The nature of the low-field magnetic phase between 6.3 K and 7.3 K is not clear, for the evidence requires that AF_2 not differ substantially in spin structure from AF_1. The phase I originates from a reorientation of the moments such that of each six sublattices, three are reversed. The Sc phase has the characteristics of a spin-flop phase. The proposed magnetic structures are pictured in Fig. 10.8. The specific heat of $NiBr_2 \cdot 2 H_2O$, which is assumed to be isostructural to the chloride, has likewise been interpreted [13] in terms of a large (negative) single ion anisotropy and ferromagnetic intrachain interaction. A theory has been presented [25] that allows spin reorientation as a function of temperature even in zero external field when large zero-field splittings are present which cause anisotropic exchange.

Although many ordered antiferromagnets appear to be simple two-sublattice systems, these materials should restrain us from drawing too many conclusions from inadequate data. The derivation of the H–T phase diagram is a necessary part of any investigation.

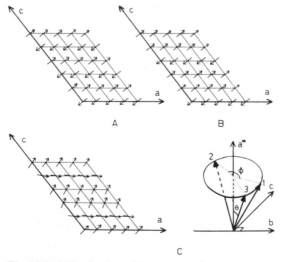

Fig. 10.8. (a) Projection of the magnetic moments on the *ac*-plane in the AF_1 phase. (b) Projection of the magnetic moments on the *ac*-plane in the intermediate phase I. The direction of the moments on three neighboring sublattices have been reversed. (c) Picture of the screw phase denoted by Sc. All moments make the same angle θ with the easy a^*-axis. From Ref. [24]

10.3 Tris(dithiocarbamates) of Iron(III)

Six-coordinate complexes of iron(III) are usually either high spin (^6S ground state) or low spin ($^2T_2$). Several factors, such as the strength of the ligand field and the covalency, determine which configuration a particular compound will assume; moreover, whatever configuration that compound has, it is usually retained irrespective of all such external influences as temperature or minor modification of the ligand. The tris(dithiocarbamates) of iron(III), $Fe(S_2CNR_2)_3$ are a most unusual set of compounds for, depending on the R group, several of them are high spin, several are low spin, and several seem to lie very close to the cross-over point between the two configurations. While these are the best-known examples of this phenomenon, a review of the subject [26] describes several other such systems. The subject of spin-crossover in iron(II) complexes has been extensively (218 references) reviewed [27]. Recent examples of other systems that can be treated formally as electronic isomers include molecules such as manganocene [28], $Fe(S_2CNR_2)_2[S_2C_2(CN)_2]$ [29] and Schiff-base complexes of cobalt(II) [30].

The dithiocarbamates and the measurements of their physical properties have been reviewed at length elsewhere [26], so that only some of the magnetic results will be discussed here. The energy level diagram is sketched in Fig. 10.9 for a situation in which the $^6A_1$ level lies E in energy above the $^2T_2$ state, where E is assumed, in at least certain cases, to be thermally accessible. The usual Zeeman splitting of the $^6A_1$ state is shown, but the splitting of the $^2T_2$ state is complicated by spin-orbit coupling effects, as indicated. The most interesting situations will occur when $E/kT \approx 1$, and the energy sublevels are intermingled.

The magnetic properties corresponding to this set of energy levels are calculated from Van Vleck's equation as

$$\chi_M = N\mu_B^2\mu_{eff}^2/3kT, \tag{10.1}$$

where

$$\mu_{eff}^2 = \frac{(\tfrac{3}{4})g^2 + 8x^{-1}(1 - e^{-3x/2}) + 105e^{-(1+E/\zeta)x}}{1 + 2e^{-3x/2} + 3e^{-(1+E/\zeta)x}} \tag{10.2}$$

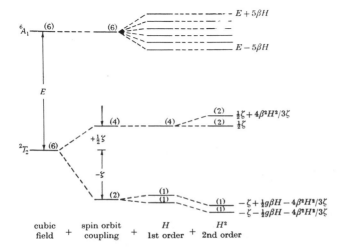

Fig. 10.9. Energy levels (not to scale) of configuration d^5 in the crossover region. From Ref. [26]

and $x = \zeta/kT$, with ζ the one-electron spin-orbit coupling constant. Usual low spin behavior corresponds to large positive E, high spin, to large negative E, so that marked deviations from this behavior occur only when $|E/\zeta| < 1$. In practice, it is difficult to apply Eq. (10.1) over a wide enough temperature interval to evaluate all the parameters, especially since the spin-orbit coupling constant must often be taken as an empirical parameter; thermal decomposition and phase changes are among the other factors which limit the application of this equation.

The exceptional magnetic behavior of the crossover systems is best seen in the temperature dependence of the reciprocal of the molar susceptibility,

$$\chi_M^{-1} = 3kT/N\mu_B^2\mu_{eff}^2 \tag{10.3}$$

and several calculated values of χ_M^{-1} are represented in Fig. 10.10. The maxima and minima have obvious diagnostic value. A fuller discussion of the behavior of these curves has been given elsewhere [26]. A selection of experimental data, along with fits to Eq. (10.1) (modified to account for different metal-ligand vibration frequencies in the two electronic states) is presented in Fig. 10.11. It is remarkable how well such a simple

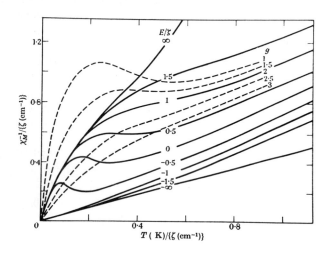

Fig. 10.10. Calculated values of χ_M^{-1}. Full lines: $g = 2$ with various values of E/ζ. Broken lines: $E/\zeta = 1$ with various values of g. From Ref. [26]

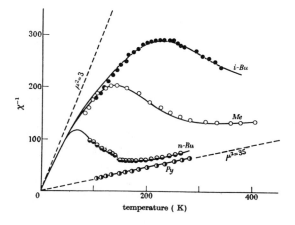

Fig. 10.11. Variation of χ^{-1} of [Fe(S$_2$CNR$_2$)$_3$] with temperature. Me: R = methyl; n-Bu: R = n-butyl; i-Bu: R = i-butyl; Py: NR$_2$ = pyrrolidyl. From Ref. [26]

model accounts for this unusual magnetic behavior. Maxima and minima are observed (when the temperature range is appropriate), and the mean magnetic moment per iron atom rises with temperature from low spin toward high spin values. However, the relative positions of the $^6A_1$ and $^2T_2$ levels cannot be estimated beyond asserting that they must lie within about 500 cm^{-1} of each other. The marked discontinuity at 145 K in the susceptibility of the di-n-butyl derivative is due to a phase change of the solid.

It should be pointed out that similar data are obtained on these materials when they are dissolved in inert solvents, which shows that the reported effects are neither intermolecular in origin nor due to some other solid state effect. Data on more than 18 compounds in this series alone have been reported, one recent example which exhibits the phenomenon being the N-methyl-N-n-butyl derivative [31].

Although the magnetic aspects of the dithiocarbamates appear to be understood on the whole, it should be pointed out that the more chemical aspects of this problem remain somewhat perplexing. For example, Mössbauer and proton NMR studies have failed to show evidence for the simultaneous population of two electronic levels [32], a result which would require a rapid crossover between the levels ($<10^{-7}$ s). An alternative explanation, based on crystal field calculations, suggests [33] that the ground state may be a mixed-spin state, the variable magnetic moment being due to a change in character of the ground state with temperature. Important vibronic contributions to the nature of the positions of the energy levels have also been suggested [34].

A further complication arises from the fact that several of these complexes crystallize with solvated solvent molecules occupying crystallographic positions in the unit cell [35]. The effect of the solvated solvent on the magnetic behavior of the ferric ion has been demonstrated in the case of the morpholyl derivative, for example, which was found to have a moment of 2.92 μ_B when benzene solvated and 5.92 μ_B when dichloromethane solvated [36].

Even though the organic group R is on the periphery of the molecule, this group determines whether a particular compound Fe(S$_2$CNR$_2$)$_3$ is high spin or low spin at a given temperature. This is a most unusual situation in coordination chemistry.

The crystallographic aspects of the problem have been reviewed [37, 38]. One expects a contraction of the FeS$_6$ core in the transition from a high-spin state to one of low-spin. Compounds which are predominantly in the high spin-state. (e.g., R=n-butyl) have a mean Fe–S distance of about 0.241 nm, while compounds predominantly low spin (e.g., R=CH$_3$, R=C$_6$H$_5$) have a mean Fe–S distance of 0.2315 nm, suggesting a contraction of about 0.008 nm in the Fe–S distances on going from the high-spin to the low-spin state. The FeS$_6$ polyhedron itself undergoes significant changes in its geometry, also.

10.4 $\mathscr{S} = \frac{3}{2}$ Iron(III)

An interesting series of compounds is provided by the halobisdithiocarbamates of iron, Fe(X)(S$_2$CNR$_2$)$_2$, for they exhibit the unusual spin state of $\frac{3}{2}$ for iron(III). In octahedral stereochemistry, as was discussed above, iron(III) must always have spin of $\frac{5}{2}$ (so-called spin-free), or $\frac{1}{2}$ (spin-paired), each with its own well-defined magnetic behavior. Since these spin states also apply to tetrahedral stereochemistry (although no $\mathscr{S} = \frac{1}{2}$

tetrahedral compounds of iron(III) have yet been reported), it is impossible to have a ground state with $\mathscr{S}=\frac{3}{2}$ in cubic coordination. But, in lower symmetry environments, such as in a distorted five-coordinate structure, the $\mathscr{S}=\frac{3}{2}$ state can in fact become the ground state, and that happens with these compounds.

The compounds are easily prepared [39] from the corresponding tris(dithiocarbamates) by the addition of an aqueous hydrohalic acid to a solution of $Fe(S_2CNR_2)_3$ in benzene. The compound with $R=C_2H_5$ and $X=Cl$ has been studied the most intensely, but a variety of compounds have been prepared and studied with, for example, $X=Cl$, Br, I or SCN, and $R=CH_3, C_2H_5, i\text{-}C_3H_7$, etc. All are monomeric and soluble in organic solvents, and retain a monomeric structure in the crystal [39–41]. The structure of a typical member of the series is illustrated in Fig. 10.12 where the distorted tetragonal pyramidal coordination that commonly occurs may be seen. Thus, the four sulfur atoms form a plane about the iron atom which lies, however, 0.063 nm above the basal plane. The Fe–X distances are 0.226 nm (Cl), 0.242 nm (Br), and 0.259 nm (I). In solution at ambient temperatures the compounds exhibit a magnetic behavior typical of a $\mathscr{S}=\frac{3}{2}$ system [39]. All available data are consistent with the fact that the orbitally non-degenerate $^4A_2$ state is the ground state. Some very similar halobis(diselenocarbamates) have also been described recently [42].

As usual, the $^4A_2$ state is split by the combined action of spin-orbit coupling and crystal field distortions [43]. Most of these pyramidal compounds exhibit large zero-field splittings, and very different values have occasionally been reported by different authors for the same compound. The major reason this has occurred lies mainly with the choice of principal axis coordinate system (see Chap. 11), for the natural choice is to assign the Fe–X axis as the z-axis. With this choice, however, values of E, the rhombic spin Hamiltonian parameter, have been found to be comparable to or larger than the axial term, D. This is not the best choice of axes, however, for it has been shown [44] that there is always a better choice of principal molecular axes whenever $|E/D|>\frac{1}{3}$.

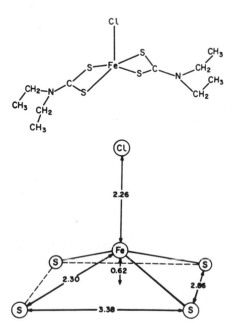

Fig. 10.12. Molecular geometry of Fe(dtc)$_2$Cl. From Ref. [47]

Ignoring for the moment the rhombic terms, the zero-field splittings 2D for these compounds vary widely in both sign and magnitude [45] and of course this determines the low-temperature magnetic behavior. Since the g-values of these compounds [46] are about 2 when considering the true spin of $\frac{3}{2}$, the situation is similar to that of tetrahedral cobalt(II) when the zero-field splittings are large. Thus, $|\pm\frac{1}{2}\rangle$ is low when D is positive, and $g'_{\parallel}=2$, $g'_{\perp}=4$; conversely, when D is negative, $|\pm\frac{3}{2}\rangle$ is low and $g'_{\parallel}=6$, $g'_{\perp}=0$. The zero-field splittings in these compounds have been determined from the measurement of the far IR spectra in the presence of a magnetic field [45] and from paramagnetic anisotropies at high temperatures [44]. Choosing the Fe–X bond as the direction of the quantization axis [47] in the ethyl derivatives, $D=+1.93\,cm^{-1}$ for the chloro compound, and $D=+7.50\,cm^{-1}$ for the bromo derivative. Large E (rhombic) terms have also been reported; writing $\delta=2(D^2+3E^2)^{1/2}$, reported zero field splittings δ for the ethyl series are $3.4\,cm^{-1}$ (Cl), $16\,cm^{-1}$ (Br), and $19.5\,cm^{-1}$ (I) [47]. On the other hand, the magnetic and Mössbauer results at lower temperatures on $Fe(Cl)(S_2CNEt_2)_2$ are more consistent if the magnetic z-axis lies in the FeS_4 plane [48]. Indeed, with this choice of axes for this molecule, and with new measurements, one finds $D/k=-3.32\,K$ and $E/k=-0.65\,K$, which leads to $\delta/k=7\,K$. Replacing the organic ligand leads to only small differences in the ZFS; for the chlorobis(N,N-di-isopropyldithiocarbamate [49], the same choice of axes leads to $D/k=-1.87\,K$, $E/k=0.47\,K$, and $\delta/k=4.14\,K$. The corresponding selenium derivative, on the other hand has $D/k=6.95\,K$, $E/k=-0.14\,K$, and $\delta/k=13.9\,K$. This ferromagnetic compound appears to display XY-like behavior.

Mössbauer and magnetization measurements [41, 50–52] first showed that $Fe(Cl)(S_2CNEt_2)_2$ orders ferromagnetically at about 2.43 K. Specific heat measurements [53] confirmed the phase transition, with a λ-anomaly appearing at 2.412 K, but interestingly, there is no clear evidence for the anticipated Schottky anomaly. This is due to the large lattice contribution, which can be evaluated in several ways [48, 53]. The susceptibility exhibits a great deal of anisotropy at low temperatures, acting like a three-dimensional Ising system. The susceptibility parallel to the easy axis, which was illustrated in Fig. 6.26, is typical of that of a demagnetization-limited ferromagnet [48].

Unraveling the physical properties of $[Fe(Br)(S_2CNEt_2)]$ has proved to be very troublesome. The compound as usually prepared has an energy level structure the reverse of that in the chloride ($|\pm\frac{1}{2}\rangle$ low rather than $|\pm\frac{3}{2}\rangle$) and with a much larger zero-field splitting. Ordering has not been found [41] at temperatures down to 0.34 K. Yet mixed crystals of the bromide with as little as 16% of the chloride present behave like the chloride, with the exchange being of comparable magnitude and the ZFS of the same sign. Resolution of this paradox appears to reside in the following [54]. The Mössbauer, EPR, susceptibility, and far infrared data which have led to the zero-field splitting parameters and ground state level structure for $Fe(Br)[S_2CN(C_2H_5)_2]_2$, have in fact been obtained on specimens for which the crystal structure is not the same as in the mixed crystals. A disintegration of single crystal specimens of $Fe(Br)[S_2CN(C_2H_5)_2]_2$ at temperatures in the neighborhood of 220 K has been observed. This does not occur for the chloride. Although the room temperature crystal structures of the chloride and the bromide are, apart from small differences in lattice constants, essentially identical, it seems evident that for liquid nitrogen temperatures and below the "natural" crystal structure of the pure bromide is no longer identical with that of the chloride. EPR and other measurements on the bromide have probed the characteristics of the natural low temperature structure. For reasons which remain unclear, it is observed that even a small fraction of $Fe(Cl)[S_2CN(C_2H_5)_2]_2$ stabilizes $Fe(Br)[S_2CN(C_2H_5)_2]_2$ in the common high temperature crystal structure of these homologous compounds. As with the pure chloride, mixed crystals at least as rich in bromide as 84% can be cooled to liquid helium temperatures and warmed to room temperature without suffering any discernible damage. In the mixed crystals then, the iron ion of the bromide species experiences a crystal field which is characteristic of the high temperature structure and which can be quite different from that of the low temperature structure of the pure bromide. The level structure of the quartet ground state can therefore differ radically from that of pure $Fe(Br)[S_2CN(C_2H_5)_2]_2$. In the mixed crystals the value of D in the spin Hamiltonian describing the quartet ground state of the iron ion in the bromide is no longer positive but rather negative, and the zero field splitting between the $|\pm\frac{1}{2}\rangle$ and $|\pm\frac{3}{2}\rangle$ Kramers doublets, $2(D^2+3E^2)^{1/2}$, is possibly not very different from that of the pure chloride. In the low temperature structure of the pure bromide, D/k has been found to be equal to 10.78 K, with the zero-field splitting equal to 21.71 K. With $D>0$ the $|\pm\frac{1}{2}\rangle$ doublet lies lowest. With $D<0$ in the mixed crystal, $|\pm\frac{3}{2}\rangle$ lies lowest, as is the case for the iron ion of the chloride species.

These suggestions of DeFotis and Cowen appear to be verified by measurements [55] on the bromide when it is recrystallized from benzene, rather than, as more commonly, from methylene chloride. A different crystal form is obtained for the compound, and it is found to order magnetically at 1.52 K.

The structure and magnetic behavior of the methyl derivatives are less well-defined [56].

10.5 Manganous Acetate Tetrahydrate

The structure [57, 58] of $Mn(CH_3COO)_2 \cdot 4H_2O$ is illustrated in Fig. 10.13, where it will be observed that the crystal consists of planes containing trimeric units of manganese atoms. The manganese atoms are inequivalent, for the central one lies on an inversion center, and is bridged to both of its nearest neighbors by two acetate groups in the same fashion as in copper acetate. However, in addition to the water molecules in the coordination spheres, there is also another acetate group present which bonds entirely differently, for one of the oxygen atoms bridges Mn_1 and Mn_2, while the second oxygen atom of the acetate group forms a longer bridging bond to an Mn_2 of an adjoining trimer unit. The net result of these structural interactions is that there is a strong AF interaction within the trimer, and a weaker interaction between the trimers.

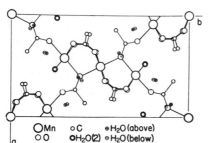

Fig. 10.13. The structure of $Mn(CH_3COO)_2 \cdot 4H_2O$. The projection is along the perpendicular c^*-direction. Two water molecules in the Mn_2 coordination octahedron which are shown superimposed are actually one above and one below the Mn plane. From Ref. [58]

There appears to be considerable short range order in the compound above the long-range ordering temperature of 3.18 K.

The inverse powder susceptibility [59] shows substantial curvature throughout the temperature region below 20 K. Between 14 and 20 K, apparent Curie-Weiss behavior is observed, $\chi = 3.19/(T+5.2)$ emu/mol, but the Curie constant is well below the value 4.375 emu-K/mol normally anticipated for $\mathscr{S} = \frac{5}{2}$ manganese. The large Curie-Weiss constant, and the curvature in χ^{-1} below 10 K, coupled with the small Curie constant, suggest that there is probably also curvature in the hydrogen region, and that in fact true Curie-Weiss behavior will be observed only at much higher temperatures. This is one piece of evidence which suggests the presence of substantial short-range order.

The zero-field heat capacity [60] is illustrated in Fig. 10.14, where the sharp peak at $T_c = 3.18$ K is only one of the prominent features. The broad, Schottky-like peak at about 0.7 K may be due to the presence of inequivalent sets of ions present in the compound; it could arise if one sublattice remains paramagnetic while the other sublattice(s) become ordered, but a quantitative fit of this portion of the heat capacity curve is not yet available. The other, unusual feature of this heat capacity curve is that the specific heat is virtually linear between 5 and 16 K. Thus, the heat capacity cannot be governed by the familiar relation, $c = aT^3 + b/T^2$, which implies that either the lattice is not varying as T^3 in this temperature region, or that the magnetic contribution is not varying as T^{-2}, or both. From the fact that less than 80% of the anticipated magnetic entropy is gained below 4 K, at the least, these facts suggest the presence of substantial short-range order.

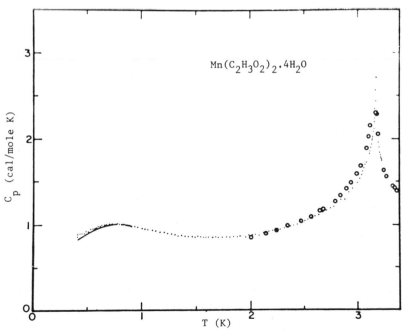

Fig. 10.14. Heat capacity c_p of powdered $Mn(CH_3COO)_2 \cdot 4H_2O$ in "zero" applied field below 3.4 K. From Ref. [60]

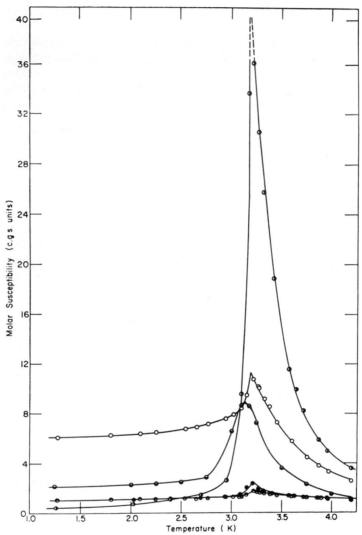

Fig. 10.15. Powder and single crystal susceptibilities of Mn(CH$_3$COO)$_2$·4H$_2$O. ○, *b*-axis; ◑ and ●, *c**-axis; ◐, *a*-axis; ◒, powder. From Ref. [59]

Unusual anisotropy was observed [59] in the single crystal susceptibilities, Fig. 10.15. The susceptibility parallel to the monoclinic *a*-axis rises sharply to an unusually large value while in the *c**-direction (*a*, *b*, *c** is a set of orthogonal axes used in the magnetic studies in the P2$_1$/a setting; the crystallographers [57, 58] prefer the P2$_1$/c setting) a similar but smaller peak is observed. The third, orthogonal susceptibility is essentially temperature-independent in this region, the bump in χ_b probably being due to misalignment. While it is clear from these data that a magnetic phase transition occurs in Mn(OAc)$_2$·4H$_2$O at 3.18 K, the behavior of χ_a alone suggests that the phase transition is not the usual paramagnetic to antiferromagnetic one.

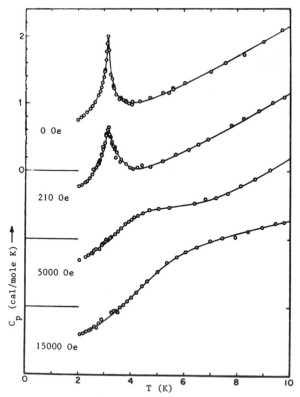

Fig. 10.16. Heat capacity c_p of powdered $Mn(CH_3COO)_2 \cdot 4\,H_2O$ in several applied magnetic fields. The zero of the vertical axis has been displaced downwards to separate different curves. The same vertical scale applies to each curve. From Ref. [60]

The compound is unusually sensitive to external magnetic fields, Fig. 10.16. The λ-peak in the specific heat broadens and shifts to higher temperature [60] in a field of as little as 210 Oe (0.0210 T); a field of 1000 Oe (0.1 T) has little effect on χ_b, but the peaks in χ_a and χ_{c*} are reduced and shifted in temperature. This weak field of the measuring coils was found to influence χ_a, and indeed a magnetic phase transition is caused by an external field of a mere 6 Oe ($6 \cdot 10^{-4}$ T) [61, 62].

The saturation magnetization of the compound [62] is only one-third as large as anticipated for a normal manganese salt. This would follow if the exchange within the Mn_2–O–Mn_1–O–Mn_2 groups is antiferromagnetic and relatively large compared to the coupling between such groups in the same plane or between planes. The neutron diffraction work [57] shows that the interactions within the Mn_2–O–Mn_1–O–Mn_2 units is AF, but ferromagnetic between different groups in the same plane and AF between adjacent planes. The zero-field magnetic structure deduced from the neutron work is illustrated in Fig. 10.17. Both the crystal and magnetic structures are consistent with the existence of substantial short range magnetic order.

NMR studies [61] are consistent with all these data. The transition at 6 Oe with the external field parallel to the a-axis is like a metamagnetic one, short-range order

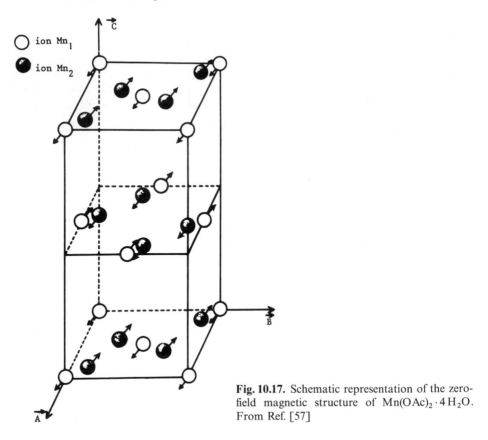

Fig. 10.17. Schematic representation of the zero-field magnetic structure of $Mn(OAc)_2 \cdot 4H_2O$. From Ref. [57]

persists above T_c, and the saturated paramagnetic state is found at about 140 Oe (0.0140 T) (at 1.1 K). No exchange constants have as yet been extracted from the data.

10.6 Polymeric NiX_2L_2

Many of the principles discussed earlier in this book come together in the series of compounds of stoichiometry MX_2L_2, where M may be Cu, Ni, Mn, Co, or Fe; X is Cl or Br; and L is pyridine (py) or pyrazole (pz). All the molecules appear to be linear chains in both structural and magnetic behavior, with *trans*-MX_4L_2 coordination spheres. The dihalo-bridged chains are formed by edge-sharing of octahedra; the structure of this isostructural series of molecules was illustrated for the Ising chain α-$CoCl_2 \cdot 2py$ in Fig. 7.1. It was pointed out in Chap. 7 that $CoCl_2 \cdot 2L$, where L is H_2O or pyridine, exhibits a large amount of short-range order, but that the effect was enhanced when the small water molecule was replaced by the larger pyridine molecule. Similarly, large amounts of short-range order as well as unusually large zero-field splittings are also exhibited by the other members of this series. The discussion will be limited to the nickel compounds.

Susceptibilities of the compounds NiX$_2$L$_2$, X=Cl, Br; L=py, pz, have been reported under several conditions [63]. Unfortunately, this series of molecules does not form single crystals with any degree of ease, and only powder measurements are available. The magnetic behavior is unusual enough to be easily observed in this fashion, however, and the results are confirmed by specific heat studies [64], so that a great deal of confidence can be placed in the work. The compounds obey the Curie-Weiss law between 30 and 120 K with relatively large positive values of θ of 7 to 18 K, depending on X and L. The Curie constants are normal for nickel(II), and since it was known that a chain structure obtains, the θ parameters were associated with the intrachain exchange constant, J. Deviations from the Curie-Weiss Law were observed below 30 K, which are due to the influence of both intrachain exchange and single-ion anisotropy due to zero-field splittings.

The analysis of the susceptibility data is hindered by the lack of suitable theoretical work for chains of $\mathscr{S}=1$ ions. Writing the Hamiltonian as

$$H = g\mu_B \boldsymbol{H} \cdot \boldsymbol{S} + D[S_z^2 - (\tfrac{1}{3})\mathscr{S}(\mathscr{S}+1)] + A\boldsymbol{S} \cdot \langle \boldsymbol{S} \rangle,$$

a molecular field term, with A=2zJ, is included to account for the intrachain interaction. The above Hamiltonian may be solved approximately for T larger than either of the parameters D/k or A/k, and, after averaging the three orthogonal susceptibilities in order to calculate the behavior of a powder, one finds [63]

$$\theta = 4zJ/3k$$

and so ferromagnetic intrachain constants of 2 to 7 K were obtained, in addition to very large zero-field splittings of the order of -25 K.

At low temperatures, and at low fields, the powder susceptibility exhibits a maximum, and then approaches zero value at 0 K. The temperatures of maximum χ, T_m, are about 3 to 7 K, and were assigned as critical temperatures; comparison with specific heat results suggests that T_c is actually slightly below T_m, as discussed earlier. A weak antiferromagnetic interchain interaction that leads to an antiferromagnetically ordered state would cause this behavior.

All these results are confirmed by the specific heat data [64]. The broad peaks in the magnetic heat capacity which are characteristic of one-dimensional ordering were observed, and transitions to long-range order, characterized by λ-like peaks, were observed; a double peak was observed for NiCl$_2 \cdot$ 2 py. The ordering temperatures are 6.05 K (NiCl$_2 \cdot$ 2 pz); 3.35 K (NiBr$_2 \cdot$ 2 pz); 6.41 and 6.750 K (NiCl$_2 \cdot$ 2 py); and 2.85 K (NiBr$_2 \cdot$ 2 py).

There are several features of these investigations that make these compounds of more than passing interest. The first is that since the compounds contain nickel, which is an $\mathscr{S}=1$ ion, zero-field splittings would be expected to, and do, complicate the specific heat behavior. In fact, these compounds exhibit some of the largest zero-field splittings yet observed being, for example, -27 K for NiCl$_2 \cdot$ 2 py, and -33 K for NiBr$_2 \cdot$ 2 pz. The parameter D/k is negative for the four compounds which places a spin-doublet as the lowest or ground state. The problem then arises of calculating the specific heat of a one-dimensional magnetic system as a function of the ratio D/J, and this was the inspiration for the work of Blöte [65]. A typical set of his results is illustrated in Fig.

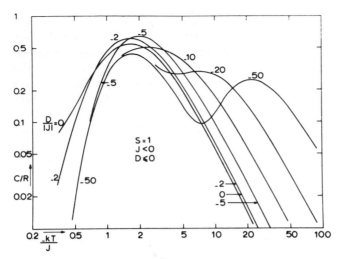

Fig. 10.18. Heat capacities of antiferromagnetic $\mathscr{S}=1$ chains with isotropic interaction and negative D terms. For large D, the extrapolated results approach the sum of a Schottky anomaly (due to the D term) and an Ising anomaly for magnetic interaction in the lower doublet. From Ref. [65]

10.18, where it will be observed that as $D/|J|$ becomes large and negative, two broad peaks are to be found in the specific heat. One is due to the magnetic chain behavior, and the second is due to the Schottky term. That these contributions are in fact additive when they are well-separated on the temperature axis (i.e., when $D/|J|$ is large) was illustrated by the data on $NiBr_2 \cdot 2\,pz$, Fig. 10.19. An equally good fit to the data was obtained by either fitting the results to the complete curve of Blöte for the Heisenberg linear chain model with uniaxial single-ion anisotropy, or by simply summing the linear chain and Schottky contributions. It should be pointed out that when $|D/J|$ is as large as it is, about 12, in this compound, that the exchange interaction occurs between ions with effective spin-doublet ground states. The situation was described in Sect. 2.5

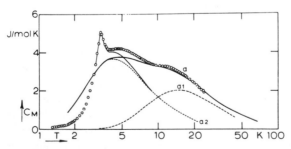

Fig. 10.19. The magnetic specific heat of $NiBr_2 \cdot 2\,pz$ as a function of temperature: 0, experimental results; $---$ al, Schottky curve for independent Ni(II) ions with single-ion anisotropy parameter $D/k = -31\,K$. $---$ a2, Ising linear-chain model $\mathscr{S} = \frac{1}{2}$; $J/k(\mathscr{S} = \frac{1}{2}) = 10.5\,K$; $---$ a, the sum curve of a1 and a2. This curve coincides for the greater part with the curve for the Heisenberg linear-chain model with uniaxial single-ion anisotropy as calculated for $\mathscr{S} = 1$ by Blöte: $J/k = 2.7\,K$, $D/k = -33\,K$. From Ref. [64]

and corresponds to an Ising $\mathscr{S}' = \frac{1}{2}$ system. Care must be used in comparing the magnetic parameters obtained by the analysis from the different points of view, since the J/k obtained from the $\mathscr{S}' = \frac{1}{2}$ formalism will be four times larger than that corresponding to the $\mathscr{S} = 1$ Hamiltonian.

The fact that the $\mathscr{S}' = \frac{1}{2}$ Ising ion has g values of $g_{\parallel} \approx 4.4$ and $g_{\perp} \approx 0$ gives rise to the other interesting feature of $NiBr_2 \cdot 2\,pz$, and that is that this Ising nature, along with the ferromagnetic intrachain interaction, causes it to be a metamagnet. The specific heat of a powdered sample was measured in a field of $5\,kOe$ ($0.5\,T$); since $g_{\perp} \approx 0$, the (unavailable) single crystal measurements were not needed. The λ-like anomaly disappeared, the maximum value of the specific heat increased and shifted to a higher temperature. Furthermore the susceptibility as a function of field [63] has a maximum at the critical field of the AF$\rightarrow$metamagnetic transition, and then maintains a constant value. Not only do these results confirm the metamagnetic behavior, but also they confirm the unusually large zero-field splittings.

10.7 Hydrated Nickel Nitrates

A series of investigations [66–69] on the hydrates of nickel nitrate illustrates practically all of the problems in magnetism. That is, within the three compounds $Ni(NO_3)_2 \cdot 2\,H_2O$, $Ni(NO_3)_2 \cdot 4\,H_2O$, and $Ni(NO_3)_2 \cdot 6\,H_2O$, zero-field splittings, anisotropic susceptibilities, both short and long-range order, and metamagnetism are all to be found.

The different hydrates may be grown from aqueous solutions of nickel nitrate, with crystals of $Ni(NO_3)_2 \cdot 6\,H_2O$ appearing when the solution is kept at room temperature, the tetrahydrate at about $70\,°C$, and the dihydrate at about $100\,°C$. The hexahydrate undergoes 5 crystallographic phase transitions as it is cooled to low temperatures, so that its crystal structure in the helium region, where the magnetic measurements have been made, is triclinic. The tetrahydrate apparently exists in two different forms, monoclinic and triclinic [70]. The dihydrate forms layers such as is illustrated in Fig. 10.20. Each nickel ion is bonded to two *trans*-water molecules, and to four nitrate oxygens that bridge to other nickel atoms.

Since these are nickel salts, of $\mathscr{S} = 1$, zero-field splittings must be considered. Ignoring the rhombic (E) crystal field term for the moment, a negative axial (D) term puts a doubly-degenerate level below the singlet. A Schottky term is anticipated in the specific heat, but also, because the lowest level has spin-degeneracy irrespective of the size of D, magnetic ordering will occur at some temperature. This is in fact the situation with the dihydrate, where $D/k = -6.50\,K$, and a λ-transition is found in the specific heat at $4.105\,K$.

But, in the hexahydrate (and apparently also the tetrahydrate) the opposite situation prevails. That is, the zero-field splitting is positive and the singlet lies lowest. In the presence of exchange interactions that are weak compared to the zero-field splitting, all spin-degeneracy, and hence all entropy, is removed as the temperature is lowered towards $0\,K$, and the system cannot undergo long-range magnetic order (in the absence of a field). The situation is precisely the same as that discussed earlier for $Cu(NO_3)_2 \cdot 2.5\,H_2O$ (Chap. 5). This argument has been put on a quantitative basis in the MF approximation by Moriya. Thus, the specific heat of $Ni(NO_3)_2 \cdot 6\,H_2O$ is

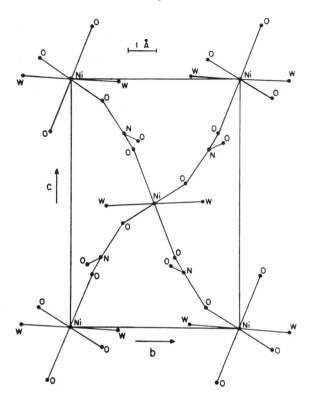

Fig. 10.20. Projection of the unit cell of nickel nitrate dihydrate on the *bc*-plane. The nickel ions form a face-centered pattern. The symbol W is used to describe a water molecule. From Ref. [66]

described in terms of a Schottky function including both D and E. It was found that D/k $=6.43$ K, and $E/k = +1.63$ K, which puts the three levels successively at 0, 4.80, and 8.06 K.

A spontaneous magnetic transition is ruled out because none was observed above 0.5 K, all the entropy anticipated for a $\mathscr{S}=1$ ion was observed at higher temperatures from the Schottky curve, and thus exchange is quite small compared to the zero-field splittings. Magnetic interactions are not necessarily zero, on the other hand, although they are not evident in zero-field heat capacity measurements. Such subcritical interactions may be observable with susceptibility measurements, however, for they may contribute to the effective magnetic field at a nickel ion when an external field is applied, even if it is only the small measuring field in a zero-external field susceptibility measurement. This appears to be the case with $Ni(NO_3)_2 \cdot 6H_2O$, where the powder susceptibility is displayed in Fig. 10.21, along with the susceptibility calculated from the parameters obtained from the heat capacity analysis. A good fit requires an antiferromagnetic molecular field constant, $A/k=0.62$ K, where $A= -2zJ$.

It has been claimed [71] that magnetic ordering can be induced in $Ni(NO_3)_2 \cdot 6H_2O$. Though the material is triclinic, the local axes of the two magnetic ions in the unit cell are parallel and the level crossing field is some 40 kOe (4 T). Indeed, magnetic ordering seems to occur at low temperatures and a phase diagram somewhat like that illustrated in Fig. 6.19 was determined, except that there are two maxima, corresponding to two crossing fields, and a minimum in between. The explanation of

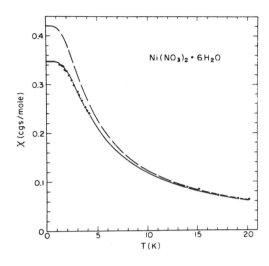

Fig. 10.21. Powder susceptibility of nickel nitrate hexahydrate as a function of temperature. Full line: theoretical curve, with $g = 2.25$, $D/k = +6.43$ K, $E/k = +1.63$ K, and $A/k = +0.62$ K. Dashed line: same, but with $A = 0$. From Ref. [69]

these results is that the major interaction between the nickel ions is pairwise, but that long-range order occurs through interpair interactions which are comparable in magnitude.

The tetrahydrate is described [66] by essentially the same zero-field splitting parameters as the hexahydrate, but it is not clear on which crystal form these measurements were performed. A number of field dependent studies have been reported [72] on β-Ni(NO$_3$)$_2 \cdot 4$H$_2$O, which is triclinic with two inequivalent molecules in the unit cell, called A and B. Complicated behavior is observed, and it is claimed that field induced ordering of the A molecules is observed when the external field is applied parallel to the local z_A-axes, and similarly when the applied field is parallel to the local z_B-axes. The phase diagrams are not equivalent, however, and it is suggested that there are two kinds of field-induced spin ordered states, corresponding to the two inequivalent nickel ion sites.

As mentioned above, the dihydrate orders spontaneously at 4.1 K. A positive Curie-Weiss constant of 2.5 K suggests that the major interactions are ferromagnetic, and a MF model based on the structurally-observed layers has been proposed. If the layers are ferromagnetically coupled, with $2z_2J_2/k = +4.02$ K, and the interlayer interaction is weak and AF with $2z_1J_1/k = -0.61$ K, then a consistent fit of the susceptibility data is obtained. Furthermore, the large single-ion anisotropy, coupled with this magnetic anisotropy suggests that Ni(NO$_3$)$_2 \cdot 2$H$_2$O should be a metamagnet, and this indeed proves to be the case. The phase diagram has been determined, and the tricritical point is at 3.85 K; the pressure dependent phase diagram has also been determined [73].

10.8 The Pyridine N-Oxide Series

The compounds [M(C$_5$H$_5$NO)$_6$]X$_2$ where the ligand is pyridine N-oxide and X may be perchlorate, nitrate, fluoborate, iodide or even bromate are of extensive interest, and several of them have been discussed earlier in the book. Several further points will be made here; an extensive review is available [74]. The compounds with M = Mn, Fe, Co,

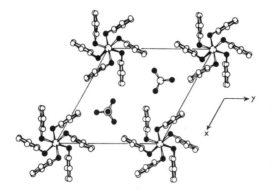

Fig. 10.22. A projection along the c-axis of one layer of the hexagonal unit cell of $[Co(C_5H_5NO)_6](ClO_4)_2$. The cobalt ions are on the corners (large open circles) and are octahedrally surrounded by the oxygens (filled circles) belonging to the C_5H_5NO groups. It should be noted that the cobalt ions shown are next nearest magnetic neighbors to one another. From Ref. [77]

Ni, Cu, and Zn are isostructural, which is especially significant because the zinc compound provides a diamagnetic host lattice for EPR and other studies, and the copper compound (at least at room temperature) presumably then does not display the distorted geometry that is common with so many other compounds. The structure [75, 76], illustrated in Fig. 10.22, is rhombohedral with but one molecule in the unit cell. The result is that each metal ion has six nearest-neighbor metal ions, which in turn causes the lattice to approximate that of a simple cubic one [77].

The metal ions in this lattice attain strict octahedral symmetry, yet they display relatively large zero-field splittings. Thus, the O–Ni–O angles are 90.3(1)° and 89.7(1)° in $[Ni(C_5H_5NO)_6](BF_4)_2$ and the O–Co–O angles are 89.97(4)° and 90.03(4)° in $[Co(C_5H_5NO)_6](ClO_4)_2$. Yet, in the nickel perchlorate compound, the zero-field splitting is very large, 6.26 K, and the parameter δ (Sect. 4.6.8) is relatively large (-500 to $-600 \, \text{cm}^{-1}$) in the cobalt compound [78]; the latter result is obtained from the highly anisotropic g-values of $[Co, Zn(C_5H_5NO)_6](ClO_4)_2$, with $g_\parallel = 2.26$, $g_\perp = 4.77$. Similarly, in the isomorphous manganese compound, the zero-field splitting parameter, D, takes [79] the very large value of 410 Oe (0.0410 T) (0.055 K).

The antiferromagnetic ordering behavior observed with the $[M(C_5H_5NO)_6](ClO_4)_2$ molecules is quite fascinating, especially as it occurs at what is, at first glance, relatively high temperatures for such large molecules. Thus, the cobalt molecule orders at 0.428 K [77]. When ClO_4^- is replaced by BF_4^- in these substances, the crystal structures remain isomorphous [75], and the magnetic behavior is consistent with this. Thus, $[Co(C_5H_5NO)_6](BF_4)_2$ orders at 0.357 K, and on a universal plot of c/R vs. $kT/|J|$, the data for both the perchlorate and fluorborate salts fall on a coincident curve, as illustrated in Fig. 10.23. The nitrate behaves similarly [80]. These molecules are the first examples of the simple cubic, $\mathscr{S} = \frac{1}{2}$, XY magnetic model. The susceptibility appears in Fig. 6.13.

The situation changes in a remarkable fashion with the copper analogs [81]. Though they are isostructural with the entire series of molecules at room temperature, the copper members distort as they are cooled. This is consistent with the usual coordination geometries found with copper, but what is especially fascinating is that the perchlorate becomes a one-dimensional magnetic system at helium temperatures, while the fluoborate behaves as a two-dimensional magnet. These phenomena result from different static Jahn-Teller distortions which set in cooperatively as the samples are cooled [82]. The different specific heat curves are illustrated in Fig. 10.24.

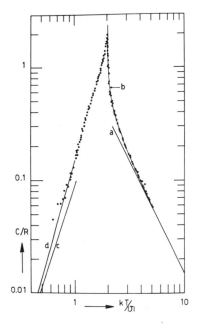

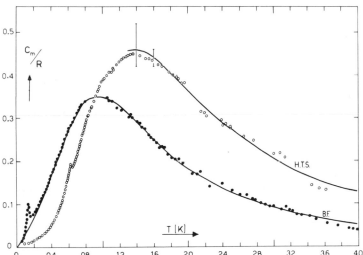

Fig. 10.23. Specific heat data for $[Co(C_5H_5NO)_6](BF_4)_2$ (open circles) and $[Co(C_5H_5NO)_6](ClO_4)_2$ (filled circles) plotted vs. relative temperature, $kT/|J|$. Curves a–d are theoretical predictions for the s.c. XY model with $\mathscr{S}=\frac{1}{2}$. From Ref. [77]

Fig. 10.24. Comparison of the magnetic specific heats found for $[Cu(C_5H_5NO)_6](ClO_4)_2$ (●) and $[Cu(C_5H_5NO)_6](BF_4)_2$ (○). Curve B–F is the Bonner-Fisher prediction for a linear chain antiferromagnet with $J/k = -1.02$ K; curve HTS is the prediction from the high-temperature series for the quadratic layer antiferromagnet with $J/k = -1.10$ K. From Ref. [77a]

The large distortions in $[Fe(C_5H_5NO)_6](ClO_4)_2$ result in Ising behavior for this $\mathscr{S}=\frac{1}{2}$ antiferromagnet. This has already been discussed in Chap. 6.

The compound $[Mn(C_5H_5NO)_6](BF_4)_2$ orders at 0.16 K [83]. Though this is not an extraordinarily low temperature, this means for this particular compound that magnetic exchange interactions are comparable in magnitude with the zero-field splittings, dipole-dipole interactions, and even nuclear hyperfine interactions. The net result is that an odd shaped specific heat peak is observed and a separation of the several contributions can only be made with difficulty.

The large zero-field splitting (6.26 K) observed with $[\text{Ni}(C_5H_5NO)_6](ClO_4)_2$ has already been discussed in Chap. 4 and field-induced ordering in this system was discussed in Chap. 6. The phase diagram was presented in Fig. 6.23. Data on the nitrate are provided in Ref. [84].

In order to determine the superexchange path in this series of molecules, it is necessary to examine the crystal structure with care. This is true, of course, in any system but especially so here, where a naive approach would suppose that the pyridine rings would effectively insulate the metal ions from one another and cause quite low ordering temperatures. Figure 10.25 illustrates several facets of the likely superexchange paths in these molecules. The rhombohedral unit cell formed by the metal ions is illustrated and closely approximates a simple cubic lattice. A reference metal ion is connected to its six nearest magnetic neighbors at a distance of about 0.96 nm by equivalent superexchange paths, consisting of a nearly collinear Co–O---O–Co bond, in which the Co–O and O---O distances are about 0.21 nm and 0.56 nm, respectively. It is important to note that the pyridine rings are diverted away from this bond so that one expects the superexchange to result from a direct overlap of the oxygen wave functions. A nice illustration of the importance of the superexchange path, or rather, the lack of a suitable superexchange path, is provided by a comparative study [85] of the hexakisimidazole- and antipyrine cobalt(II) complexes. The shortest direct metal-metal distances are long, being, for example, 0.868 nm in $[\text{Co}(Iz)_6](NO_3)_2$; furthermore, while effective paths such as a Co–O---O–Co link can be observed in the antipyrine compound, the angles are sharper than in the pyridine N-oxide compound, and the O---O distance is also longer (0.76 nm vs. 0.56 nm for the C_5H_5NO salt). Neither compound orders magnetically above 30 mK.

The molecule $[\text{Co}(DMSO)_6](ClO_4)_2$, where DMSO is $(CH_3)_2SO$, behaves similarly magnetically; the coordination sphere is similar to that found with the pyridine N-oxides. However, it does not order above 35 mK. This must be due to the stereochemistry of the sulfoxide ligands, which do not provide a suitable superexchange path.

Another series of compounds which contain pyridine N-oxide has recently [85] been investigated. The parent compound is of stoichiometry $Co(C_5H_5NO)_3Cl_2$, but crystal structure analysis showed that it should be formulated as

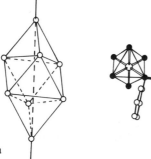
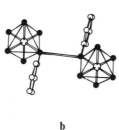

a   **b**

Fig. 10.25. (a) The rhombohedral cell formed by the cobalt ions in $[\text{Co}(C_5H_5NO)_6](ClO_4)_2$. (b) The superexchange path connecting cobalt ions that are nearest magnetic neighbors. From Ref. (77)

$[Co(C_5H_5NO)_6](CoCl_4)$. The compounds therefore contain two magnetic subsystems, that of the octahedral cations and that of the tetrahedral anions. Since the crystal field splittings of the cobalt in the two sites is substantially different, one might expect rather complex magnetic ordering. The compound $[Co(C_5H_5NO)_6](ZnCl_4)$ could also be prepared, and it was found to be isostructural to the Co/Co compound. This is important because the magnetic study of the Co/Zn system helped in deciphering the magnetic properties of the Co/Co compound. The measurements showed that the Zn atoms entered the lattice highly preferentially, residing only on the tetrahedral positions. Finally, the analogous bromide compounds $[Co(C_5H_5NO)_6](CoBr_4)$ and $[Co(C_5H_5NO)_6](ZnBr_4)$ were also prepared and shown to be isostructural to the chloride analogs. The compounds behave similarly, which gives further credence to the analysis of the chloride salts.

The compounds are monoclinic and belong to the space group Co. Superexchange paths of the familiar kind, Co–O---O–Co, are found in each of the three crystallographic directions, but no efficient Co–Cl---Cl–Co paths are evident in the crystal structure analysis. A view of the structure is presented in Fig. 10.26.

Specific heat data on $[Co(C_5H_5NO)_6](CoCl_4)$ and $[Co(C_5H_5NO)_6](ZnCl_4)$ are displayed in Fig. 10.27. The sharp anomalies occurring near 1 K may be attributed to a magnetic ordering of the octahedral cobalt ions, since that is the only magnetic contribution observed for the Co/Zn compound. The transition temperature, 0.95 K, is the same for both compounds, and thus there are two uncoupled magnetic subsystems in $[Co(C_5H_5NO)_6](CoCl_4)$. The Co/Co system exhibits additional paramagnetic contributions at low temperatures which are absent in the zinc analog, and these have been attributed to the cobalt ions at the tetrahedral sites. These are Schottky anomalies, due to the zero-field splitting of the ground state. Similar results were obtained for the bromide analogs.

These interpretations are confirmed by the susceptibility measurements, which are displayed in Fig. 10.28 for $[Co(C_5H_5NO)_6](CoCl_4)$. The most prominent feature in the data is the large spike in the b-axis data at 0.94 K, which is indicative of weak ferromagnetism. The susceptibilities of $[Co(C_5H_5NO)_6](ZnCl_4)$, Fig. 10.29, are somewhat similar, weak ferromagnetism once again being observed parallel to the b-axis. The

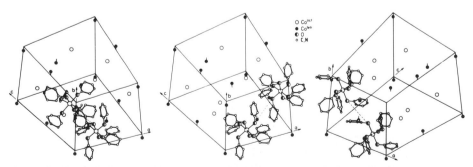

Fig. 10.26. Positions of the tetrahedrally coordinated and octahedrally coordinated Co^{2+} ions in the unit cell of $[Co(C_5H_5NO)_6](CoCl_4)$. For clarity the octahedral surroundings consisting of C_5H_5NO-groups are drawn for two nearest neighbors only. In each figure the Co–O---O–Co superexchange path for a different (independent) direction is indicated. The Cl^- ions surrounding the tetrahedrally coordinated Co^{2+} ions are omitted. From Ref. [85]

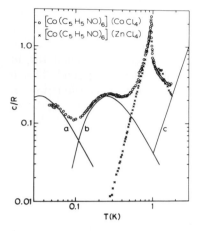

Fig. 10.27. Specific heat of powder samples of $[Co(C_5H_5NO)_6](CoCl_4)$ (o) and $[Co(C_5H_5NO)_6](ZnCl_4)$ (×). The Schottky curves a and b are discussed in the text. Curve c is the estimated lattice contribution. A small impurity contribution can be seen at 1.3 K in the data on $[Co(C_5H_5NO)_6](ZnCl_4)$. From Ref. [85]

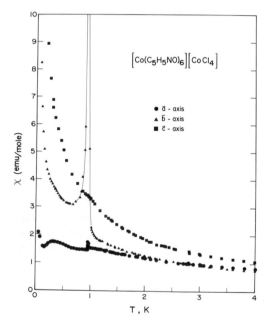

Fig. 10.28. Susceptibility data parallel to the three crystallographic axes of $[Co(C_5H_5NO)_6](CoCl_4)$. The maximum value of χ as measured in the b-direction at $T = T_c$ equals ~ 100 emu/mol. From Ref. [85]

peak occurs at the same temperature, 0.94 K. The preferred axis of spin alignment is the a-axis, while the c-axis is a perpendicular axis. These features are more readily apparent in the data on the Co/Zn compound than in those on the Co/Co one because in the latter there is a strong paramagnetic contribution from the tetrahedral $(CoCl_4)$ ions. The fact that the ordering anomalies in the specific heat curves in the Co/Zn compound are nearly as sharp as in the Co/Co salt and occur at precisely the same transition temperature provides strong evidence that in the Co/Zn system the preference of the zinc atoms for the tetrahedral sites is nearly 100%.

The general features of the susceptibilities of the bromide analogs are similar, with the striking exception that weak ferromagnetism is not observed. The data are not as clean as with the chlorides, which suggests that the discrimination of the zinc for

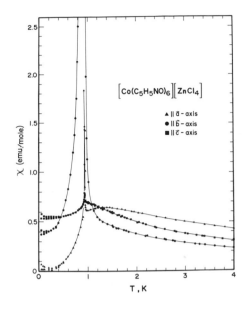

$[Co(C_5H_5NO)_6][ZnCl_4]$

▲ || $\bar{a}$ - axis
● || $\bar{b}$ - axis
■ || $\bar{c}$ - axis

Fig. 10.29. Susceptibility data parallel to the three crystallographic axes of $[Co(C_5H_5NO)_6]$ $(ZnCl_4)$. Note the change in the vertical scale compared to Fig. 10.28. The maximum value of χ as measured in the b-direction at $T=T_c$ equals ~ 100 emu/mol. From Ref. [85]

occupation of the tetrahedral sites is not as great as with the chlorides. The ordering temperatures are 0.65 K.

The subsystems involving the tetrahedral cobalt ions do not undergo a magnetic phase transition at temperatures down to 40 mK.

10.9 The $A_2[FeX_5(H_2O)]$ Series of Antiferromagnets

It has long been known that one can easily prepare salts of the $[FeX_5(H_2O)]^{2-}$ anion from aqueous solutions. Slow evaporation of stoichiometric mixtures of metal halide and ferric halide in acidic solution yields crystals of the desired $A_2[FeX_5(H_2O)]$ salt; only the monohydrate seems to grow under these conditions [86]. These substances have recently been found to undergo magnetic ordering at easily accessible temperatures. The compounds are antiferromagnets, and the exchange interactions are much stronger than is usually observed with hydrated double salts of the other transition metal ions.

All the compounds but one are orthorhombic although they are not isomorphic; the cesium compound $Cs_2[FeCl_5(H_2O)]$ belongs to the space group Cmcm [87, 88] and is isomorphic to the analogous ruthenium(III) material [89]. The remaining compounds except for monoclinic [90] $K_2[FeF_5(H_2O)]$ belong to the space group Pnma. A sketch of the structure of $Cs_2[FeCl_5(H_2O)]$ is presented in Fig. 10.30. The compounds $(NH_4)_2[FeCl_5(H_2O)]$ and $(NH_4)_2[InCl_5(H_2O)]$ are isomorphous, which is useful because the indium compound thus furnishes a colorless, diamagnetic analog. This is one of the largest series of structurally-related antiferromagnets of a given metal ion available.

The crystals contain discrete $[FeX_5(H_2O)]^{2-}$ octahedra. Several exhaustive discussions of the structures and the way they affect magnetic properties have appeared

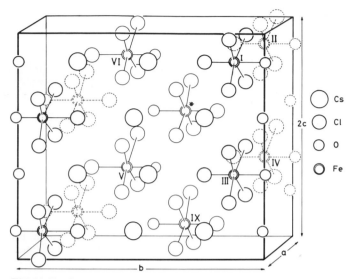

Fig. 10.30. Crystal structure of $Cs_2[FeCl_5(H_2O)]$. The unit cell is doubled along the c-axis. For clarity, only the Cs atoms in the frontal bc-plane have been included. The differences in thickness of the atoms indicate different bc-planes. From Ref. [92]

[91–93]. The essential feature is that there are no bridging groups between the metal ions, such as an Fe–O–Fe or Fe–Cl–Fe linkage. All the superexchange paths are of the sort Fe–Cl---Cl–Fe, Fe–Cl---H–O(H)–Fe or variants thereof. In some cases there are two such paths between any two given metal ions. What is unusual is how well these relatively long paths transmit the superexchange interaction. The chloride and bromide compounds span ordering temperatures from 6.54 K to 22.9 K [86, 94]. These relatively high transition temperatures may be compared with those for such other hydrated halides as $Cs_2CrCl_5 \cdot 4 H_2O$ ($T_c = 0.185$ K; Ref. [95]) or $Cs_2MnCl_4 \cdot 2 H_2O$ ($T_c = 1.8$ K; Ref. [96]). These substances all exhibit weak anisotropy, and are therefore good examples of the Heisenberg magnetic model.

The cesium chloride material is perhaps the compound that has been most thoroughly studied. Its susceptibility is displayed in Fig. 10.31, and the reader will note that three data sets, measured along each of the crystallographic directions, are pictured there. Above the maximum which occurs at about 7 K, the data sets are coincident, within experimental error. This is a graphic illustration of what is meant by the term magnetic isotropy, and the kind of behavior to which we assign the name, Heisenberg magnetic model compound. The same data are plotted with different (reduced) scales in Fig. 10.32 in order to facilitate comparison with theory. The fit illustrated [97] is to a high-temperature-series expansion for the susceptibility of the simple-cubic, spin $\mathscr{S} = \frac{5}{2}$ Heisenberg antiferromagnet. The exchange constant, which is the fitting parameter, in this case is $J/k = -0.310$ K. As the temperature is decreased, long range order sets in and the susceptibility behaves like that of a normal anitferromagnet, with the susceptibility parallel to the axis of preferred spin alignment dropping to zero at low temperatures. The a-axis is the easy axis, as it is with all the chloride and bromide analogs.

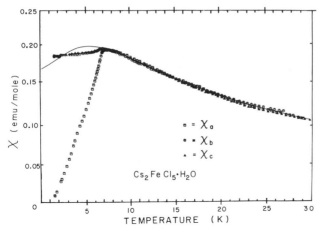

Fig. 10.31. Magnetic susceptibility of Cs$_2$[FeCl$_5$(H$_2$O)] measured along the three crystal axes. From Ref. [86]

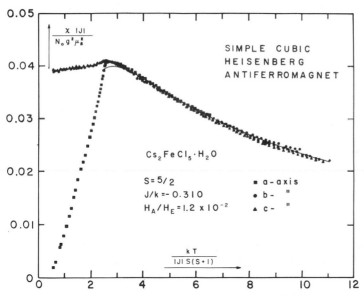

Fig. 10.32. The magnetic susceptibilities of Fig. 10.31, plotted on reduced scales in order to emphasize the isotropy above the critical temperature. The solid curve is the calculated high temperature series expansion. From Ref. [97]

The specific heat of Cs$_2$[FeCl$_5$(H$_2$O)] has also been measured, [86, 92] and is illustrated in Fig. 10.33. The transition temperature is a relatively low one, 6.54 K. This is advantageous because the magnetic contribution to the specific heat is the more easily determined the lower the magnetic transition temperature. The lattice contribution was evaluated empirically in this case. The resulting data have been analyzed in detail [92], the analysis being consistent with a three-dimensional character to the lattice. The entropy parameter $(S_\infty - S_c)/R$ takes the value 0.42 for infinite spin $\mathscr{S}$ on a

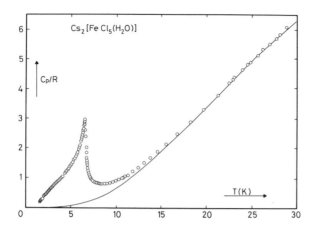

Fig. 10.33. Specific heat of $Cs_2[FeCl_5(H_2O)]$. The points are experimental and the curve is the estimated lattice contribution. From Ref. [92]

simple-cubic lattice (there are no calculated values for $\mathscr{S}=\frac{5}{2}$, but the values should be close), and the experimental value is 0.4. The agreement is good and is one of the strongest arguments in favor of the three-dimensional nature of the ordering.

The susceptibility of $Rb_2[FeCl_5(H_2O)]$ has been measured twice, [86, 87] as has the specific heat [93]. The susceptibility data were initially interpreted in terms of a one-dimensional chain with an important molecular field (interchain) correction. Both sets of data have since been analyzed in terms of magnetic lattice dimensionality crossover theory [93].

In the case of interchain interactions one must consider the Heisenberg hamiltonian with two different types of neighbors

$$H = -2J \overset{(d)}{\underset{i,j}{\sum}} S_i \cdot S_j - 2J' \overset{(3-d)}{\underset{i,j}{\sum}} S_i \cdot S_j,$$

where the first summation runs over nearest neighbors in d lattice directions and the second is along the other $(3-d)$ directions. In the present case, $d=1$. The high temperature series expansion calculated for this model, in the classical limit of $\mathscr{S}=\infty$, has been analyzed recently for several values of $R=(J'/J)$ [93, 98]. The case of a lattice dimensionality crossover from a linear chain system to a simple cubic lattice was studied by calculating the values of the reduced susceptibility $\chi|J|/Ng^2\mu_B^2$ and temperature $kT/|J|\mathscr{S}(\mathscr{S}+1)$ for different values and signs J and J'. The smaller the value of R, the more a material resembles a magnetic linear chain. For $Rb_2[FeCl_5(H_2O)]$ the result is $R=0.15$ and $2J/k=-1.45\,K$, when the results are scaled to the true spin of $\mathscr{S}=\frac{5}{2}$. Clearly, while there is some low-dimensional character the material cannot be considered, even to a first approximation, to be a linear chain.

The potassium salt, $K_2[FeCl_5(H_2O)]$, with a transition temperature of 14.06 K has one of the highest T_c's ever reported for a hydrated transition metal chloride double salt. This is clearly indicative of unusually extensive exchange interactions. The susceptibility behavior resembles that of $Cs_2[FeCl_5(H_2O)]$, except that the data are shifted to higher temperatures, consistent with the higher ordering temperature. The specific heat [91] exhibits the usual λ-feature, but the relatively high transition temperature makes difficult the separation of the lattice and magnetic contributions.

Just the same, it has been estimated that about 85% of the total expected magnetic entropy has already been gained below the transition temperature. This is consistent with the primarily three-dimensional character of the magnetic lattice. The data have been reanalyzed [93] according to the same model used to analyze $Rb_2[FeCl_5(H_2O)]$. An exchange constant $2J/k = -1.44 K$ ($\mathscr{S} = \frac{5}{2}$) was estimated, with a crossover parameter $R = 0.20$–0.35.

Two bromides have also been investigated [94], $Cs_2[FeBr_2(H_2O)]$ and $Rb_2[FeBr_5(H_2O)]$. They have high ordering temperatures, 14.2 and 22.9 K, respectively.

The transition temperatures vary as expected. For a given alkali ion, T_c for a bromide compound is always higher than that of the chloride analogue. This is a common phenomenon. There is also a general increase in transition temperature with decreasing radius of the alkali ion. Superexchange interactions have approximately an r^{-12} dependence between metal ion centers.

The phase diagrams for $Cs_2[FeCl_5(H_2O)]$, $Rb_2[FeCl_5(H_2O)]$, and $K_2[FeCl_5(H_2O)]$ have all been determined and that for the cesium compound is illustrated in Fig. 10.34. The diagrams are quite similar to one another, as well as to the schematic diagram in Fig. 6.14.

The first phase diagram of the $A_2[FeX_5(H_2O)]$ series of materials that was examined was [97] that of $Cs_2[FeCl_5(H_2O)]$. The bicritical point was found at $T_b = 6.3 K$ and $H_b = 14.7 kOe$ (0.147 T). Unfortunately, $H_c(0)$ is estimated to be about 150 kOe (0.15 T) for $Cs_2[FeCl_5(H_2O)]$, a value difficult to reach. In cases such as this, one can also make use of the relationship that the zero-field susceptibility perpendicular to the easy axis extrapolated to 0 K takes the value $\chi_\perp(0) = 2M_s/(2H_E + H_A)$, where $M_s = Ng\mu_B\mathscr{S}/2$ is the saturation magnetization of one antiferromagnetic sublattice. Since $\chi_\perp(0)$ is relatively easy to measure, this relationship in combination with that for $H_{SF}(0)$ can be used to determine H_A and H_E. The anisotropy field was estimated as $H_A = 0.88 kOe$ (0.088 T) and the exchange field $H_E = 75.9 kOe$ (7.59 T). The parameter α is then 1.2×10^{-2}. One interesting feature is that the antiferromagnetic-paramagnetic phase boundary is quite vertical $[T_b/T_c(0) = 0.96]$ as has been observed in several other low-anisotropy antiferromagnets.

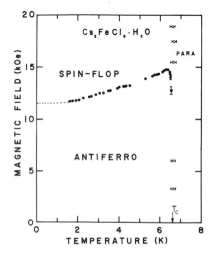

Fig. 10.34. Magnetic phase diagram of $Cs_2[FeCl_5(H_2O)]$. From Ref. [97]

The phase diagram of $Rb_2[FeCl_5(H_2O)]$ is quite similar [87, 99]. The bicritical point is found at 9.75 K and 17.8 kOe (1.78 T). With H_A of only 0.6 kOe (0.06 T) and an H_E increased to 172 kOe (17.2 T), the anisotropy appears to be even less in this compound; α is 3.4×10^{-3}. Therefore this salt is also a good example of the Heisenberg magnetic model system.

The last compound in this series whose phase diagram has been determined is $K_2[FeCl_5(H_2O)]$ [99]. The bicritical point occurs at 13.6 K and 34.1 kOe (3.41 T), and the spin-flop field extrapolated to zero temperature is about 27 kOe (2.7 T). One can therefore calculate an anisotropy field of 1.7 kOe (0.17 T) and an exchange field of 199 kOe (19.9 T). The anisotropy parameter α is 8.5×10^{-3}.

The anisotropy field appears to be anomalously small in all of these crystals. Perhaps the dipole-dipole interactions, which should be small, act in opposition to the other contributors to the anisotropy field. The iron systems are compared in Table 10.3 with several other compounds which are good examples of the Heisenberg magnetic model. Among the spin $\mathscr{S} = \frac{5}{2}$ systems, it will be seen that the temperature of the critical point is more convenient for future studies of, for example, critical phenomena.

The fluoride, $K_2[FeF_5(H_2O)]$ [100], is considered separately because it is the lone $A_2[FeX_5(H_2O)]$ compound which clearly exhibits what might be called classical antiferromagnetic linear chain behavior. That is, it shows the characteristic broad maximum in the susceptibility which identifies quasi-one-dimensional systems; the specific heat, which should likewise show a broad maximum, has not yet been measured. This difference in behavior from the other systems described follows from the structural features of this (monoclinic) compound. Though there are still discrete octahedra in the crystal lattice, they are connected by hydrogen bonds; it is well-known that fluoride ion can hydrogen-bond more strongly than either chloride or bromide. Thus, although fluoride usually does not provide as efficient an exchange path as chloride or bromide, in this case strong hydrogen bonding provides a more directional character to the exchange. The material indeed appears to provide the first good example of a one-dimensional antiferromagnet of iron(III).

The three susceptibility data sets exhibit broad maxima at about 3.4 K; a portion of the data is illustrated in Fig. 10.35. The system undergoes long range antiferromagnetic ordering at $T_c = 0.80$ K, with the preferred direction of spin alignment in the ac-plane.

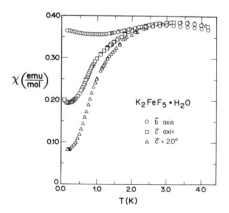

Fig. 10.35. Zero field magnetic susceptibilities of $K_2[FeF_5(H_2O)]$ below 4.2 K. The b-axis is the hard one, and the others c and $c + 20°$ form angles of $50°$ and $30°$ respectively with the easy axis of the magnetization. From Ref. [100]

Table 10.3. Some antiferromagnetic Heisenberg 3D magnetic model systems.

	Crystal lattice	$\mathscr{S}$	T_c, K	H_A/H_E
$CuCl_2 \cdot 2H_2O$	Orthorhombic	$\frac{1}{2}$	4.36	10^{-2}–10^{-3}
$KNiF_3$	Cubic (perovskite)	1	246	Small
Cr_2O_3	Rhombohedral	$\frac{3}{2}$	307.0	2.9×10^{-4}
$RbMnF_3$	Cubic (perovskite)	$\frac{5}{2}$	83.0	5×10^{-6}
MnF_2	Tetragonal (rutile)	$\frac{5}{2}$	67.33	1.6×10^{-2}
$Cs_2FeCl_5 \cdot H_2O$	Orthorhombic	$\frac{5}{2}$	6.54	1.2×10^{-2}
$Rb_2FeCl_5 \cdot H_2O$	Orthorhombic	$\frac{5}{2}$	10.00	3.4×10^{-3}
$K_2FeCl_5 \cdot H_2O$	Orthorhombic	$\frac{5}{2}$	14.06	8.5×10^{-3}

A consideration of the transition temperature following the calculations of Oguchi [101] yields a value for $R = |J'/J| = 1.4 \times 10^{-2}$. The low dimensional magnetic character is not ideal. Weak interactions of the metal ions with their next-nearest-neighbors become important below 10 K and the correlation between magnetic moments on different chains produces a lattice dimensionality crossover from a linear-chain system to an anisotropic simple cubic lattice.

The application of the classical approximation for real systems is known to be reliable for large values of the spin such as in this case, with $\mathscr{S} = \frac{5}{2}$. In the limit of a pure linear chain ($J' = 0$) the exact solution has been calculated by Fisher and the susceptibility is given by Eq. (7.6),

$$\chi = \chi_C \frac{1 + u(K)}{1 - u(K)},$$

with

$$u(K) = (\coth K) - \left(\frac{1}{K}\right), \qquad \chi_C = \frac{Ng^2\mu_B^2\mathscr{S}(\mathscr{S}+1)}{3kT}$$

and

$$K = \frac{2J\mathscr{S}(\mathscr{S}+1)}{kT}.$$

The curve has been plotted in Fig. 10.36 with $J/k = -0.40$ K, together with the experimental points.

One of the most significant problems yet to be worked on with these salts concerns the properties of mixed and diluted compounds. As was pointed out above, the indium and iron salts are isomorphous and since the indium materials are diamagnetic, it would be interesting to determine the phase diagram of the mixed systems. The transition temperatures will decrease with dilution, but the detailed behavior needs to be mapped out. A more extensive review of these compounds is given elsewhere [102].

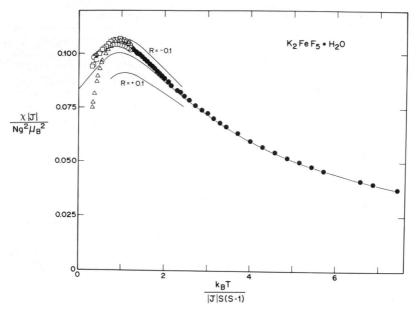

Fig. 10.36. Zero field susceptibilities of $K_2[FeF_5(H_2O)]$ in the paramagnetic region. The full symbols are data points which have been taken on powders, while the open ones represent the measurements in the three principal directions. The continuous line is the solution for the classical Heisenberg linear chain and the values of R give the calculations for three-dimensional crossover with ferro- and antiferromagnetic interchain constants. From Ref. [100]

10.10 Some Dilution Experiments

Random or disordered magnetic systems have attracted a great deal of interest recently. This is a large field, including such subjects as metallic spin glasses, amorphous magnets and random alloys of two magnetic species. We concentrate here on a few site-diluted magnets, that is, systems in which magnetic atoms are replaced by non-magnetic atoms in increasing amounts. The magnetic ordering temperature T_c is thereby decreased and for sufficiently large impurity doping T_c reduces to zero at a critical concentration p_c of magnetic atoms. The value of p_c depends on the number of magnetic neighbors and on the lattice structure. We follow the review article of de Jongh [103] closely throughout this section.

Let $T_c(p)$ be the long-range ordering temperature for the concentration p of magnetic ions. It is assumed that an isostructural diamagnetic material is available and that the exchange constant does not vary with dilution. The molecular field prediction is simply that the ratio $T_c(p)/T_c(1)$ should increase linearly from zero, for $p=0$, to 1, at $p=1$. It should not surprise that the ratio changes when the Heisenberg, Ising or XY models are used instead, and one then finds results such as are illustrated in Fig. 10.37 for the fcc lattice. Note that $p_c \simeq 0.3$; in general, p_c is larger for 2D lattices than for 3D lattices. The Ising and XY models have downward curvature over the range of p values, but the Heisenberg model exhibits an inflection point.

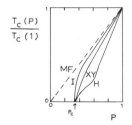

Fig. 10.37. The qualitative dependence on p of the magnetic ordering temperature of 3D ferromagnets. From Ref. [103]

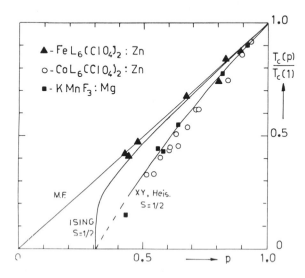

Fig. 10.38. Experimental variation of $T_c(p)$ for simple cubic antiferromagnets, compared with theory for the s.c. Ising and XY, $\mathscr{S}=\frac{1}{2}$ models. From Ref. [103]

Some data on $T_c(p)$ vs. p are illustrated in Fig. 10.38 for three simple cubic antiferromagnets. The systems are the $\mathscr{S}=\frac{5}{2}$ Heisenberg compound $KMnF_3:Mg$ [104], the $\mathscr{S}=\frac{1}{2}$, XY system $[Co(C_5H_5NO)_6](ClO_4)_2:Zn$ measured by Algra [105] and the Ising, $\mathscr{S}=\frac{1}{2}$ system $[Fe(C_5H_5NO)_6](ClO_4)_2:Zn$ measured by Mennenga [106]. The data are compared with the respective theoretical calculations, and the agreement is reasonable. The predictions for s.c. Heisenberg and XY models coincide only because of the scale of the figure. Since the extreme anisotropy inherent to either the Ising or XY models will always be incompletely realized in the experimental compounds, one may expect data on Ising or XY materials to be shifted somewhat in the direction of the Heisenberg prediction. All the data may be extrapolated to the critical value $p_c \simeq 0.31$ expected for the s.c. lattice. The pyridine N-oxide systems are particularly suitable for studies of this sort since the lattice parameters for the three compounds $[M(C_5H_5NO)_6](ClO_4)_2$, $M = Co$, Fe, Zn are equal within about 0.1%. Accordingly no variation of the exchange with dilution has been observed.

As would be expected intuitively, the susceptibility of antiferromagnetic systems increases sharply upon dilution. This is shown in data of Breed [104] on $KMnF_3:Mg$, in Fig. 10.39. One imagines that there will be extra paramagnetic contributions arising from finite isolated clusters containing an odd number of manganese spins. The density of spins in isolated clusters however decreases very rapidly above p_c so that this cannot be the only reason for the divergence. The explanation, due to Harris and Kirkpatrick, is that the increase is caused by local fluctuations in the total magnetic moment which

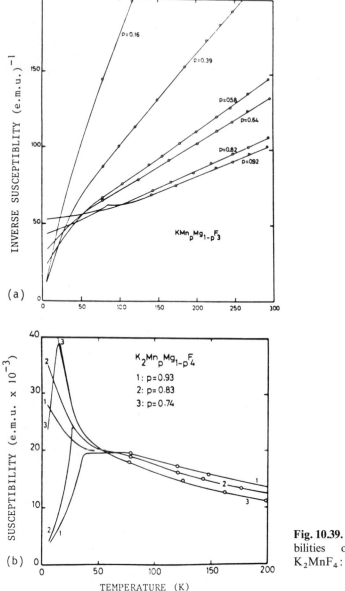

INVERSE SUSCEPTIBLITY (e.m.u.)$^{-1}$

p=0.16

p=0.39

p=0.58

p=0.64

p=0.82

p=0.92

$KMn_pMg_{1-p}F_3$

(a)

SUSCEPTIBILITY (e.m.u. × 10^{-3})

$K_2Mn_pMg_{1-p}F_4$

1: p=0.93
2: p=0.83
3: p=0.74

(b)

TEMPERATURE (K)

Fig. 10.39. Experimental suscepti-bilities of $KMnF_3$:Mg and K_2MnF_4:Mg. From Ref. [103]

arise because, in the neighborhood of a nonmagnetic impurity, the balance of the two antiferromagnetic sublattices is destroyed. Similar data on single-crystals of $[Co(C_5H_5NO)_6](NO_3)_2$:Zn have also been obtained [107].

Some interesting results come out of specific heat studies of diluted antifer-romagnets. Data on the 3D Ising compound Cs_3CoCl_5:Zn, as measured by Lagendijk [108], are illustrated in Fig. 10.40. The anomaly at T_c is not appreciably broadened as p decreases, the peak at p=0.5 being nearly as sharp as the one at p=1. The shape of the

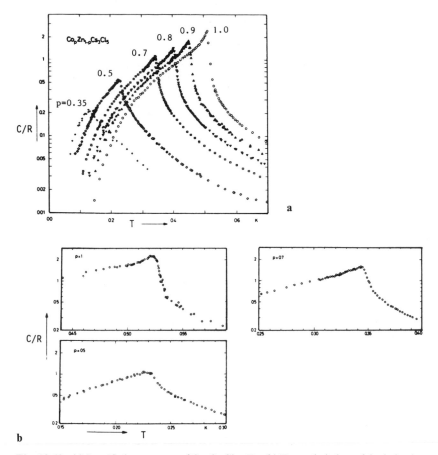

Fig. 10.40. (a) Specific heat curves of Cs_3CoCl_5 : Zn. (b) Expanded view of the behavior near $T_c(p)$

anomaly is also not affected dramatically, although the entropy change above T_c becomes more important with decreasing T_c. The entropy change above T_c is about 15% for $p = 1$, increases to about 34% at $p = 0.5$, and is about 50% for $p = 0.35$. Thus there is a high degree of short-range order which develops above $T_c(p)$ in the infinite magnetic clusters present for $p > p_c$. This is evidenced in some cases by a broad, Schottky-like maximum at a temperature of the order of the exchange constant.

This is clearly apparent in the data for $[Co(C_5H_5NO)_6](ClO_4)_2$: Zn as shown in Fig. 10.41. A broad maximum appears already for $p \leq 0.7$; the fraction of magnetic spins in isolated finite clusters is negligible at these concentrations, so that the broad anomaly must be associated with the infinite cluster of spins. Indeed, this behavior resembles to a degree that of a quasi-lower-dimensional magnetic system. The shape of the maximum may be seen to depend on the degree of dilution (Fig. 10.41b). Interestingly, it was found that, just for $p = p_c$, the experimental curve could be described by the prediction for a $\mathscr{S} = \frac{1}{2}$ linear chain XY magnet, as shown in Fig. 10.41b. The exchange constant needed for the fit is just equal to the value of that for the pure (three-dimensional) compound.

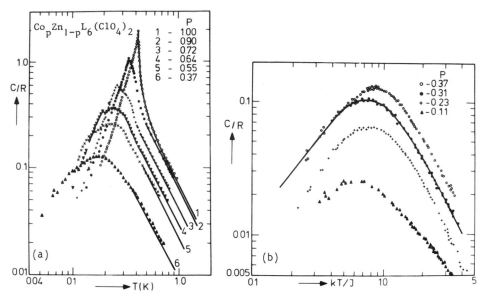

Fig. 41. (a) Specific heat data for $[Co(C_5H_5NO)_6](ClO_4)_2 : Zn$. Solid curves are theoretical predictions for the limiting high-temperature behavior of the s.c., $\mathscr{S} = \frac{1}{2}$ magnet, assuming random dilution and no dependence of the exchange on p. (b) Specific heat behavior for p values close to $p_c = 0.31$. The solid curve is the prediction for the XY, $\mathscr{S} = \frac{1}{2}$ linear chain, using the exchange constant of the pure compound. From Ref. [103]

This leads us to the effect of dilution on the transition to 3D magnetic order of a quasi one-dimensional system. Doping such a system with nonmagnetic ions will have a drastic effect on T_c since this will break up the chains into finite segments and very effectively reduce the magnetic correlations between spins on the same chain. Indeed, it has been shown that a few percent of nonmagnetic impurities may be sufficient to reduce T_c by a factor of five or even more. This was observed with $Cs_2CoCl_4 : Zn$ [109].

Finally, there are many recent studies in which, instead of diluting a magnetic system with nonmagnetic ions, a random mixture is considered of two magnetic species with competing interactions and/or competing spin anisotropies. An example, composed of isostructural Ising and XY systems, would be a sample of

$$[Fe_{1-p}Co_p(C_5H_5NO)_6](ClO_4)_2 .$$

Because of the anisotropies, the Fe moments will tend to align along the z-axis, while the Co moments will order preferentially in the xy-plane perpendicular to z. This very material has been studied by Mennenga [106]. Here the two interpenetrating magnetic systems may order independently of one another. The anisotropy energies are so much stronger than the exchange energies that there is an almost complete decoupling between the two subsystems.

A different result is found when the exchange energies are comparable to the anisotropy energies. The coupling between the two subsystems is enhanced and the neighboring Ising and XY moments will then divert away from the z-axis and the

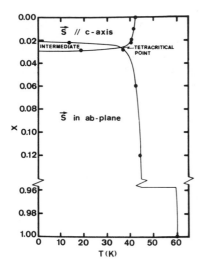

Fig. 10.42. Phase diagram for $K_2Mn_pFe_{1-p}F_4$. From Ref. [110]

xy-plane, respectively. Very small concentrations of one component can be sufficient to polarize the whole system, and there will be a shift of both $T_c(p)$ curves. An experimental phase diagram that results is illustrated in Fig. 10.42 for the system $K_2Mn_pFe_{1-p}F_4$ [110]. The Ising anisotropy for K_2MnF_4 is of dipolar origin and thus very small compared to the strong crystal field XY anisotropy in K_2FeF_4. The result is a crossing of the two phase boundaries at a concentration of only about 2.6% of Fe(II) impurities in K_2MnF_4. The point where the curves cross is called a tetracritical point.

Many other such systems have been studied lately [103]. We mention but two more, $Mn_pFe_{1-p}Cl_2 \cdot 2H_2O$ [111] and $Mn(Cl_pBr_{1-p})_2 \cdot 4H_2O$ [112]. In the first case, the components are structurally isomorphous and consist of chemical linear chains. The intrachain exchange is antiferromagnetic in the manganese salt, while it is ferromagnetic in the iron salt. These competing exchange interactions, along with the competing orthogonal spin anisotropies, lead to a rather complex phase diagram.

In the second example, the exchange field differs by only about 20% between the pure components, $MnCl_2 \cdot 4H_2O$ and $MnBr_2 \cdot 4H_2O$. But the anisotropy between them varies by about a factor of four, the bromide being more anisotropic. The bromide orders at a higher temperature (2.12 K, vs. 1.62 K) and this is apparently due to the greater anisotropy. By mixing the two isomorphous salts, one is able to tune the anisotropy by varying p, and as a consequence the variation of T_c with p allows the study of the effect of the change in anisotropy on the magnetic properties.

10.11 Biomagnetochemistry of Cobalt(II)

We turn here to a subject that illustrates how the kind of research described in this book may lead to unanticipated new advances, and that is the application of magnetic properties to problems in biochemistry.

There has been much interest recently in cobalt-containing enzymes. We note studies on the spectral properties of cobalt carboxypeptidase A [113], on the spectroscopy at low temperatures of the intermediates in the reaction of the same enzyme with dansyl oligopeptides [114], as well as EPR studies of single crystals of both cobalt and copper carboxypeptidase A [115]. Furthermore, there are reports on the X-ray absorption spectra of cobalt carbonic anhydrase [116] as well as on the crystal structures of nickel and cobalt carboxypeptidase A [117]. The role of cobalt as a probe of the structure and function of carbonic anhydrase has been reviewed [118]. All this activity, and these cobalt-containing enzymes do not even occur naturally! These efforts suggest that the study of cobalt-containing enzymes must be useful for an understanding of the natural (zinc-containing) enzymes. The emphasis here will be on simple model compounds of cobalt.

Cobalt(II) is a particularly useful ion for magnetic studies because its magnetic properties are very sensitive to changes in its environment (Ref. [119] and Sect. 6.13). Though both octahedral and tetrahedral cobalt(II) act as effective spin $\mathscr{S} = \frac{1}{2}$ ions at low temperatures, the g-values are quite different for the two geometries. The g-values of six-coordinate or quasi-octahedral cobalt also exhibit great variability from compound to compound. All of this is well-understood and allows an assignment of stereochemistry. Fewer empirical data are available concerning the electronic structure of five-coordinate cobalt. The diversity of magnetic behavior exhibited by cobalt(II) is illustrated by the selection of results that were presented in Table 6.7. The crystal structure of every one of those materials is known. The variation of the crystal field parameters is much larger than that found with such a common ion in biochemical systems as copper(II), which always exhibits g-values of about 2.1 to 2.2. The zero-field splitting is an irrelevant parameter for Cu(II). Among the other metals, zinc is important in biological systems, but it is diamagnetic and colorless.

Measurements at low temperatures are required in order to observe the interesting magnetic properties of cobalt. That is, when the excited states become depopulated and the ion acts in whatever geometry it may have as a spin $\mathscr{S} = \frac{1}{2}$ ion. EPR absorption in cobalt compounds is scarcely ever observed above 4.2 K because of fast spin-lattice relaxation. In a recent compilation [119] of magnetic and spectroscopic data on cobalt(II), 61 tetrahedral compounds are listed. Of these, the EPR spectra of only 5 of them appear to have been reported. Of 42 five-coordinate compounds, the EPR spectra of about three of them have been reported. The only tetrahedral compounds of cobalt which have been investigated as extensively as those listed in Table 6.7 are the A_3CoX_5 (A = Cs, Rb; X = Cl, Br) series of materials.

Cobalt(II) has proved to be particularly useful as a reporter ion in biochemical systems because it can be studied by a variety of physical techniques. Most cobalt-containing materials such as enzymes (with the native zinc replaced by cobalt, a reaction which in many cases allows the system to retain most of its natural biological activity [120]) are too magnetically dilute for direct susceptibility measurements. A number of interesting model systems have been described in the literature, however, and we describe some of these below.

One of the central themes in the research on the magnetochemistry of cobalt in biochemical systems has been the attempt to identify and define the coordination environment of the metal. The geometry is not well-defined, because cobalt may be four-, five- or six-coordinated, and the different polyhedra are flexible and may be distorted as well. The definition of coordination sphere and its influence on the concomitant magnetic properties is a central question. Some natural materials have been studied, such as cobalt carbonic anhydrase [121]. However most of the published research has concerned either model compounds or natural compounds in which the cobalt has been introduced, usually by replacement of zinc. The magnetic susceptibility of several compounds has been examined, but by and large the research had used such tools as electron paramagnetic resonance (EPR), optical spectroscopy, or magneto circular dichroism.

A report [121] on the electronic and EPR spectra and the susceptibility of cobalt human carbonic anhydrase concentrated of defining the spin state of the sample and its change with the addition of cyanide.

A study [122] of the magnetic susceptibility of cobalt carboxypeptidase A over an extended temperature range (30–120 K) led to an assignment of the metal stereochemistry as five-coordinate; this was done on the basis of empirical correlations of the values of the effective magnetic moments of a number of simpler compounds. A similar study [123] assigned a distorted tetrahedral structure to cobalt(II) stellacyanin. In both cases, the assignments were made in conjunction with spectroscopic measurements, but in neither case was the nature of the zero field splitting (ZFS) determined.

Several model compounds have recently been reported that mimic some of the properties of carbonic anhydrase. They are $[Co(H_2O)L](ClO_4)_2$, where L is tris[(3,5-dimethyl-1-pyrazolyl)methyl]amine [124], and {tris[(4,5-dimethyl-2-imidazolyl)-methyl]phosphine oxide}Co(II) [125]. Though a variety of spectroscopic data have been reported, there are no magnetic measurements reported as yet on either of these compounds. The former compound was designed to favor five-coordination, and that has indeed been found [126].

Makinen and co-workers [127–130] have recently worked out an EPR method for determining the ZFS in biological materials, and used it to assign the coordination geometry in such samples as the active-site-specific Co^{2+}-reconstituted enzyme, liver alcohol dehydrogenase. The method is an indirect one, depending on the temperature dependence of the CW microwave saturation behavior (the Orbach process); zero-field splittings have been determined for a number of materials, but the sign of the ZFS is not available from these measurements. (Recall for example that the $CoCl_4^{2-}$ ion has approximately the same magnitude of ZFS in both Cs_3CoCl_5 and Cs_2CoCl_4, but the sign of the ZFS changes between the two compounds.) Assignments of coordination geometry were based on an empirical correlation of the magnitude of the ZFS in several compounds; the ZFS has been determined directly in just a few compounds, and in no five-coordinate cobalt compounds. A number of new results are included in the most recent version of this correlation [130]. This new method has two special limitations. First, the zero-field splitting of those compounds whose spin-lattice relaxation rates remain too fast in the liquid helium region for saturation to occur may not be measured, and secondly, paramagnetic impurities may perturb the temperature dependence of the relaxation of the cobalt.

Many other compounds have been studied by EPR, but only for the measurement of the g-values. This has been done [131] for compounds such as $[Co(C_2H_5COO)_2(imidazole)_2]$, which has g-values of 2.06, 3.43, and 5.37. This compound, along with several similar ones [132, 133], has been proposed as a model for zinc metalloenzymes. The crystal structures of these compounds have been reported, but the ZFS remains unknown. A minor change in the nature of the compound results in an important change in the observed magnetic properties: the compound $[Co(CH_3COO)_2(imidazole)_2]$ has g-values of 2.16, 4.23, and 4.49. Both the acetate and the propionate exhibit a somewhat distorted tetrahedral coordination sphere; the values of all the angles at the Zn(Co) are quite similar, but the sense of the distortion is not identical. The ZFS of the acetate has been reported [128] as 4.8 cm^{-1} (of unspecified sign) but there is as yet no report on the ZFS of the propionate.

Makinen [128, 130] has reported a ZFS of 25.3 cm^{-1} for [Co(2-picoline N-oxide)$_5$] (ClO$_4$)$_2$, an example of the uncommon coordination number five for cobalt. The molecule is important because he has used it in a correlation of ZFS parameters which has led him to suggest that the hydrolysis of esters by carboxypeptidase A requires a penta-coordinate metal ion [128, 129]. The EPR spectrum of [Co(2-CH$_3$–C$_5$H$_4$NO)$_5$] (ClO$_4$)$_2$ has been reported [134] but there is no direct measurement of the ZFS nor of the low temperature magnetic properties of this compound. Indeed, there is no such report on any penta-coordinate cobalt(II) compound.

The system $[Co(Et_4dien)Cl_2]$ is another penta-coordinate compound [135] which has also been studied by Makinen and Yim [128]. They suggest a ZFS of 42–56 cm^{-1}, which is a very large value.

Yet another five-coordinate cobalt compound which may have interesting magnetic properties and of which there is very little known is the aforementioned $[Co(H_2O)L]$ (ClO$_4$)$_2$ [124, 126]. It appears to be high-spin, and therefore its magnetic ground state may be different from the two compounds mentioned above. The compound seems to be an interesting model for carbonic anhydrase.

Horrocks and co-workers [131–133] have studied a number of compounds, $[Co(RCOO)_2(2-X-Im)_2]$, which are interesting because, while the ligands are all similar, small differences have led to different crystal structures. They have compared them with the analogous zinc materials. The compounds reported are:

$[Co(CH_3COO)_2(Im)_2]$	4-coordinate
$[Co(C_2H_5COO)_2(Im)_2]$	4-coordinate
$[Co(CH_3COO)_2(2-Me-Im)_2]$	6-coordinate
$[Co(C_2H_5COO)_2(2-Me-Im)_2]$	4-coordinate
$[Co(n-C_3H_7COO)_2(2-Me-Im)_2]$	6-coordinate
$[Co(CH_3COO)_2(2-Et-Im)_2]$	4-coordinate,

where Me is CH_3 and Et is C_2H_5, and Im is imidazole. If these compounds have any value as model compounds for metalloenzymes, then the small differences among them should provide a wealth of information that will be useful for the further development of empirical relations between magnetic parameters and biological structure and reactivity.

Little is known about magnetic interactions between cobalt ions in biological molecules. A prerequisite for the existence of such interactions of course is that two (or more) magnetic ions be close enough to interact, and this is not a common phenomenon in biochemical systems or even the model systems which have been reported. Nevertheless, we note that the compound $[Co-H_2ATP-2,2'-bipyridyl]_2 \cdot 4H_2O$ has been studied as a model for the ATP transport mechanism. That is because ternary complexes such as this exhibit high stability towards hydrolysis. The crystal structure of this material [136] shows that it is dimeric, with the cobalt atoms bridged by two –O–P–O– moities from the triphosphate system. This is just the kind of structure which should provide a strong superexchange path, and a magnetic interaction. The zinc compound is isostructural.

10.12 References

1. Ballhausen C.J., (1962) Introduction to Ligand Field Theory, McGraw-Hill, New York
2. McGarvey B.R., (1966) in: Transition Metal Chemistry, Vol. 3, p. 90, Ed.: Carlin R.L., Marcel Dekker, Inc., New York
3. Cotton F.A. and Wilkinson G., (1980) 4th Ed.: Advanced Inorganic Chemistry, J. Wiley and Sons, New York
4. Abragam A. and Bleaney B., (1970) Electron Paramagnetic Resonance of Transition Ions, Oxford University Press, Oxford
5. Orton J.W., (1968) Electron Paramagnetic Resonance, Iliffe Books, Ltd., London
6. a) Mabbs F.E. and Machin D.J. (1973) Magnetism and Transition Metal Complexes, Chapman and Hall, London b) Annual reviews of the literature are provided by the Annual Reports of the Chemical Society, London, and by the Specialist Periodical Reports, also published by the Chemical Society, in the series: Electronic Structure and Magnetism of Inorganic Compounds
7. Gmelin: Handbuch Anorg. Chem., Vol. 57, Part B, Sect. 2, p. 553 (Fig. 213), Springer (1966) Berlin, Heidelberg, New York
8. Robinson W.K. and Friedberg S.A., Phys. Rev. 117, 402 (1960)
9. McElearney J.N., Losee D.B., Merchant S., and Carlin R.L., Phys. Rev. B 7, 3314 (1973)
10. Polgar L.G., Herweijer A., and de Jonge W.J.M., Phys. Rev. B 5, 1957 (1972)
11. Bizette H., Compt. rend. 243, 1295 (1956)
12. Bhatia S.N., Carlin R.L., and Paduan-Filho A., Physica 92 B, 330 (1977)
13. Kopinga K. and de Jonge W.J.M., Phys. Lett. 43 A, 415 (1973)
14. Van Uitert L.G., Williams H.J., Sherwood R.C., and Rubin J.J., J. Appl. Phys. 36, 1029 (1965)
15. Haseda T., Kobayashi H., and Date M., J. Phys. Soc. Japan 14, 1724 (1959)
16. Mizuno J., J. Phys. Soc. Japan 16, 1574 (1961)
17. Hamburger A.I. and Friedberg S.A., Physica 69, 67 (1973)
18. Kleinberg R., J. Chem. Phys. 50, 4690 (1969)
19. Kleinberg R., J. Appl. Phys. 38, 1453 (1967)
20. Paduan-Filho A., Becerra C.C., and Oliveira, Jr. N.F., Phys. Lett. 50 A, 51 (1974)
21. Figueiredo W., Salinas S.R., Becerra C.C., Oliveira, Jr. N.F., and Paduan-Filho A., J. Phys. C 15, L-115 (1982)
22. Vianna S.S. and Becerra C.C., Phys. Rev. B 28, 2816 (1983)
23. Oliveira, Jr. N.F., Paduan-Filho A., Salinas S.R., and Becerra C.C., Phys. Rev. B 18, 6165 (1978)
24. Swüste C.H.W., Botterman A.C., Millenaar J., and de Jonge W.J.M., J. Chem. Phys. 66, 5021 (1977)
25. de Neef T. and de Jonge W.J.M., Phys. Rev. B 10, 1059 (1974)
26. Martin R.L. and White A.H., Trans. Met. Chem. 4, 113 (1968)
27. Gütlich P., Struct. Bonding 44, 83 (1981)
28. Switzer M.E., Wang R., Rettig M.F., and Maki A.H., J. Am. Chem. Soc. 96, 7669 (1974)
29. Pignolet L.H., Patterson G.S., Weiher J.F., and Holm R.H., Inorg. Chem. 13, 1263 (1974)
30. Kennedy B.J., Fallon G.D., Gatehouse B.M.K.C., and Murray K.S., Inorg. Chem. 23, 580 (1984); Zarembowitch J. and Kahn O., Inorg. Chem. 23, 589 (1984)
31. Terzis A., Filippakis S., Mentzafos D., Petrouleas V., and Malliaris A., Inorg. Chem. 23, 334 (1984)
32. Merrithew P.B. and Rasmussen P.G., Inorg. Chem. 11, 325 (1972)

33. Harris G., Theor. Chim. Acta **5**, 379 (1966); ibid., **10**, 119 (1968)

34. Hall G.R. and Hendrickson D.N., Inorg. Chem. **15**, 607 (1976)

35. Healy P.C. and Sinn E., Inorg. Chem. **14**, 109 (1975); Butcher R.J. and Sinn E., J. Am. Chem. Soc. **98**, 2440, 5139 (1976); Sinn E., Inorg. Chem. **15**, 369 (1976); Cuckauskas E.J., Deaver B.S., and Sinn E., J. Chem. Phys. **67**, 1257 (1977)

36. Malliaris A. and Papaefthimious V., Inorg. Chem. **21**, 770 (1982)

37. Leipoldt J.G. and Coppens P., Inorg. Chem. **12**, 2269 (1973)

38. Hoskins B.F. and Pannan C.D., Inorg. Nucl. Chem. Lett. **11**, 409 (1975)

39. Martin R.L. and White A.H., Inorg. Chem. **6**, 712 (1967); Hoskins B.F. and White A.H., J. Chem. Soc. **A 1970**, 1668

40. Healey P.C., White A.H., and Hoskins B.F., J. Chem. Soc. **D 1972**, 1369

41. Chapps G.E., McCann S.W., Wickman H.H., and Sherwood R.C., J. Chem. Phys. **60**, 990 (1974)

42. DeFotis G.C., Failon B.K., Wells F.V., and Wickman H.H., Phys. Rev. B **29**, 3795 (1984)

43. Carlin R.L., Science, **227**, 1291 (1985)

44. Wickman H.H., Klein M.P., and Shirley D.A., J. Chem. Phys. **42**, 2113 (1965)

45. Bracket G.C., Richards P.L., and Caughey W.S., J. Chem. Phys. **54**, 4383 (1971)

46. Wickman H.H. and Merritt F.R., Chem. Phys. Lett. **1**, 117 (1967)

47. Ganguli P., Marathe V.R., and Mitra S., Inorg. Chem. **14**, 970 (1975)

48. DeFotis G.C., Palacio F., and Carlin R.L., Phys. Rev. B **20**, 2945 (1979)

49. Mitra S., Figgis B.N., Raston C.L., Skelton B.W., and White A.H., J. Chem. Soc. Dalton **1979**, 753

50. Wickman H.H., Trozzolo A.M., Williams H.J., Hull G.W., and Merritt F.R., Phys. Rev. **155**, 563 (1967); ibid., **163**, 526 (1967)

51. Wickman H.H. and Trozzolo A.M., Inorg. Chem. **7**, 63 (1968)

52. Wickman H.H. and Wagner C.F., J. Chem. Phys. **51**, 435 (1969); Wickman H.H., J. Chem. Phys. **56**, 976 (1972)

53. Arai N., Sorai M., Suga H., and Seki S., J. Phys. Chem. Solids **38**, 1341 (1977)

54. DeFotis G.C. and Cowen J.A., J. Chem. Phys. **73**, 2120 (1980)

55. Decurtins S., Wells F.V., Sun K.C.-P., and Wickman H.H., Chem. Phys. Lett. **89**, 79 (1982)

56. Yoshikawa M., Sorai M., and Suga H., J. Phys. Chem. Solids **45**, 753 (1984)

57. Burlet P., Burlet P., and Bertaut E.F., Solid State Comm. **14**, 665 (1974)

58. Bertaut E.F., Duc T.Q., Burlet P., Burlet P., Thomas M., and Moreau J.M., Acta Cryst. B **30**, 2234 (1974)

59. Flippen R.B. and Friedberg S.A., Phys. Rev. **121**, 1591 (1961)

60. Schelleng J.H., Raquet C.A., and Friedberg S.A., Phys. Rev. **176**, 708 (1968)

61. Spence R.D., J. Chem. Phys. **62**, 3659 (1975)

62. Schmidt V.A. and Friedberg S.A., Phys. Rev. **188**, 809 (1969)

63. Witteveen H.T., Rutten W.L.C., and Reedijk J., J. Inorg. Nucl. Chem. **37**, 913 (1975)

64. Klaaijsen F.W., Thesis, (1974) Leiden; Klaaijsen F.W., Blöte H.W.J., and Dokoupil Z., Solid State Comm. **14**, 607 (1974); Klaaijsen F.W., Dokoupil Z., and Huiskamp W.J., Physica **79 B**, 547 (1975). Similar studies on the manganese analogs are published by Klaaijsen F.W., Blöte H.W.J., and Dokoupil Z., Physica **81 B**, 1 (1976)

65. Blöte H.W.J., Physica **79 B**, 427 (1975)

66. Berger L. and Friedberg S.A., Phys. Rev. **136**, A 158 (1964)

67. Schmidt V.A. and Friedberg S.A., Phys. Rev. B **1**, 2250 (1970)

68. Polgar L.G. and Friedberg S.A., Phys. Rev. B **4**, 3110 (1971)

69. Herweijer A. and Friedberg S.A., Phys. Rev. B **4**, 4009 (1971)

70. Morosin B. and Haseda T., Acta Cryst. B **35**, 2856 (1979)

71. Wada N., Matsumoto K., Amaya K., and Haseda T., J. Phys. Soc. Japan **47**, 1061 (1979)

72. Wada N., J. Phys. Soc. Japan **49**, 1747 (1980); Wada N. and Haseda T., J. Phys. Soc. Japan **50**, 779 (1981); Wada N., Kashima Y., Haseda T., and Morosin B., J. Phys. Soc. Japan **50**, 3876 (1981)

73. Salem-Sugui, Jr. S., Ortiz W.A., Paduan-Filho A., and Missell F.P., J. Appl. Phys. **53**, 7945 (1982)

74. Carlin R.L. and de Jongh L.J., Chem. Revs. to be published
75. van Ingen Schenau A.D., Verschoor G.C., and Romers C., Acta Cryst. **B 30**, 1686 (1974)
76. Bergendahl T.J. and Wood J.S., Inorg. Chem. **14**, 338 (1975)
77. a) Algra H.A., de Jongh L.J., and Carlin R.L., Physica, **93 B**, 24 (1978)
    b) Algra H.A., de Jongh L.J., Huiskamp W.J., and Carlin R.L., Physica **83 B**, 71 (1976)
78. Carlin R.L., O'Connor C.J., and Bhatia S.N., J. Am. Chem. Soc. **98**, 685 (1976)
79. O'Connor C.J. and Carlin R.L., Inorg. Chem. **14**, 291 (1975)
80. Bartolome J., Algra H.A., de Jongh L.J., and Carlin R.L., Physica **B 94**, 60 (1978)
81. Algra H.A., de Jongh L.J., and Carlin R.L., Physica **93 B**, 24 (1978)
82. Reinen D. and Krause S., Solid State Comm. **29**, 691 (1979)
83. Algra H.A., de Jongh L.J., and Carlin R.L., Physica **92 B**, 258 (1977)
84. Bos W.G., Klaassen T.O., Poulis N.J., and Carlin R.L., J. Mag. Mag. Mat. **15**, 464 (1980)
85. Mennenga G., de Jongh L.J., Huiskamp W.J., Sinn E., Lambrecht A., Burriel R., and Carlin R.L., J. Mag. Mag. Mat. **44**, 77 (1984)
86. Carlin R.L., Bhatia S.N., and O'Connor C.J., J. Am. Chem. Soc. **99**, 7728 (1977)
87. O'Connor C.J., Deaver, Jr. B.S., and Sinn E., J. Chem. Phys. **70**, 5161 (1979)
88. Greedan J.E., Hewitt D.C., Faggiani R., and Brown I.D., Acta Cryst. **B 36**, 1927 (1980)
89. Hopkins T.E., Zalkin A., Templeton D.H., and Adamson M.G., Inorg. Chem. **5**, 1431 (1966)
90. Edwards A.J., J. Chem. Soc. Dalton **1972**, 816
91. McElearney J.N. and Merchant S., Inorg. Chem. **17**, 1207 (1978)
92. Puertolas J.A., Navarro R., Palacio F., Bartolome J., Gonzalez D., and Carlin R.L., Phys. Rev. **B 26**, 395 (1982)
93. Puertolas J.A., Navarro R., Palacio F., Bartolome J., Gonzalez D., and Carlin R.L., Phys. Rev. **B 31**, 516 (1985)
94. Carlin R.L. and O'Connor C.J., Chem. Phys. Lett. **78**, 528 (1981)
95. Carlin R.L. and Burriel R., Phys. Rev. **B 27**, 3012 (1983)
96. Smith T. and Friedberg S.A., Phys. Rev. **177**, 1012 (1969)
97. Paduan-Filho A., Palacio F., and Carlin R.L., J. Phys. (Paris) **39**, L-279 (1978)
98. Puertolas J.A., Navarro R., Palacio F., Bartolome J., Gonzalez D., and Carlin R.L., J. Mag. Mag. Mat. **31–34**, 1243 (1983)
99. Palacio F., Paduan-Filho A., and Carlin R.L., Phys. Rev. **B 21**, 296 (1980)
100. Carlin R.L., Burriel R., Rojo J., and Palacio F., Inorg. Chem. **23**, 2213 (1984), and to be published
101. Oguchi T., Phys. Rev. **133**, A 1098 (1964)
102. Carlin R.L. and Palacio F., Coord. Chem. Revs. **65**, 141 (1985)
103. de Jongh L.J., Proceedings of the European Physical Society Summer School on "Magnetic Phase Transitions," Ettore Majorana Centre, Erice, Italy, July, 1983, eds. Elliott R.J. and Ausloos M., Springer-Verlag, N.Y., 1983
104. Breed D.J., Gilijamse K., Sterkenburg J.W.E., and Miedema A.R., Physica **68**, 303 (1973)
105. Algra H.A., de Jongh L.J., Huiskamp W.J., and Reedijk J., Physica B **86–88**, 737 (1977); Algra H.A., de Jongh L.J., and Reedijk J., Phys. Rev. Lett. **42**, 606 (1979)
106. Mennenga G., Thesis (1983) Leiden; Mennenga G., de Jongh L.J., Huiskamp W.J., and Reedijk J., J. Mag. Mag. Mat., **43**, 3, 13 (1981)
107. Carlin R.L., Lambrecht A., and Burriel R., unpublished
108. Lagendijk E. and Huiskamp W.J., Physica **62**, 444 (1972)
109. Bartolome J., Algra H.A., de Jongh L.J., and Huiskamp W.J., Physica B **86–88**, 707 (1977)
110. Bevaart L., Frikkee E., Lebesque J.V., and de Jongh L.J., Phys. Rev. B **18**, 3376 (1978)
111. DeFotis G.C., Pohl C., Pugh S.A., and Sinn E., J. Chem. Phys. **80**, 2079 (1984)
112. Westphal C.H. and Becerra C.C., J. Phys. C **13**, L 527 (1980)
113. Geoghegan K.F., Holmquist B., Spilburg C.A., and Vallee B.L., Biochemistry **22**, 1847 (1983)
114. Geoghegan K.F., Galdes A., Martinelli R.A., Holmquist B., Auld D.S., and Vallee B.L., Biochemistry **22**, 2255 (1983)
115. Dickinson L.C. and Chien J.C.W., J. Am. Chem. Soc. **105**, 6481 (1983)
116. Yachandra V., Powers L., and Spiro T.G., J. Am. Chem. Soc. **105**, 6596 (1983)

117. Hardman K.F. and Lipscomb W.N., J. Am. Chem. Soc. **106**, 463 (1984)

118. Bertini I. and Luchinat C., Acc. Chem. Res. **16**, 272 (1983)

119. Banci L., Bencini A., Benelli C., Gatteschi D., and Zanchini C., Struct. Bond. **52**, 37 (1982). See also the related review: Bencini A. and Gatteschi D., Trans. Met. Chem. (N.Y.) **8**, 1 (1982)

120. Vallee B.L., Riordan J.F., Johansen J.T., and Livingston D.M., Cold Spring Harbor Symp. Quant. Biol. **36**, 517 (1971)

121. Haffner P.H. and Coleman J.E., J. Biol. Chem. **248**, 6630 (1973)

122. Rosenberg R.C., Root C.A., and Gray H.B., J. Am. Chem. Soc. **97**, 21 (1975)

123. Solomon E.I., Wang R.-H., McMillin D.R., and Gray H.B., Biochem. Biophys. Res. Comm. **69**, 1039 (1976)

124. Bertini I., Canti G., Luchinat C., and Mani F., Inorg. Chem. **20**, 1670 (1981)

125. Brown R.S., Salmon D., Curtis N.J., and Kusuma S., J. Am. Chem. Soc. **104**, 3188 (1982)

126. Benelli C., Bertini I., DiVaira M., and Mani F., Inorg. Chem. **23**, 1422 (1984)

127. Yim M.B., Kuo L.C., and Makinen M.W., J. Magn. Res. **46**, 247 (1982)

128. Makinen M.W. and Yim M.B., Proc. Natl. Acad. Sci. USA **78**, 6221 (1981)

129. Kuo L.C. and Makinen M.W., J. Biol. Chem. **257**, 24 (1982)

130. Makinen M.W., Kuo L.C., Yim M.P., Wells G.B., Fukuyama J.M., and Kim J.E., J. Am. Chem. Soc. **107**, 5245 (1985)

131. Horrocks, Jr. W. De W., Ishley J.N., Holmquist B., and Thompson J.S., J. Inorg. Biochem. **12**, 131 (1980)

132. Horrocks, Jr. W. De W., Ishley J.N., and Whittle R.R., Inorg. Chem. **21**, 3265 (1982)

133. Horrocks, Jr. W. De W., Ishley J.N., and Whittle R.R., Inorg. Chem. **21**, 3270 (1982)

134. Bencini A., Benelli C., Gatteschi D., and Zanchini C., Inorg. Chem. **19**, 3839 (1980)

135. Dori Z., Eisenberg R., and Gray H.B., Inorg. Chem. **6**, 483 (1967)

136. Cini R. and Orioli P., J. Inorg. Biochem. **14**, 95 (1981); Orioli P., Cini R., Donati D., and Mangani S., J. Am. Chem. Soc. **103**, 4446 (1981)

11. Some Experimental Techniques

11.1 Introduction

We conclude with a brief sketch of some experimental procedures for magnetic measurements. In keeping with the rest of the book, the discussion will be limited to methods for measuring magnetic susceptibilities and specific heats. Since local requirements and experience usually determine the nature of each apparatus, relatively few details will be given; a number of commercial instruments are available for measuring susceptibilities, and these are not discussed.

11.2 Specific Heat Measurements

We have assumed an elementary knowledge about heat capacities (the change in the temperature of a system for the transfer of a given amount of heat) and of specific heats (the heat capacity of a system per unit mass). Specific heats of interest in this book have referred to conditions of constant magnetization or constant applied field. These can be measured in either an adiabatic or isothermal calorimeter [1].

An electrical method of measurement is always used [1–3]. A resistance wire is usually wound around the specimen, and a known current is passed through the wire for a measured time period. The electrical energy dissipated by the wire and which flows into the sample may be interpreted as heat. In general, each calorimeter, the heating coils, and the thermometer depend on the nature of the material to be studied and the temperature range desired. Each apparatus is essentially a unique installation.

In calorimetry of solids at low temperatures, the sample is suspended in a highly-evacuated space according to one method of measurement [4]. A single-crystal sample is preferable in this case, so that a heating coil may be wound around the sample. A thermometer, usually a germanium resistance diode, may be glued to the sample for the measurement of its temperature. The sample may be cooled by employing helium exchange gas to allow thermal contact between the sample and the bath of liquid helium. This method suffers from the disadvantage that all the exchange gas must then be removed before measurements are made. This frequently cannot be done with a high degree of confidence.

Under adiabatic conditions, heat is not transferred from the sample to its surroundings. The temperature of the sample is measured as a function of time, as in classical calorimetry. In order to take account of background heat leaks, the following procedure is used [2, 3, 5].

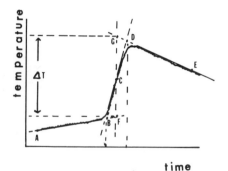

Fig. 11.1. A sample recorder chart indicating the temperature vs. time for a heat capacity measurement

A typical temperature vs. time graph is illustrated in Fig. 11.1. The line AB represents the background change in resistance (i.e., temperature) before the current is turned on. At the time corresponding to point B, the current is established for the time t (usually a matter of a few seconds). Then the temperature is continuously monitored as a function of time, resulting in the new segment labeled DE. Line FG is constructed midway between points B and D, and the two lines AB and DE are extended as shown, giving the points F and G. The molar heat capacity c at the temperature corresponding to point C is then given by

$$c = \frac{MW\varepsilon It}{(wt)\Delta T},$$

where (wt)/MW is the number of models of the sample, ΔT is the temperature change indicated in the figure, ε is the measured voltage across the sample, and I is the current. The resolution of the experiment is measured by ΔT, which is frequently as small as 0.01 K. The quality of the data depends on many factors, such as the time required for the sample to attain internal thermal equilibrium.

An alternative method [6] for the measurement of powders involves compressing the sample hydraulically in a calorimeter can, a procedure which increases the area of thermal contact between the sample and the sample holder, and also increases the filling factor without requiring exchange gas. This method requires that corrections be made for the "addenda," which is a generic term meaning the sample can, the residual exchange gas, any grease mixed with the sample to increase thermal contact, and so on. This is usually accomplished by running the experiment as a blank, without a sample present. The lower the measuring temperature, the smaller will be the correction for the addenda, for its heat capacity also decreases with temperature, as does that of the lattice. The magnetic contribution will be more evident, the lower the temperature at which it occurs.

All the rest is electronics, and vacuum and cryogenics technology. The thermometer must be calibrated and its resistance measured with precision, the sample needs to be cooled to the desired initial temperature, the voltage and the current must be measured accurately, and the thermal isolation of the sample must be as great as possible.

A calorimetric method has been described [7] in which the sample is cooled by adiabatic demagnetization. The sample is cooled by contact with a cooling salt, after which it must be isolated by means of a superconducting switch.

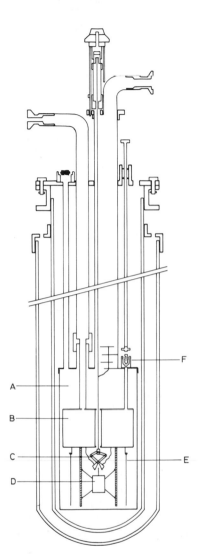

Fig. 11.2. A standard cryostat with a calorimetric insert. A) vacuum chamber; B) liquid helium pot; C) heat switch; D) calorimetric vessel; E) adiabatic shield; and F) needle valve. From Ref. [8]

A typical calorimeter [8] is illustrated in Fig. 11.2. The inner vacuum can is immersed in liquid helium, while the liquid helium pot B is thermally isolated from the main helium bath. By pumping on this pot through a minute hole, it is possible to obtain a starting temperature of about 1 K in the pot. The thermal contact between the calorimeter D and the pot B is established with the aid of a mechanical heat switch that can be externally operated.

11.3 Gouy and Faraday Balances

We begin the discussion about susceptibilities with a brief discussion of the classical force method. These are now rarely used for measurements below 80 K, but they are

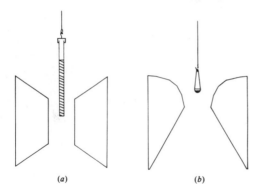

Fig. 11.3. Sample configuration for force measurement of magnetic susceptibility using a) the Gouy method, and b) the Faraday method. From Ref. [11]

conceptually easier to appreciate than some of the more sophisticated methods. They are described in detail elsewhere [9–11], as well as in all the classical texts on magnetism. These methods are sensitive and straightforward, but they always require a substantial applied magnetic field.

Both methods require that the magnetic material be placed in an inhomogeneous magnetic field. A displacement force is then exerted on the sample, drawing it into a region of higher field. The displacement force depends on both the magnetization and the field gradient so that measurement of the force gives direct information on the magnetic susceptibility of the sample [11].

The Gouy method is the simplest of all. In this method the sample is placed in a long cylindrical tube which is suspended from an analytical balance. The sample tube is positioned between the poles of a magnet such that one end of the tube is in a region of homogeneous field and the other is in the region of zero-field. This is illustrated in Fig. 11.3, where it will be seen that this arrangement results in an inhomogeneous field at the sample. When the sample is positioned in this fashion, a paramagnetic material experiences an increase in weight while a diamagnetic substance experiences a decrease in weight as a result of the displacement force exerted on the sample. A large amount of material is required and the uniformity of packing of the sample is important.

The Faraday method requires specially designed pole faces, as shown in Fig. 11.3, to place a small uniform sample in a region where the product of the field times the field gradient is constant. The force is then independent of the packing of the sample and depends only on the total mass of the material present. The method is sensitive and highly reproducible, single crystals can be measured, but the weight changes are small and the suspension devices are usually fragile. The force methods are not suitable for the measurement of canted antiferromagnets nor of ferromagnets because of the required applied field. This is also true with regard to the popular vibrating magnetometers which are available commercially.

O'Connor describes these and other methods further [11].

11.4 Susceptibilities in Alternating Fields

The static susceptibility that was introduced in deriving the Curie law in Chapter I was defined as M/H. In a number of experiments one actually measures the dynamic or differential susceptibility, dM/dH, by means of applying an oscillating magnetic field. A system of magnetic ions may not always be capable of following immediately the changes of this external magnetic field. That is, the redistribution of the magnetic spins over the energy levels proceeds via a relaxation process characterized by a time constant τ. This method owes much to the pioneering work of Gorter and his colleagues at the Kamerlingh Onnes Laboratorium, Leiden.

Let the sample be placed in a magnetic field H(t) that varies sinusoidally

$$H(t) = H_0 + h \cdot \exp(i\omega t).$$

The field H_0 is constant and may be zero, while $\omega = 2\pi\nu$ is the angular frequency of the ac field. The magnetic field H(t) induces a time variation of the magnetization

$$M(t) = M_0 + m(\omega)\exp(i\omega t)$$

and there can be a phase shift between M(t) and H(t) due to relaxation effects. Thus, $m(\omega)$ is a complex quantity. The differential susceptibility $\chi(\omega)$ is equal to $m(\omega)/h$; it is a complex quantity and is written as

$$\chi(\omega) = \chi'(\omega) - i\chi''(\omega). \tag{11.1}$$

The real part $\chi'(\omega)$ is called the dispersion while $\chi''(\omega)$ is referred to as the absorption.

The influence of the frequency ω of the oscillating field on the measured differential susceptibility is directly related to the relaxation time, τ. At the low-frequency limit, when the spin system remains in equilibrium with the lattice, then $\omega\tau \ll 1$ and the measured susceptibility is identical to the low-field static susceptibility. This low frequency limit of the differential susceptibility is called the isothermal susceptibility, χ_T, thus expressing the fact that the spins maintain thermal equilibrium with the surroundings.

The other, high frequency, limit is obtained if the differential susceptibility is measured with an oscillating field of high frequency. Then the magnetic ions are incapable of redistributing themselves in accordance with the applied field. This case corresponds to $\omega\tau \gg 1$, and the assembly of spins is found to be uncoupled from its surroundings. The susceptibility in this case is the so-called adiabatic susceptibility, χ_S, which is strongly field-dependent. Indeed, at strong fields $\chi_S \to 0$. The susceptibilities are defined as $\chi_T = (\partial M/\partial H)_T$ and $\chi_S = (\partial M/\partial H)_S$.

One finds [12] that the two susceptibilities defined above are related as

$$\chi = \frac{\chi_T - \chi_S}{1 + i\omega\tau} + \chi_S \tag{11.2}$$

or, since $\chi = \chi' - i\chi''$, then

$$\chi'(\omega) = \frac{\chi_T - \chi_S}{1 + \omega^2\tau^2} + \chi_S \tag{11.3}$$

and

$$\chi''(\omega) = \frac{\omega\tau(\chi_T - \chi_S)}{1 + \omega^2\tau^2}. \tag{11.4}$$

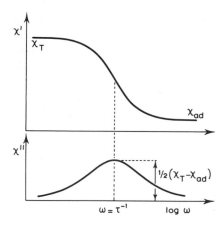

Fig. 11.4. Frequency dependence of χ' and χ'', according to Eqs. (11.3) and (11.4). (χ_{ad} is χ_S).

The frequency dependency of χ' and χ'' is drawn schematically in Fig. 11.4. The in-phase component χ' is χ_T when $\omega \ll \tau^{-1}$, while $\chi' = \chi_S$ if $\omega \gg \tau^{-1}$. The out-of-phase component χ'' approaches zero at both of these limits, showing a maximum around the frequency $\omega = \tau^{-1}$. This figure demonstrates the difficulty of comparing a differential susceptibility with the simple models presented earlier. It is not enough to determine whether a susceptibility possesses an out-of-phase component or not, for one may still be measuring χ_S rather than χ_T.

One frequently plots $\chi''(\omega)$ vs. $\chi'(\omega)$ in a so-called Argand diagram, such as is represented in Fig. 11.5. The isothermal susceptibility χ_T may be obtained from the right-hand side of the semi-circular figure ($\omega \ll \tau^{-1}$); in the limit of high frequencies ($\omega \gg \tau^{-1}$) the adiabatic susceptibility is obtained at the left-hand side of the Argand diagram. The relaxation time τ may be obtained, at least in ideal cases such as illustrated, from the angular frequency at the top of the diagram, when $\omega = \tau^{-1}$.

The ratio between the adiabatic susceptibility and the isothermal one was shown in Eq. (3.12) to be equal to the ratio between the specific heats, c_M and c_H. The specific heat referred to in Eq. (3.21) is c_M, and we now write it, in the high-temperature limit, as

$$c_M = b/T^2 .$$

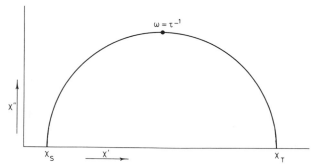

Fig. 11.5. Ac susceptibilities as a function of frequency, ω; this is called an Argand diagram. The situation illustrated corresponds to a relaxation process with a single time constant τ

Furthermore, c_H was found in Eq. (3.15) as CH^2/T^2; more generally, when there are interactions between the magnetic ions, one obtains,

$$c_H = (b + CH^2)/T^2.$$

Thus the ratio c_M/c_H and thus also χ_S/χ_T become

$$\chi_S/\chi_T = c_M/c_H = b/(b + CH^2). \tag{11.5}$$

This ratio is valid only in the asymptotic T^{-2} limit for the specific heat and when the magnetic field is not too strong ($g\mu_B H < 2kT$) so that saturation effects on χ_T can be omitted. In the limit of H going to zero, one sees from Eq. (11.5) that χ_S/χ_T becomes 1, or that $\chi_S = \chi_T$. Thus relaxation effects can be neglected in this limit.

In strong external fields, χ_S/χ_T becomes zero. In this case, the differential susceptibility may have any value between χ_T and 0.

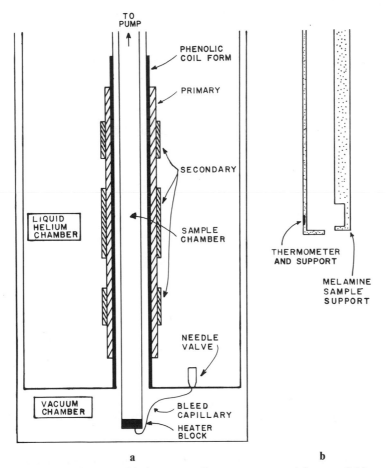

Fig. 11.6. A dewar and inductance coil arrangement used for zero-field *ac* mutual inductance measurements, 1.5–20 K. From Ref. [4]

The ac mutual inductance method is perhaps the best and most widely-used method for differential susceptibility measurements at low temperatures [13]. It is not generally as sensitive as the force methods but it has the advantages that crystal susceptibilities may be measured directly, that zero-applied field measurements may be made, and that relaxation effects are easily observed.

The sample is placed in an inductance coil system such as is illustrated in Fig. 11.6 [4, 14]. The sample is usually placed inside a system consisting of a primary coil and two secondary coils. The secondary coils are equally but oppositely wound on top of the primary and are connected in series. Moving a magnetic sample from the center of one secondary coil to the center of the other causes a change in the voltage over the secondary coils, proportional to the ac susceptibility of the sample. The coils are wound uniformly of fine magnet wire. The homogeneous measuring field depends on the current of the ac signal as well as the geometry of the coil system and, if, necessary, can be reduced to as little as 1 Oe (10^{-4} T) or less. In general, the coils are immersed in liquid helium. Since the applied field is coincident with the axis of the coils, a crystal can be readily oriented for the measurement of the desired susceptibility.

Lock-in amplifiers can be used to vary the frequency and to allow measurement of both components of the complex susceptibility, χ' and χ''. In the ac susceptibility χ' is the in-phase inductive component, and χ'' is the out-of-phase or resistive component. The resistive component is proportional to the losses in the sample, which are due to relaxation effects. These may indicate spin-lattice relaxation effects or absorption due to the presence of permanent ferromagnetic moments.

The method is versatile. It has been extended to measurements over the frequency interval 0.2 Hz to 60 kHz and to temperatures up to 200 K [15], to measurements in an

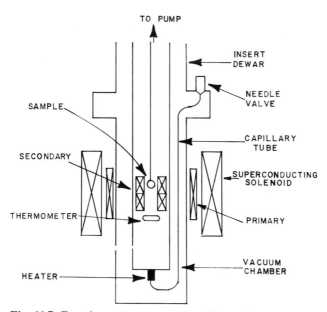

Fig. 11.7. Experimental arrangement of insert dewar for variable temperature susceptibility measurements using the *ac* mutual inductance method, along with an applied magnetic field. From Ref. [16]

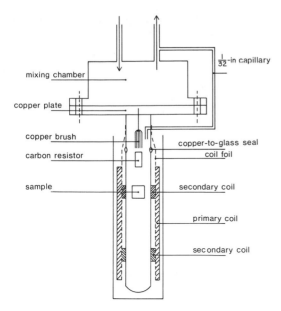

mixing chamber

$\frac{1}{32}$-in capillary

copper plate

copper brush

carbon resistor

copper-to-glass seal
coil foil

sample

secondary coil

primary coil

secondary coil

Fig. 11.8. Experimental coil arrangement for mutual inductance susceptibility measurements in a dilution refrigerator. From Ref. [17]

applied field (for example, Ref. [16]), and it has been used in a dilution refrigerator for measurements down to 40 mK [17]. An apparatus for field-dependent measurements is shown in Fig. 11.7. In this case the coil system is placed inside a cryostat which contains a superconducting coil. Measurements can be made with this particular system over the temperature interval 1.5–30 K and in fields up to 6 T.

In the experiments at very low temperatures [17] a different coil system is required because the sample cannot be moved, but the principle of the method is the same. The system is illustrated in Fig. 11.8. The main problem here is to maintain thermal equilibrium between the sample and the ^{3}He–^{4}He bath in the mixing chamber.

11.5 Anisotropic Susceptibilities

We conclude this chapter with some comments on the relationships between magnetic anisotropy and crystal symmetry [18–20]. We follow Ref. [19] closely here.

In homogeneous but anisotropic substances, the magnetization M is a vector quantity and depends on the direction as well as the magnitude of the applied field H. In such cases the susceptibility χ is a tensor and

$$M = \chi H$$

or

$$\begin{bmatrix} M_1 \\ M_2 \\ M_3 \end{bmatrix} = \begin{bmatrix} \chi_{11} & \chi_{12} & \chi_{13} \\ \chi_{12} & \chi_{22} & \chi_{23} \\ \chi_{13} & \chi_{23} & \chi_{33} \end{bmatrix} \begin{bmatrix} H_1 \\ H_2 \\ H_3 \end{bmatrix},$$

where the vector components M_1, M_2, M_3, H_1, H_2, H_3 refer to an orthogonal coordinate system fixed in the crystal. The tensor χ is then symmetric with six independent elements. The intensity of magnetization in the i-th direction ($i = 1, 2, 3$) is given by

$$M_i = \chi_{i1}H_1 + \chi_{i2}H_2 + \chi_{i3}H_3 .$$

It is always possible to find an orthogonal coordinate system in which the only non-vanishing elements of χ are along the diagonal.

$$\begin{bmatrix} \chi_1 & 0 & 0 \\ 0 & \chi_2 & 0 \\ 0 & 0 & \chi_3 \end{bmatrix} = \chi_p .$$

The elements χ_1, χ_2, and χ_3 are known as the principal susceptibilities and their directions (i.e., the coordinate axes) are the principal magnetic axes.

The directions of the principal magnetic axes depend on the symmetry of the crystal. Cubic crystals are necessarily isotropic. In the axial crystal systems (tetragonal, rhombohedral, hexagonal) one principal axis ($\chi_{\parallel}$) coincides with the fourfold or trigonal symmetry axis of the crystal lattice. The other two ($\chi_{\perp}$) are perpendicular to this and may be arbitrarily specified. The principal susceptibilities of orthorhombic crystals are in general all different and their directions coincide with the a-, b-, and c-crystallographic axes. Monoclinic crystals have low symmetry; the direction of the unique twofold or b-axis defines one principal susceptibility axis. The other two axes lie in the ac-plane, but their directions are arbitrary and must be determined by experiment. In the triclinic case there is only translational symmetry and none of the principal magnetic axes are crystallographically determined. The tensor χ must be determined by experiment and then diagonalized by a transformation of the coordinate system. The details of this procedure are discussed by Mitra [18] and Horrocks and Hall [19].

In order to apply all the theoretical arguments presented earlier in the book, one must then find the principal molecular susceptibilities. These are the susceptibilities of the individual molecules or complex ions in the crystal, referred to a local coordinate system. The subscripts x, y, z are used for these principal molecular susceptibilities, χ_x, χ_y, and χ_z. Incidently, when there are axial (D) and rhombic (E) terms in the spin-Hamiltonian, it is usual [21] to choose the local axes such that the ratio $|E/D|$ is less than $\frac{1}{3}$.

For a triclinic system with but one molecule in the unit cell, the principal molecular and crystal axes will coincide. However, these must be determined by experiment, for the axes may have any orientation in the crystal. Furthermore, there are no necessary relationships between the axes systems and the molecular structure; these must be determined by structural analysis.

In monoclinic symmetry, it is common to consider an orthogonal coordinate system such as a, b, and c^*. Relative to these axes the direction cosines of the principal molecular susceptibility axes χ_x, χ_y, and χ_z are:

$$
\begin{array}{c}
\,\begin{array}{ccc} a & b & c^* \end{array}\\
\begin{array}{c}\chi_x\\\chi_y\\\chi_z\end{array}
\begin{bmatrix}
\alpha_x & \beta_x & \gamma_x\\
\alpha_y & \beta_y & \gamma_y\\
\alpha_z & \beta_z & \gamma_z
\end{bmatrix}.
\end{array}
$$

This coordinate system is also appropriate in orthorhombic symmetry except that c^* is replaced by c (which is now orthogonal to a and b). The relationship between principal molecular and crystal susceptibilities is expressed as

$$\chi_1 = \chi_a = \sum (\chi_x \alpha_x^2 + \chi_y \alpha_y^2 + \chi_z \alpha_z^2)$$
$$\chi_2 = \chi_b = \sum (\chi_x \beta_x^2 + \chi_y \beta_y^2 + \chi_z \beta_z^2)$$
$$\chi_3 = \chi_c = \sum (\chi_x \gamma_x^2 + \chi_3 \gamma_y^2 + \chi_3 \gamma_3^2),$$

where the summation is taken over all independently oriented molecules.

Axial crystals allow only the measurement of $\chi_{\parallel}$ and $\chi_{\perp}$; if the molecular unit has less than axial symmetry there is insufficient information available for the determination of χ_x, χ_y, and χ_z. Usually, however, this does not happen and the principal symmetry axis of the molecule is aligned parallel to that of the crystalline lattice. Then,

$$\chi_{\parallel}^{cryst} = \chi_{\parallel}^{mol} \quad \text{and} \quad \chi_{\perp}^{cryst} = \chi_{\perp}^{mol}.$$

Traditional methods such as the angle-flip or critical-couple method [18–20] measure anisotropies, $\Delta\chi$, alone. These have now been supplanted by ac methods, particularly those using SQUIDS [22], which measure each susceptibility individually. Another recent development [23] is the construction of an ac apparatus with horizontal coils. This allows rotation diagrams to be made of the susceptibility in the xz-plane.

11.6 References

1. Stout J.W., (1968) in: Experimental Thermodynamics, Vol. 1, Chap. 6; (McCullough J.P. and Scott D.W., Eds.), Butterworths, London; Westrum, Jr. E.F., Furakawa G.T. and McCullough J.P., ibid., Chap. 5; Hill R.W., Martin D.L., and Osborne D.W., ibid., Chap. 7. These articles, which appear to be the most thorough available on traditional experimental procedures, are, respectively, on isothermal calorimetry, adiabatic calorimetry, and on calorimetry below 20 K. More recent, but also shorter, reviews are provided by Gmelin E., Thermochim. Acta **29**, 1 (1979); and Lakshmikumar S.T. and Gopal E.S.R., J. Indian Inst. Sci. A **63**, 277 (1981). Measurements of small samples and in a high magnetic field are reviewed by Stewart G.R. Rev. Sci. Instrum. **54**, 1 (1983)
2. Zemansky M.W., (1968) 5th Ed.: Heat and Thermodynamics, McGraw-Hill, New York
3. Rives J.E., (1972) in: Transition Metal Chemistry, Vol. 7, p. 1. (Ed. Carlin R.L.) M. Dekker, New York
4. McElearney J.N., Losee D.B., Merchant S., and Carlin R.L., Phys. Rev. B **7**, 3314 (1973)
5. Love N.D., Thesis, (1967) Michigan State University

6. Klaaijsen F.W., Thesis, (1974) Leiden
7. Algra H.A., de Jongh L.J., Huiskamp W.J., and Carlin R.L., Physica **92B**, 187 (1977)
8. Puertolas J.A., Navarro R., Palacio F., Bartolome J., Gonzalez D., and Carlin R.L., Phys. Rev. B **26**, 395 (1982)
9. Figgis B.N. and Lewis J., (1965) in: Techniques of Inorganic Chemistry, Vol. 4, pp. 137–248. (Ed.: Jonassen H.B. and Weissberger A.), Interscience, New York
10. Earnshaw A., (1968) Introduction to Magnetochemistry, Academic Press, New York
11. O'Connor C.J., Prog. Inorg. Chem. **29**, 203 (1982)
12. Carlin R.L. and van Duyneveldt A.J., (1977) Magnetic Properties of Transition Metal Compounds, Springer-Verlag, New York. See also the thesis of Groenendijk H.A., (1981) Leiden
13. Pillinger W.L., Jastram P.S., and Daunt J.G., Rev. Sci. Instrum **29**, 159 (1958)
14. Bhatia S.N., Carlin R.L., and Paduan-Filho A., Physica **92B**, 330 (1977)
15. Groenendijk H.A., van Duyneveldt A.J., and Willett R.D., Physica B **101**, 320 (1980). The physics group at Zaragoza has been able to extend their measurements to 300 K.: Burriel R. and Palacio F., private communication
16. Carlin R.L., Joung K.O., Paduan-Filho A., O'Connor C.J., and Sinn E., J. Phys. C **12**, 293 (1979)
17. van der Bilt A., Joung K.O., Carlin R.L., and de Jongh L.J., Phys. Rev. B **22**, 1259 (1980)
18. Mitra S., (1973) in: Transition Metal Chemistry, Vol. 7, p. 183, (Ed. Carlin R.L.) M. Dekker, New York
19. Horrocks, Jr. W. De W. and Hall D. De W., Coord. Chem. Revs. **6**, 147 (1971)
20. Mitra S., Prog. Inorg. Chem. **22**, 309 (1976)
21. Wickman H.H., Klein M.P., and Shirley D.A., J. Chem. Phys. **42**, 2113 (1965)
22. van Duyneveldt A.J., J. Appl. Phys. **53**, 8006 (1982)
23. Burriel R. and Carlin R.L., unpublished

Formula Index

Subject Index